Hailstorms

Prediction, Control and Damage Assessment

Second Edition

Hailstorms

Prediction, Control and Damage Assessment

Second Edition

P. Kumar
Professor and Head,
Department of Core Engineering and Engineering Sciences,
MAEER's MIT College of Engineering,
Pune, India.

CRC Press is an imprint of the
Taylor & Francis Group, an **informa** business

Hailstorms: Prediction, Control and Damage Assessment, *Second Edition*
by P. Kumar

Published by

BS Publications
A unit of BSP Books Pvt. Ltd.
4-4-309/316, Giriraj Lane, Sultan Bazar,
Hyderabad - 500 095, India.

For

CRC Press
Taylor & Francis Group, an informa business
6000 Broken Sound Parkway, NW
Suite 300, Boca Raton, FL 33487
711 Third Avenue
New York, NY 10017
2 Park Square, Milton Park
Abigdon, Oxon OX14 4RN, UK
www.taylorandfrancisgroup.com

For distribution in rest of the world other than the India, Pakistan, Nepal, Myanmar (Burma), Bhutan, Bangladesh and Sri Lanka.

ISBN: 978-1-138-04777-8

British Library Cataloguing in Publication Data
A Catalogue record for this book is available from the British Library

Printed at: Sanat Printers, Kundli, Haryana, India.

Dedicated to my

wife Mrs. Kiran Kumar

and daughter Dr. Urvashi Shrivastava

Contents

Chapter 11

Damage Due to Hail and Risk Cover

Appendix - A

Appendix - B

डॉ. लक्ष्मण सिंह राठोड़
मौसम विज्ञान विभाग के महानिदेशक
एवं
विश्व मौसम संगठन में भारत के स्थाई प्रतिनिधि

Dr. L.S. Rathore
Director General of Meteorology &
Permanent Representative of India with W.M.O.
and Member Executive Council of W.M.O.

भारत सरकार
भारत मौसम विज्ञान विभाग
पृथ्वी विज्ञान मंत्रालय
मौसम भवन, लोधी रोड़
नई दिल्ली–110003
Government of India
India Meteorological Department
Ministry of Earth Sciences
Mausam Bhawan, Lodhi Road
New Delhi - 110003

Foreword

Hailstorms is disastrous weather phenomenon which causes enormous loss to life and property all over the world. India holds highest number of loss of human life in a single hailstorm in 1888 in Moradabad. This compilation uniquely encompasses climatology, microphysics, synoptic studies, measurements, observations, predictions, suppression and risk assessment, altogether in one document for India and neighbourhood.

The hailstorm climatogical data and research, carried out by several tropical meteorologists with case-studies, given in this book would provide not only theoretical back-up but also practical guide for the researchers, weather forecasters, decision makers and insurance agencies. For the prediction of hailstorms over Indian region in particular and SE Asia in general, the book contains regional and seasonal cases so as to get stochastic rules for prediction of this complex weather feature – whose perfect numerical simulation is still awaited. This book also addresses the shortcomings in the current hail mitigation efforts and rejuvenates hope for future for such experiments.

I firmly believe that this book would serve as useful reference material, as well as textbook for M.Sc., M.Phil and Ph.D. students of Universities and several training institutions for Weather forecasters, Pilots and Navigators not only in India and SAARC countries but all over world.

I compliment Dr. P. Kumar and pleased to write the foreword for a very useful book on a subject which will help assisting disaster risk reduction.

LSRathore
10.05.2016

(Laxman Singh Rathore)

Preface to Second Edition

Each year damage due to hailstorm to human life, livestock, and property and agricultural is appalling in nature. Even then awareness towards such disaster is relatively much less than several other natural calamities e.g. earthquakes, Tropical Storms and Tsunami etc. Meteorologists, all over the world, seem to be over obsessed with modeling the predictions of thunderstorms and tornados but the efforts being made towards its control is very rare. On an average during the last decade 28 persons died each year (standard deviation = 38) in India. 1997 was the worst year, which recorded 130 deaths by hailstone hit in any single year during the last decade. However human activities have by and large taken it as part of living. This book is aimed to assist researchers, scientists, insurance agencies and statutory bodies not only improve accurate prediction of hailstorm but also to control its damage by various weather modification and safety measures. Insurance agencies should find a few approaches described in the book to be innovative while estimating the premium for the damage against hailstorms – may it be household property or agricultural crops. Author wishes to revitalize the interest of all to help mitigate this catastrophe of nature.

Broadly two problems enter into the study of hailstorms e.g. the cloud physics problem of the growth of hail stones and the synoptic problem which relates to the thermodynamic investigation of the conditions that produce hailstorms. Only a few thunderstorms have hails reaching the ground and not many of them have it, even in the most suitable parts of the clouds. With the help of Doppler Radar or high power land based or airborne Radars more recent efforts have been made as how better to recognize a thunderstorm which carries hailstones. Further it is extremely important to know the properties of hailstones, the average distribution of hailfall over the earth surface and the relevant parameters of hail-damage which include the region of hailfall, the area suffering hail damage, the velocity of hail precipitation front on the ground, the thickness of the layer of fallen hail, the sizes, forms, density and structures of hail and its spectrum, the duration of hail fall, the rates of which hail damage occurs in different region as dependent on the seasons and time of day etc. Despite several studies in the past, our present knowledge on the synoptic aerological conditions under which hail falls and physical properties of hailstorms are still not very clearly understood.

This book is divided into five main parts which deals with the climatology, physical properties of hail and its measurements, prediction, its control and damage assessment. In the second edition of the book Chapter I has been expanded to include global and regional climatology and Chapters II, III, IV and V have been revised to include recent researches on hail and convective clouds. Chapter X discusses candid answers for the causes of unsatisfactory results in hail-mitigation operations and Chapter XI has been made more comprehensive to describe types of damages by hailstorms. For ready reference of meteorological definitions Appendix – B has been revised. Recent advances in cloud physics with respect to aerosol properties required for condensation and ice nucleation process has been described. As the synoptic and topographical conditions widely vary across the Indian subcontinent hence for the purpose of prediction Indian region has been divided in four parts as NE, NW, Central and South Indian region. Separate chapters discuss the typical features of hailstorm formation. Forecasters in any part of the world could find fair order of similarity in the synoptic models discussed amongst the four regions of India, with their regions, too. This book, therefore, should serve weather forecasters, worldwide.

Author is thankful to Dr. Susan C. Van den Heever and Dr Russ S. Schumacher, Department of Atmospheric Science, Colorado State University and also to Dr. T.N. Krishnamurti, University of Chicago, U.S.A. for offering their valuable remarks to help improve the contents of this book.

It is hoped that the book would rejuvenate the waning efforts in hail mitigation experiments worldwide and would provide useful text material to researchers in the field of cloud physics, meteorology, agriculture scientists, weather forecasters and also a reference material to insurance companies which cover the crop and property damage insurance.

Dr. P. Kumar

About the Author

Dr. P. Kumar was conferred Ph.D. (Mathematics) by Banaras Hindu University, Varanasi in 1973 for his researches in Fluid Mechanics and Magnetohydrodynamics (MHD). As post-doctoral research fellow of Council of Scientific and Industrial Research (CSIR), India he initiated innovative researches in artificial rain making through MHD.

He was commissioned in Meteorological Branch of Indian Air Force; wherein he served for 22 years. His glorious IAF tenure included Operational Weather Forecasting at various regions of India. He carried out theoretical as well as applied researches and modelling in Mountain Waves at Indian Institute of Technology, Delhi. He was also Senior Instructor, and guide to M.Phil and Ph.D in Meteorology at Air Administrative College, Coimbatore affiliated to Bharathiar University, Coimbatore, India. Presently he is recognized Ph.D. guide from Department of Mathematics and Department of Atmospheric and Space Sciences at Savitribai Phule University, Pune, India.

Beside the current book he is also author of book entitled "Hydraulic Machines: Fundamentals of Hydraulic Power Systems" jointly published by this publisher and CRC press, U.K. He has published 86 research papers in Indian and foreign journals and has six Indian patents to his credit related to ingenious hail control technology developed by him under the research project sanctioned by Indian Council of Agriculture Research, Ministry of Agriculture, Government of India.

He has been recipient of President of India Medal-1973, for Meritorious Services; Best Application Paper Award in 17th National System Conference held at Indian Institute of Technology, Kanpur-1993; Best Teacher Award-2011; Bharat Jyoti Award-2012; Mother Teressa Award of Excellence-2012; Best Citizens of India Award-2013; Nelson Mandela Peace award-2014.

Beside several Indian Universities he had also been guest speaker abroad at University of Melbourn, Australia; CSIRO, Melbourn, Australia; Estov University, Budapest, Hungary; Taiwan National University, Taipei and National Central University, Taiwanvon the invitation of Taiwan Academy of Sciences.

Key areas of his researches include Mountain Waves, Air Pollution, Thunderstorm, Tornado, Duststorm and Hailstorm, Tropical Storm and Climate Dynamics.

CHAPTER 1

Climatology of Hailstorm

Almost everyone at one time or another has seen hailstones. They are large particles of ice which fall out of some thunderstorms. By convention, hail has a diameter of at least 5 mm. on the ground. For aviation purposes, the WMO's technical recommendations define the code GR (from the French word: grêle) for significant hail with diameters of 5 mm or greater, notably to distinguish from other forms of solid precipitation including snow pellets, ice crystals and pellets, snow grains and snow, all related to processes such as freezing of rain or successive melting rather than severe convection (WMO, 2008). In many parts of the world they appear rarely, but in some areas they occur with discouraging regularity. Prominent amongst the unlucky places are the north India and Himalayan region, equatorial Africa, Nigeria, Kenya, eastern and northwestern plains of United states and Canada, southern Europe e.g. north Italy, Southern Germany, France, Spanish plateau, region east of Adratic sea, the Caucasus region, southern Brazile, Northern Argentina and southeastern Australia. In these areas Hailstorms not only occur frequently but also with a violence that beats the crop into ground, strips fruits off trees and causes widespread damage to buildings. Hailstorms are great aviation hazards and in several cases have killed humans on ground.

Hail is the solid precipitation in the form of hard pellets of ice or lumps of ice which fall from cumulonimbus clouds. The pellets are spherical, conical or irregular in shape and often have a structure of concentric layers of alternatively clear and opaque ice. They are variable in size, usually a few millimeters in diameter but sometimes very much bigger. However, among the early documented hailstone accounts of giants, the official U.S record is held by one that fell in Potter, Nebraska, in 1928. It measured 17 inches in circumference and 5.4 inches in diameter and weighed one and half pound. Now there are several recorded evidences of much larger hails. During the storm that produced this giant, hailstones fell with such speed impact that many of them were buried in the ground.

1.1 Geographical Distribution of Hailstorm

Broadly latitudinal zones of hailfalls extend over the globe from equator to about 65^0 lat N and 60^0 lat S; albeit south of 55^0 S very few hailstorm are reported. The farthest polward storm in both the hemespheres is reported at Igarka, Russia, near 67.3^0 N; 90^0 E on afternoon of 28 July 2002 (Cecil and Blankenship, 2012). The wide zone of hailstorm occurrence in both the hemespheres is further modified by continents over which suitable instability is generated in moist warm airmasses containing not too large number of condensation nuclei. Chepovskaya (1966) has summed up the results of many investigations into the occurrence of hail in various regions of the globe. Over Kericho, a hill station in Kenya (east Africa) which is located within 50km from the equator has hailstorms frequency of about once in 3 days almost throughout the year (Sansom - 1971). At stations situated near large water bodies, the frequency of days with hail is approximately half (1/2) to two-third (2/3) to that at stations situated some distance from coast or within 500 km from the coast over oceans. The lesser hail activity near the large water basins and over seas and oceans is attributed to two factors. First one is comparatively meager convective activity due to lower temperature of underlying surface (Lemons 1942), Mason (1957), Pastukh and Sokhrina (1957). Second factor is the presence of considerable number of condensation nuclei (CCN) over the oceanic and coastal regions due to the bursting of sea water bubbles and their release into atmosphere (Sulakvelidze-1967). These condensation nuclie below the 0^0 C isotherm level cause shedding out of most of the water content as rain droplets below the freezing level by condensation and coalescence processes. Brook et al (2003) and Cecil and Blankenship (2012) produced climatology of large hails based on reanalysis of data and passive microwave images respectively. Their results are discussed in separate section in 1.2 for severe hailstorms.

Other than India hail fall is significantly observed over eastern and northwestern plains of USA, Mexico, Canada, western mountains of south American continent, southern Brazile and north Argentina southern Europe, north Italy,southern Germany, France, Spanish plateau, Caucasus region, regions east of Adriatic sea, northern planes of China and central Tibet(Zhang et al, 2008) central, western and south central Africa, Kenya, Negeria, Medagaskar, Armenia, Georgia, Azerbadzhan, Moldavia, Turkmenistan, Ukraine, North Krasnodar (Kabardino), Balkanian SSR, North Osselian SSR, Krasnodar and Stavnopol. Transcaucasia, North Caucasus and also south Kazakhstan, southeastern Australia are places with high frequency of hailfall. Large number of hailfall are reported from the northern region of Indian subcontinent, particularly northeastern India, Bangladesh, northwest India, north Pakistan and north Afghanistan. Annual frequency in this region is highest in the western Himalayan and Tibet region and next highest in the eastern Himalaya region and the mountainous tracts of North-West Frontier Province of Pakistan. It decreases rapidly towards adjoining plains. The coastal tracts of the peninsula are regions where hailstorms are comparatively less.

Surface-based climatology of hailstorms are, however, limited by inconsistencies in observational networks and reporting practices. These generally vary from country to country and also vary with population density within individual countries. Several previous studies have examined climatology from individual countries or groups of countries. Frisby and Sansom (1967), Williams (1973), and Barnes (2001) provide reviews of hailstorm climatology from many countries, but the disparities in reporting between those countries are obvious.

For practical reasons, hail frequency is often quantified in terms of hail days per year for a certain region instead of counting the number of all individual storms. In this definition, a hail day is considered a day when hail occurred anywhere within the reference area. In contrast to hailstorm counts, the specified number of hail days will not increase linearly with the extent of the reference area, as long as the latter is smaller or similar to the extent of hailstreak (refer sec. 1.6 for definition). This makes comparisons of hail day counts for regions of different sizes a difficult task. Due to the variable sizes and shapes of the reference areas considered in the literature, the count of hail days per year at a small reference area like a hailpad seems to be the most convenient measure.

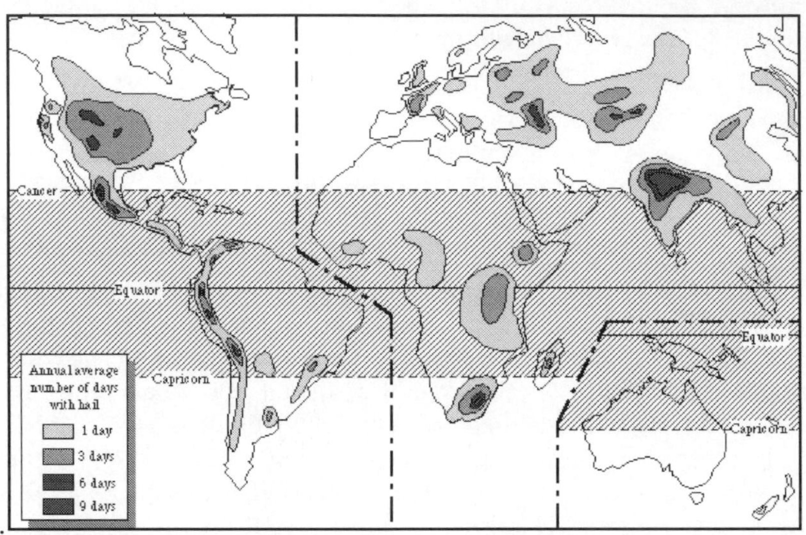

Fig. 1.1 Global map of annual hail days, from Williams (1973), based largely on Frisby and Sansom (1967) based on surface observations. (Colour plates Pg. No. 351)

Fig. 1.1, by Williams (1973) shows a composite global map based on surface hail data largely by Frisby and Sansom (1967). It highlights mountainous areas that are more prone to graupel or small hail and showers of the large hail. Data collected, was from surface weather stations or hail pads and could be dominated by cases with hailstones ~1 cm in diameter or smaller.

Due to inconsistancies in surface observational network Hand and Cappelluti (2011) prepared global hail climatology based on the 1 Jan 2004 to 31 Dec 2008 data by Convection Diagnostic Procedure (CDP) of U.K. Met Office. Refer Fig. 1.2.

Seasonal hail day densities were calculated by dividing the number of hail days by the total number of possible daily diagnoses. Although CDP has limitation of not being able to dignose large hailstones (≥ 19 mm) it generally exhibits good result on comparing with ground observations.It rightly indicates high hail days over southeast Africa and over Andies during summers and pre summers of southern hemisphere. Absence of hail over India during southwest monsoon months Fig 1.2(c) and during this period high hail days over mid latitudes over Asia and Europe and south west Canada are compatible with the ground observations..

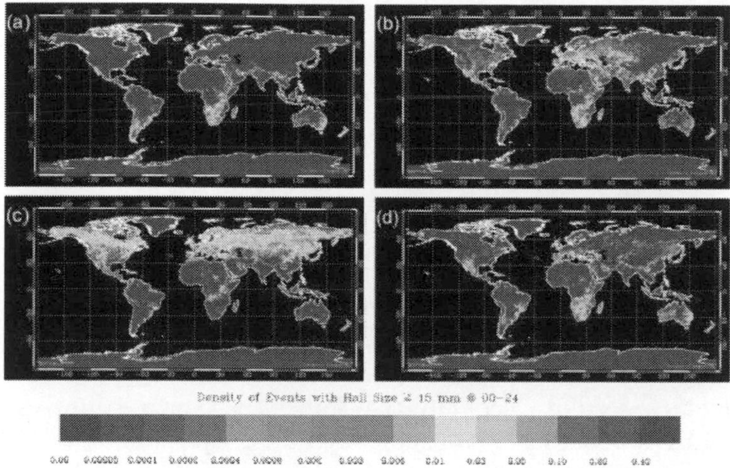

Fig. 1.2 CDP global distribution of seasonwise hail day density for hail size ≥ 15 mm (00 to 24UTC) in 1° × 1° squares. Fig. (a) December/January/February, (b) March/April/May, (c) June/July/August and (d) September/October/November. (Colour Plates Pg. No. 352)

1.2 Global Climatology of Severe Hailstorm

Brook et al (2003) used reanalysis dataset that cover the globe with special grid spacing of 200 km and temporal spacing of six hour from National Centre of Atmospheric Research (NCAR) and Unites States National Centre of Environment Protection (NCEP) for the period of three years (1997 to 1999). The thermodynamic criteria developed by them in eastern United States for large hail (≥ 5 cm in diameter), wind gust (at least 120 kmh^{-1}) or Tornado of at least F2 damage(refer Appendix-B) was applied to globe to estimate the frequency of favourable condition for severe thunderstorm. Albiet the study provided a glimps of global climatology of severe hailstorms but merely three year's data could not be considered enough for the purpose.

Satellite-borne passive microwave radiometers record brightness temperature (refer Appendix-B) depressions due to the scattering of upwelling radiation by large ice hydrometeors (graupel, hail). Cecil and Blankenship (2012) prepared global climatology (Fig.1.3) of storms producing large hail (1-in. diameter, or ~2.5 cm) on eight years data (2003 to 2010) from satellite measurements using Advanced Microwave

Scanning Radiometer for Earth Observing System (AMSR-E) instrument aboard the Aqua mission, by 36-GHz Polarized Corrected Brightness Temperature (PCT). Their method was compared with ground observations in United States and then it was extended to cover globe with assuming similar conditions.

Although this method provided a giant step towards unique, consistent comparison between regions that cannot be consistently compared using ground-based records because of varying data collection standards but regional variability in the thermodynamic profiles of regions out of U.S. may not be same with those in U.S. Hence a hailstone might be more likely to melt before reaching the surface in tropical Africa or India than in the United States. Also prediction of hailstorm over oceans by satellite borne observation is matter of severe scrutiny. The wavelength of 36 GHz is 8 mm. Half wavelength particle size (e.g. 4 mm) resonates best in this. Hence instead of a very large hailstones an ensemble of many 4 mm hails are likely to present even colder brightness temperatures (BT) which is likely to be mistaken as large hails over oceans. Hence satellite interpretation of hailstorm over ocean or over land where primarily oceanic air mass prevails in any particular season may give incorrect information with respect to occurrence of hailstorms.

Over India thunderstorms do occure but hailstorms are very few during Southwest monsoon season (July to September). This is mainly because oceanic air mass extends over the Indian subcontinent during the southwest monsoon months from July to September. As the southwest monsoon gradually advances from south east to northwest region of south Asia hail observations also incessantly shift westward. Hailstorm activity again increases over Indian region after the withdrawal of southwest monsoon i.e. post monsoon season over India.

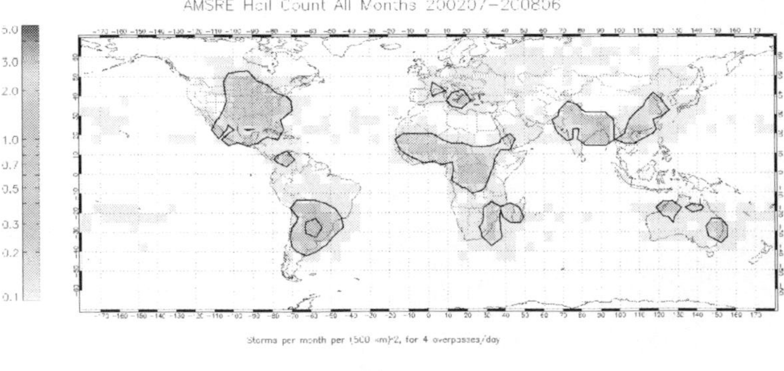

(a)

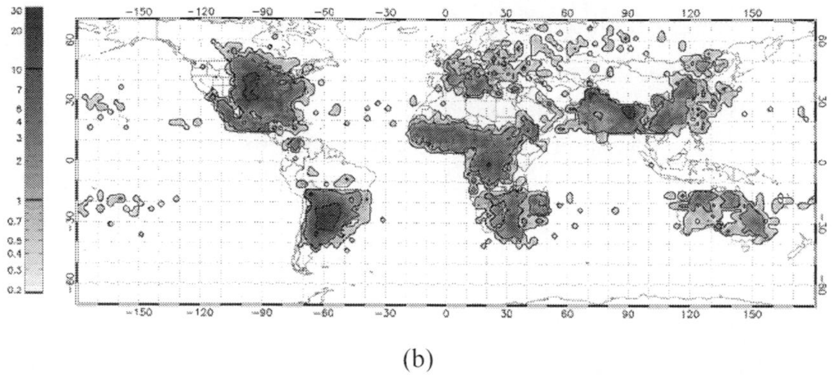

(b)

Fig. 1.3 Global climatology of storms producing large hail (1-in. diameter, or ~2.5 cm) on eight years data (2003 to 2010) from satellite measurements using Advanced Microwave Scanning Radiometer for Earth Observing System (AMSR-E) 36 GHz PCT; 500 km^2 per 4 satellite over passes in a day. Scale on left shows (a) storm per month (b) storm per year per.

In Fig. 1.3 severe (hail 1-in. diameter, or ~2.5 cm) hailstorms are indicated most often in a broad region of northern Argentina and southern Paraguay and a smaller region in Bangladesh and eastern India. Numerous hailstorms are also estimated in the central and south-eastern United States, northern Pakistan and north-western India, central and western Africa, and south-eastern Africa (and adjacent waters). Fewer hailstorms are estimated for other regions over land and scattered across subtropical oceans. Very few are estimated in the deep tropics other than in Africa.

The vast majority of potential hailstorms over the oceans are located within a few hundred kilometers of continents. A few are scattered across the open oceans, mostly between ±15° and 35° latitude in the Pacific and North Atlantic. Others are found over regions extending several hundred kilometers east from continents in the ±15°–40° latitude belts (offshore from the United States, Brazil, Uruguay, Argentina, South Africa, Mozambique, Australia, and China). Extension of Hailstorm off shore over the oceans is yet to be regionally validated by ground observation.

1.2.1 Seasonality in Severe Hailstorms

As per Cecil and Blankenship (2012) most continental regions show seasonality with hailstorms peaking in late spring or summer. The southwest monsoon alters the hailstorm climatology around the Indian subcontinent. About 75% of the hailstorms on the eastern side (around Bangladesh) occur from April through June, generally before monsoon onset.

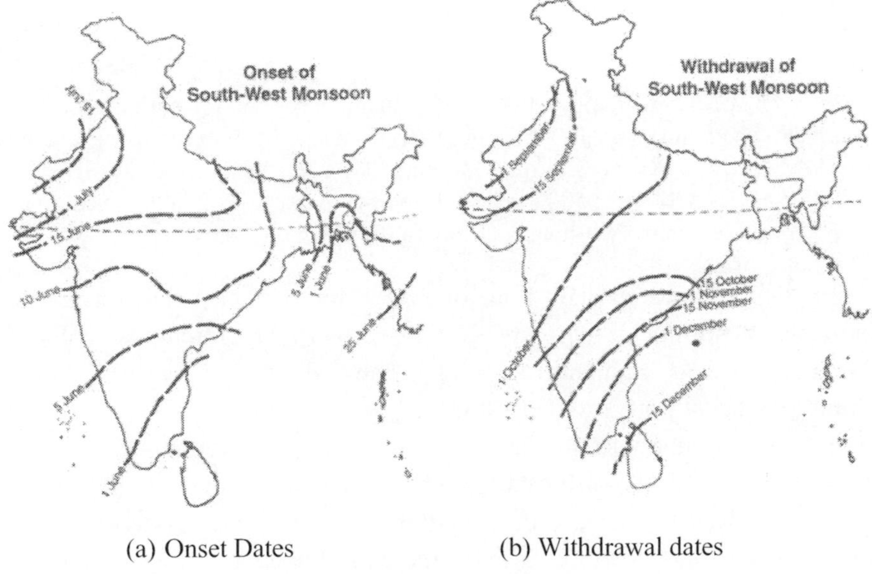

(a) Onset Dates

(b) Withdrawal dates

Fig. 1.4 Normal onset and withdrawal dates of southwest monsoon over Indian subcontinent.

Hailstorm activity shifts towards northwestward over northern India as the monsoon gradually advances over Indian region and engulfs entire continent by 15 August. Refer Fig. 1.4(a) for the normal dates of the onset over India. An arc along the foothills in northern Pakistan becomes particularly active from mid-June through mid-August. Details are discussed in sec. 1.4.

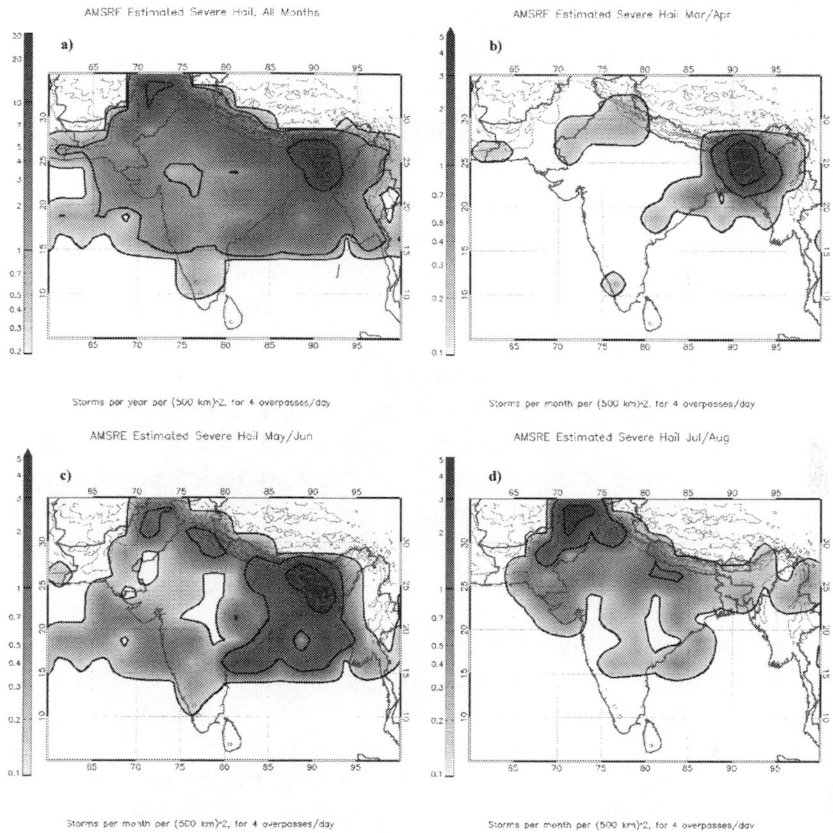

Fig. 1.5 AMSR-E hail climatology near India for (a) annual, (b) March–April, (c) May–June, and (d) July–August. Light contours are elevation, contoured at 1-km intervals.

The South Asian monsoon plays an important role in the distribution of hailstorms from Pakistan to east India. This region has two centers of peak activity (Fig. 1.5a), with the greatest concentrations in Bangladesh and northern Pakistan. During March and April, hailstorm locations are tightly concentrated over Bangladesh and east India (Fig. 1.5b). In May and June (Fig. 1.5c), this region expands to include the northern Bay of Bengal and more of eastern India, including the coasts of Orissa and Andhra Pradesh. Hailstorms begin to occur near the foothills of the Himalayas in northern India and northern Pakistan in late June. By July and August (Fig. 1.5d), northern Pakistan has a large number of hailstorms. This is in conformity with the expected view that hail forms over the region where oceanic salt rich aerosols containing airmass is diluted of its sea salt aerosols by the continental airmass. As during August entire Indian region is having oceanic airmass hail activity is sporadic only. Occasional hailstorms are spread across southeastern Pakistan and the northern half of India. Romatschke et al. (2010) showed a similar shift of deep convective cores from east in the premonsoon season to west during the monsoon.

In most individual years, hailstorm occurrence in and near Bangladesh, begins very near 1 April. The season peaks around 20 May, but it varies by a few weeks from year to year. Storm counts dramatically decrease after about 20 June, presumably with the onset of the monsoon (refer fig. 1.4 for onset dates over India). About 75% of the retrieved hailstorms in and near Bangladesh occur between 1 April and 20 June. Most years have a few hailstorms derived from AMSR-E data in September, but they are rare during July and August. Chowdhury and Banerjee (1983) show a sharp peak in April but with very few hailstorms after May. Frisby and Sansom (1967) list about 90% of the hail days in this region in 1951–60 as occurring during March–May, with a few more in June, and a few more in August–October. With the shift of storm locations from east to west, the region near the foothills in northern India has more storms during June and July. Some years have a concentrated period of 1-3 weeks with several storms. For the entire dataset, about one-third of hailstorms in the region occur between 20 June and 9 July. Almost half occur between 20 June and 29 July. Brightness temperature of ensamble of small hails may give spurious proxy for large solid hails hence is likely to introduce errors in the identification of hailstorm in place of thunderstorm.

In the global, annual hailstorm climatology (Fig. 1.3b), an area in northern Pakistan (30°–35°N, 70°–75°E) has a higher concentration of storms than any place other than Bangladesh, the central United States, and southeastern South America.

Storm locations show a preferencc for the foothills, in arcs of foothills from norththwest to north to norththeast. A large number of hailstorms are restricted to a short period, usually from late June to mid-August (plus or minus a couple weeks from year to year). Only 1 out of 79 storms in the AMSR-E database for this region occurred earlier than 9 June. About 75% occurred between 20 June and 18 August. Houze et al. (2007) and Romatschke et al. (2010) describe the meteorological conditions as favoring deep, intense convection in this region: moist southwesterly flow from the Arabian Sea warms while crossing the desert and builds instability under a cap of warm, dry continental air advected from the

Afghan plateau. It is lifted by the terrain, and the concave indentation of the mountain barrier may add low-level convergence.

Figure 1.6 shows the global distribution of satellite-inferred severe hailstorms on a 2.5° grid in bimonthly intervals.

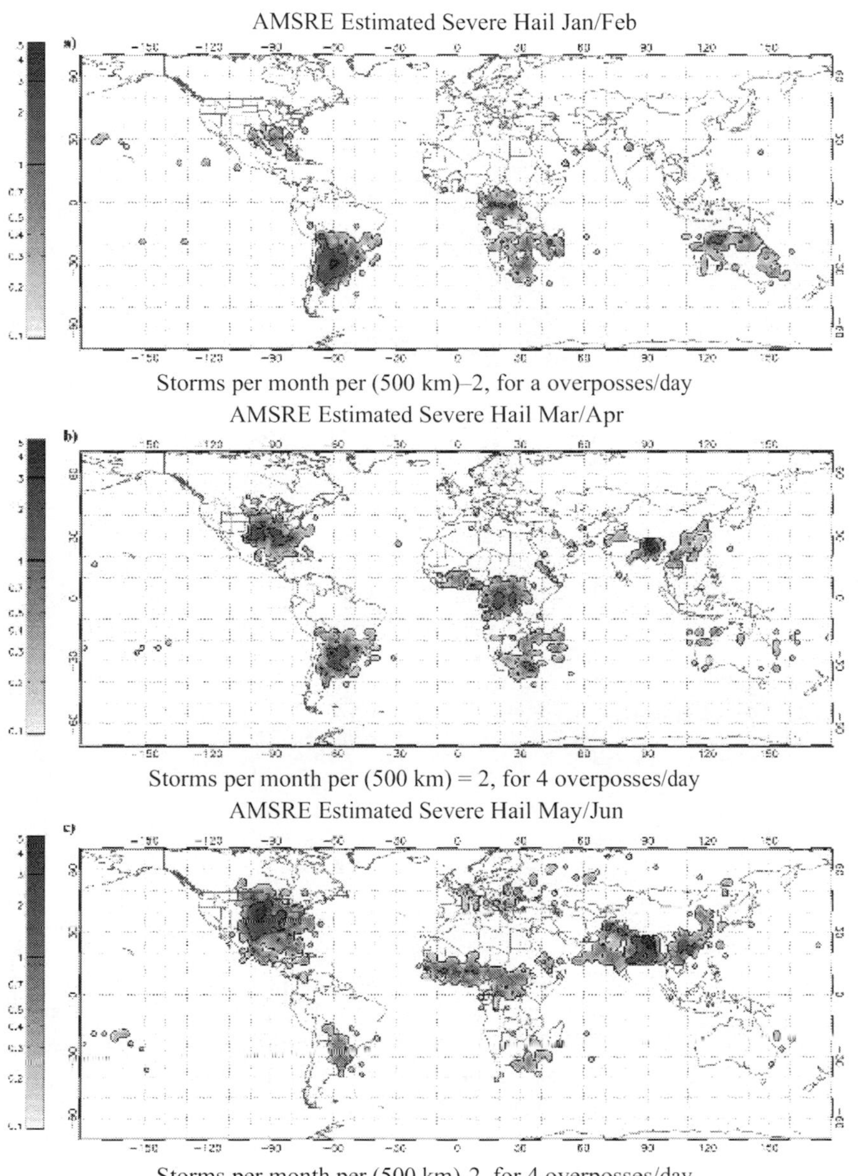

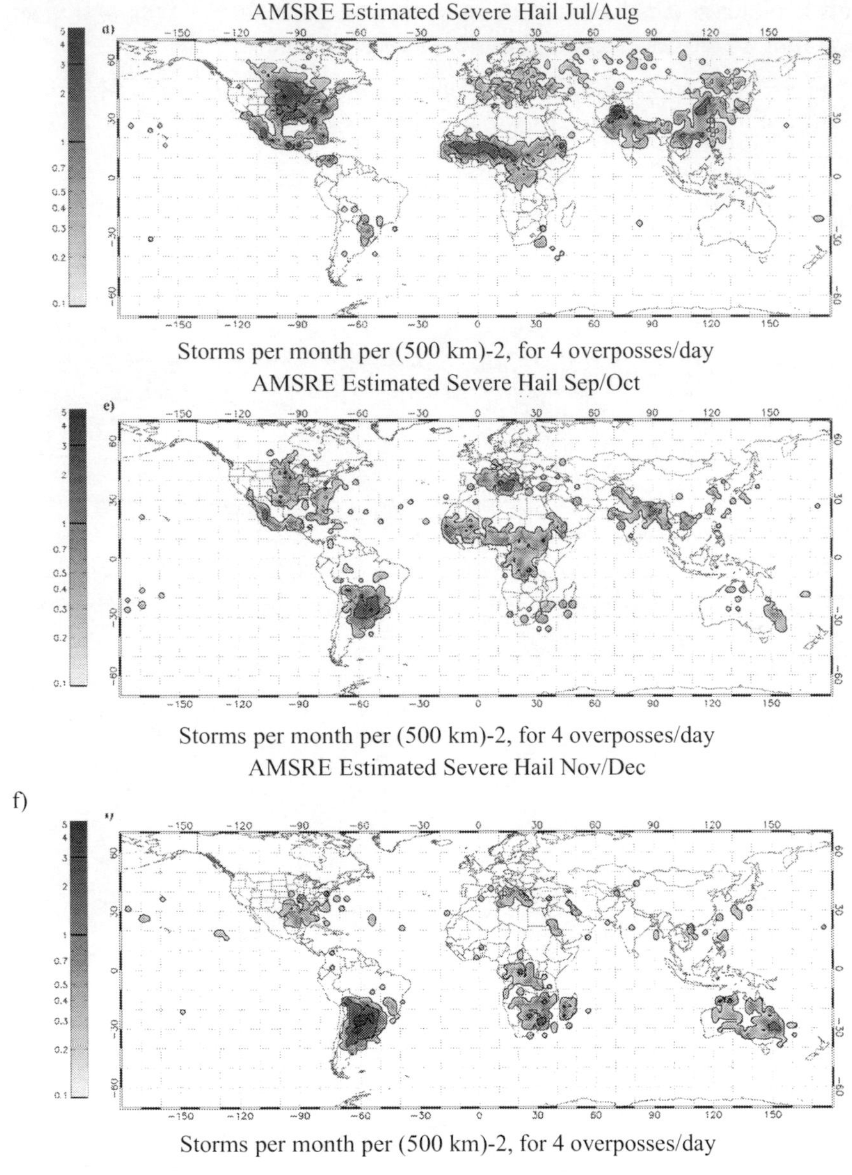

Fig. 1.6 Bimonthly AMSR-E hail climatology for (a) January–February, (b) March–April, (c) May–June, (d) July–August, (e) September–October, and (f) November–December.

The satellite-inferred hailstorms are most common in late spring and summer in particular locations. Tropical Africa has evidence of some hailstorms throughout the year, with the location of the maximum moving north and south following the noontime sun. This leads to seasonal peaks in Northern Hemisphere summer in West Africa and the Sahel Fig. 1.6d), and peaks near the equinoxes in equatorial Africa (Figs. 1.6 b, e). The seasonal peak for southeastern Africa is less pronounced (Figs. 1.6a, b, f), with storms over the offshore waters during much of the year. European activity generally peaks in summer (Fig. 1.6d), but storm counts are greater in autumn for the Mediterranean (Fig. 1.6e). This probably has to do with the Mediterranean Sea remaining warm while cooler air advances from Europe.

Over Australian region hailstorms inferred by AMSR-E are almost entirely during late spring and summer (Figs. 1.6 f, a). The few occurring in early spring tend to be near the east coast (roughly from Brisbane to Sydney), consistent with the result of Schuster et al. (2005), that the hail season extends from October to February in New South Wales and peaks in November–December. The AMSR-E-derived storms in summer cover a broader area, with a peak in the northwest. In East Asia. AMSR-E estimates that springtime hailstorms are concentrated from Thailand to Beijing, with peak activity between Hanoi and Hong Kong (Figs. 1.6b, c). Frisby and Sansom (1967) list several stations in Thailand with hail reports, primarily in March and April. They describe hail observations as rare in Vietnam and Hong Kong. In summer, the AMSR-E-derived storms extend through all of eastern and northeastern China, and into far southeastern Russia. Peak concentration is in the lower Yellow River basin (Fig. 1.6d). Very few storms meet hail criteria in this region after August. Xie et al. (2010) show a strong summer peak in large hail occurrence for northern parts of China, with a spring peak in southwestern China (Guizhou Province). Their data included hail sizes for only four regions in China, with only 6% of hail observations reaching the 2-cm diameter threshold they used to define “severe” hail. The percentage was about twice as large in Guizhou Province as in the northern regions, making it difficult to infer a climatology of large hail from the more general climatology of hail observations in China (Zhang et al. 2008).

North of 40°N across Europe and Asia, the hailstorms are almost entirely during the summer months (Figs. 1.6c,d), except for the aforementioned autumn storms near the Mediterranean. The summer peak in hailstorm frequency at high latitudes is expected and consistent with ground-based studies (e.g., Webb et al. 2001 for Britain; Tuovinen et al. 2009 for Finland). Not much seasonality is seen from the few storms scattered across open ocean regions, well removed from land. There may be a slight preference for autumn (South Pacific in Figs. 1.6b,c; North

Pacific in Fig. 1.6f) and winter (North Pacific in Fig. 1.6a). When vigorous extratropical troughs penetrate the subtropics and tropics in these seasons, they occasionally provide a favorable combination of dynamic support and thermodynamic instability over the relatively warm waters.

Southeastern South America has a broader region of frequent hailstorms, centered on northeastern Argentina and southern Paraguay. The peak in the concentration from AMSR-E is in the Chaco state of northern Argentina. Matsudo and Salio (2010) place the hail maximum farther southwest, based on reports at conventional surface weather stations in Argentina from 2000 to 2005. In early spring, the common storm locations according to AMSR-E center on southeastern Paraguay and extend into adjoining parts of southern Brazil and northeastern Argentina. Late spring and early summer have the most hailstorms, and the most active region extends from Bolivia and southern Brazil southward across Paraguay, Uruguay, and northern Argentina to about Santa Rosa in the state of La Pampa. No particular location has as high a concentration of storms as premonsoon Bangladesh, but the South American active region covers a much broader area. Further into summer, the center of activity shifts a bit southwestward and fewer storms are found in Brazil. Storm counts decrease in autumn and winter, with locations shifting toward the eastern side of the La Plata basin (Uruguay to eastern Paraguay) by midwinter (Fig. 1.6d).

In North America, winter (Fig. 1.6a) and early spring hailstorms are most common over the Gulf of Mexico and Gulf Coast states. By late spring, the hailstorm region extends into southern Canada and covers most of the United States east of the Rocky Mountains and south or west of New England. The highest concentrations in our AMSR-E sample are from Oklahoma to Iowa. A broad maximum centered on Iowa covers the midwestern United States by summer. Storm counts decrease in autumn (Fig. 1.6e) and are mostly limited to the southeastern United States for late autumn and winter (Figs. 1.6f,a).

1.2.2 Diurnal Variation Severe Hailstorms

Fig. 1.7 shows diurnal variation in severe hailstorm_for five geographical regions e.g United States, southeast South America, central Africa, Pakistan, and Bangladesh. All five regions have minima between sunrise and 1200 LST and have maxima between 1500 and 2100 LST. Regionwise details of diurnal variability in India is described in separate chapters devoted for the Indian regions in this book. The southeastern South American (broadly defined in this subsection as 18°–38°S, 72°–52°W) grouping actually has a broad peak between 1500 and 2400 LST, with a little more than half its storms during those 9 h. The other regions have a rapid decrease in storm counts after 9:00 p.m. A disproportionate share of the southeastern South American storms are horizontally extensive, mature MCSs, lasting well into the night.

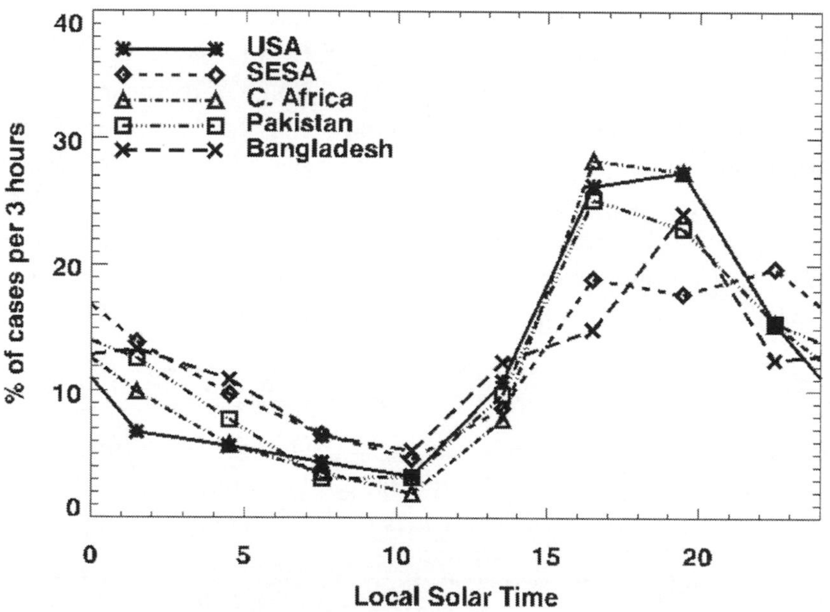

Fig. 1.7 Percentage of TMI-estimated hail cases in 3-h increments of LST for south central/southeastern United States, southeast South America, central Africa, Pakistan, and Bangladesh.

Ground-based hail reports from Chowdhury and Banerjee (1983) show a sharper afternoon peak for Bangladesh, between 1400 and 1800 LST. The diurnal cycle of Bangladesh's severe thunderstorms from Yamane et al. (2010) is more consistent with a peak occurrence between 2000 and 2100 and several still occurring after 2400 LST.

The diurnal curves for the United States (30°–38°N, 108°–80°W), central and West Africa (10°S–18°N, 20°W–40°E), and Pakistan (22°–38°N, 64°–76°E) are very similar to each other, with more than

50% of storms occurring between 1500 and 2100 LST and less than 10% between 0600 and 1200 LST. Pakistan has a smaller fraction during late afternoon than the United States and central and West Africa, and more between midnight and sunrise.

To put the AMSR-E observations into a diurnal context, (Cecil and Blankenship, 2012) also gridded and mapped the percentage of Tropical Rainfall Measuring Mission (TRMM) Microwave Imager (TMI) derived hailstorms occurring between 0000 and 0300 and between 1200 and 1500 LST (not shown). As per them analysis, along with Fig.1.7, suggests that AMSR-E likely overestimates the number of hailstorms in the Bangladesh, Pakistan, and southeastern South American regions compared to the United States and central and West Africa. AMSR-E's diurnal sampling may underestimate the number of severe hailstorms in the central United States by ~20%–50%. That being said, a global (tropics and subtropics) climatology derived from TMI instead of AMSR-E also suggests Bangladesh and southeastern South America have more hailstorms per year than the other regions (equatorward of 38°).

1.3 Regional Climatology of Hailstorms

For regional hail climatology a few regions on globe are selected which represent diverse climate and geography. Hail climatology of Indian regions are covered in detail in other chapers of this book.

1.3.1 Canada

Average number of hail days per year, based on the 1951-1980 climate normals (Environment Canada, 1987) are shown in Fig. 1.8. The highest hail frequencies occur in interior British Columbia and Alberta. In British Columbia, the largest hail frequencies occur just northwest and northeast of Williams Lake, most likely as a result of a combination of daytime heating of the interior valley combined with upslope flow along the mountains adjacent to the valley.

In Alberta, the largest frequencies occur just east of the Rockies. A series of maxima and minima appear well correlated with topography, with the minima along the river valleys and the maxima over higher terrain.

Fig. 1.8 Average number of hail days per year, based on the 1951 – 1980 climate normals (Environment Canada, 1987). The contours were hand drawn, based primarily upon about 350 weather stations. Some secondary climate station data was included in a subjective way, but not documented (Etkin, D and Brun, 2001).

1.3.2 United States

Average annual number of hail days based on data from 1901 to 1994 are shown in fig 1.9. Major high hail frequency areas are in the Rocky Mountains and the northwest Pacific Coast, and the lowest frequencies are in the nation's southwest and southeast regions.The key aspect of the nation's hail pattern is the considerable spatial variability across the nation and in most states. The national frequency varies from less than one hail day per year along most of the East Coast and parts of the desert southwest, to more than five hail days annually in the mountains of Colorado and Wyoming and along the Pacific North west coast.

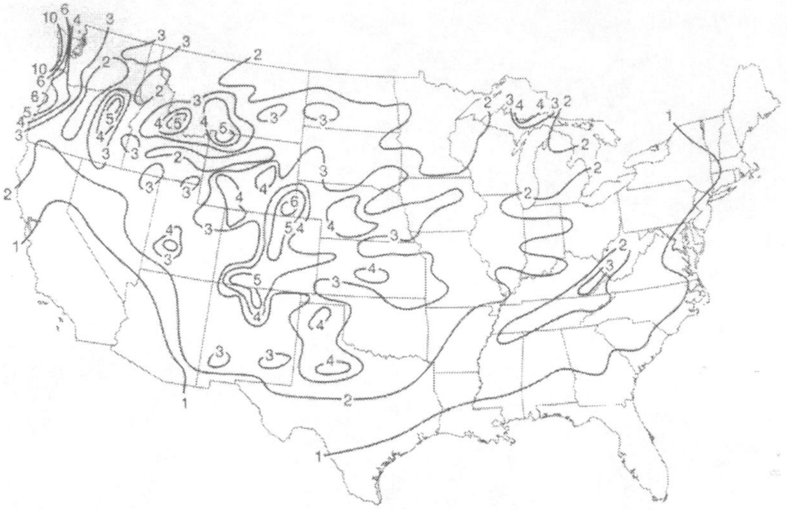

Fig.1.9 Average annual number of hail days based on data from 1901 to 1994 (from thunderstorms across the nation: An atlas of storms, hail and their damages in 20th century Copyright 2001 Stanley A, Changnon).

Several relatively high incidence areas are in and along the Rocky Mountains due to orographic effects. The Deep South and Gulf Coast have many thunderstorms but there is very little hail at the surface there because the descending hailstones melt in the high air temperatures below the cloud base and become rain drops. The highest frequencies are in areas extending from eastern Colorado into the Midwest, and in scattered locations in the northern Rockies and in the Pacific Northwest. In summer (June- August) the national pattern shows that the highest hail averages are in the Rockies, extending from Montana south into New Mexico. Orographically-induced hailstorms often develop along the front range of the Rocky Mountains.Due to the shape of the mountains, some areas are favored for storm development, and these result in paths of hailstorms that stretch eastward into the Dakotas, Nebraska, and Kansas. Summer hail is most frequent in June, decreasing in July, and becoming the least frequent in August. Hail in summer is very infrequent (0.1, or once in 10 years) in the Deep South and along the West Coast. The fall average hail-day pattern shows a few small moderate high hail days along the front range of the Rockies. The largest hail values exist along the Pacific Coast in Washington and Oregon, and in areas downwind of the Great Lakes. Pacific storms in the fall and winter frequently create small hailstones, leading to a 10 hail day annual average along the Washington coast.

1.3.3 South America

Hailstorms are generally observed over Southeastern South America mostly centered over northeastern Argentina and southern Paraguay. Argentina is, relatively highest risk zone of hailfall in south America.Therefore, a detailed climatology of hailstorm over Argentina is being discussed hereunder.

Based on data during the 1960–2008 regions lying between 30° and 40°S as well as those dominated by mountains present the highest hail frequencies in Argentina. Refer Fig. 1.10. The eastern and coastal areas of the country experience hail events mainly during springtime but they may start in late winter and continue through the beginning of summer. Events in western and central Argentina also predominate in spring but the maximum frequencies are observed during summer months.

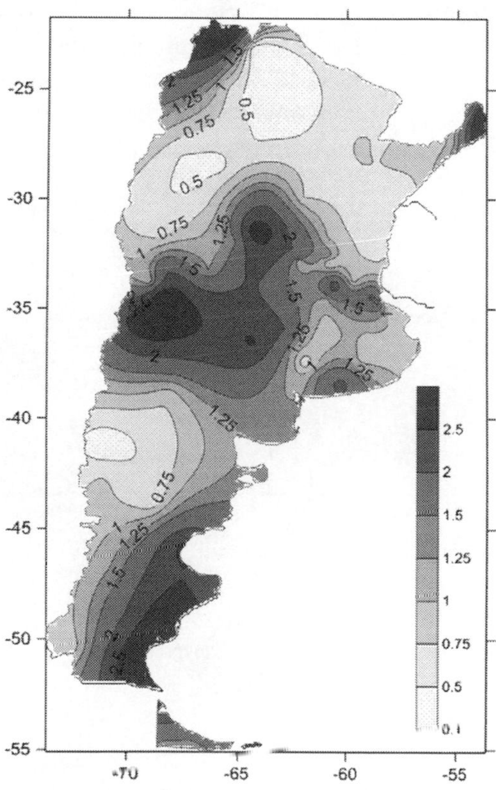

Fig.1.10 Mean annual hail events.

Trends in the annual number of hail events calculated for each region indicate that events in northwestern and northeastern Argentina have been increasing as well as in southern Patagonia. On the other hand, in central Argentina, southern Buenos Aires–La Pampa, northern Buenos Aires–Litoral and northern Patagonia trends are negative and statistically significant in the first two regions, basically by the decrease of events during spring and summer.

1.3.4 Africa

Mean annual hail frequency over Africa is shown in Fig. 1.11. In general hailstorms are observed from 10^0 N to farsouth over the African continent but high frequency (> 5 per year) are observed over central Ethopia, Kenya, Rawandi, Burundi, Tanzania in west central Africa and and over eastern south Africa and Lesotho region in the mainland. Central Medagasear Island also records (> 5 per year) hail frequency. All these high frequency zone fall over high topographic region of eastern and south eastern Africa. Over Kericho, a hill station in Kenya (east Africa) which is located within 50km from the equator has hailstorms frequency of about once in 3 days almost throughout the year (Sansom 1971).

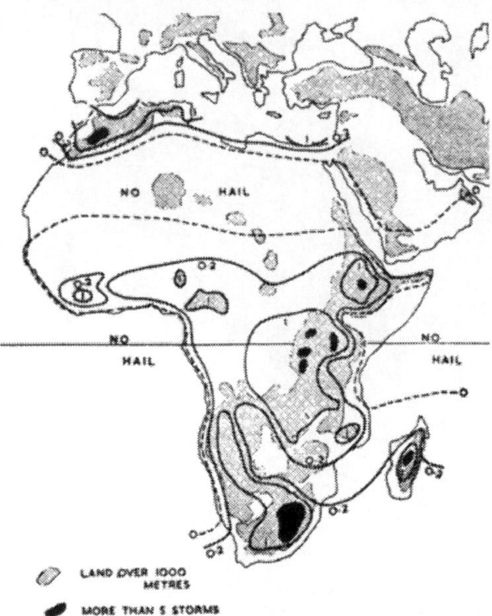

Fig. 1.11 Mean annual hail frequency over africa (sansom 1966).

(Narayan R. Gokhale. Hailstorms and Hailstone Growth, SUNY Press, 01-Jan-1975 - Hail - 465 pages)

More recently Le Roux and Oliver (1996) compiled hailday frequency based on 113 recording stations in South Africa, Lesotho and Swaziland as shown in Fig. 1.12.

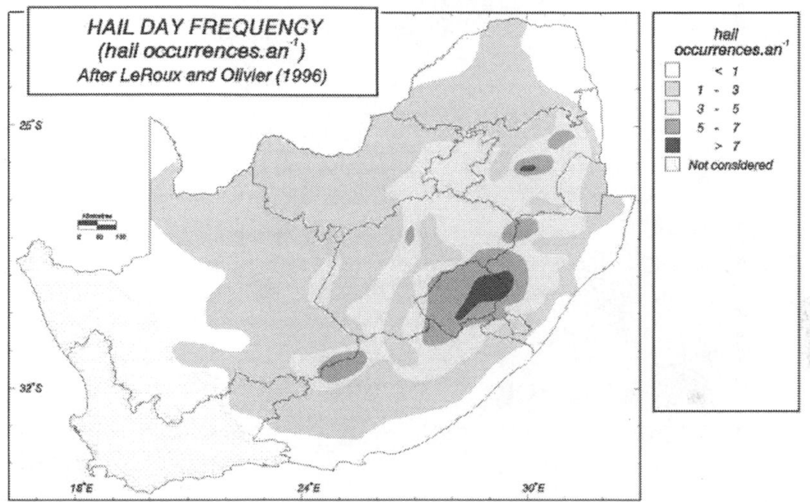

Fig. 1.12 Hail days per annum based on South African Weather Weather Bureau, 1986 from 113 recording stations in South Africa, Lesotho and Swaziland.(Le Roux and Oliver, 1996)

It indicates that at higher latitudes (> 32°S), HDF (Hail Days Frequency) increases with altitude. Between 26° and 32°S, HDF appears to increase rapidly up to altitudes of 600 m and then again between 1000-1500 m. In the subtropical regions at latitudes below 26 °S, the increase in HDF is again more or less linear with altitude, but it decreases above 1700 m.

1.3.5 Europe

In the central Eurpe (Germany, Switzerland, and Austria, it can be assumed that hail frequency decreases from west to east and from south to north. Western Europe is mainly influenced by the proximity to the Atlantic Ocean, which leads to higher static stability due to the damped diurnal and seasonal temperature amplitude. Thus, hail frequency varies substantially in Western Europe (Punge et al., 2014). Refer Fig. 1.13. France is frequently affected by severe hailstorms, hail occurs rarely in Benelux (Belgium, the Netherlands, and Luxembourg) and the British Isles.

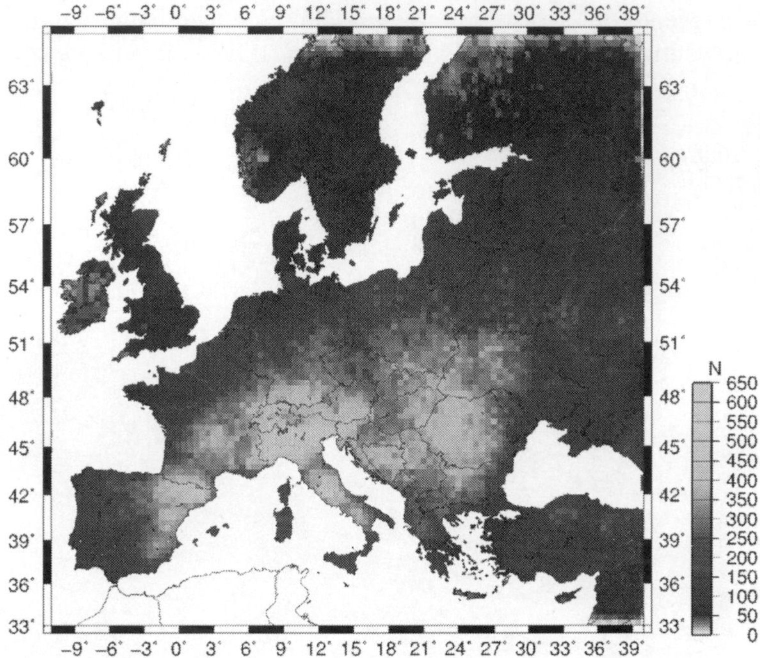

Fig. 1.13 Hail event frequency as estimated by Punge et al. (2014). (Colour plates Pg. No. 351)

In contrast, these latter regions are characterized by a high density of low pressure systems and associated fronts, and hence convective storms generating graupel or small hail occur quite frequently and throughout the year. The climate of Southern Europe including Italy and the Iberian Peninsular is dominated by the high insolation and proximity to the Mediterranean, where warm and moist air masses are advected from S to W directions. A few regions, such as northern Italy, feature some of the highest hail frequency in Europe (e.g., Punge et al., 2014). In Southeastern Europe, including the countries of Romania, Moldova, Bulgaria, Greece, Turkey, Cyprus, and those of the Balkan Peninsular, the general climate is strongly influenced by the Mediterranean and – for the eastern part – by the Black Sea. Hail occurs frequently at several hot spots, for example in the NWof Romania. However, only limited scientific literature is available, mainly based on hail observations at meteorological stations, which may also include ice pellets or graupel. In Northern Europe, hail is less common compared to most other parts of Europe, mainly due to the prevailing colder climate. In addition, the proximity to the seas, especially to the cool North Atlantic, inhibits strong convective activity. This applies especially for large hail. By contrast, ice pellets or graupel occur quite frequently and throughout the whole year, dominating in several cases the statistics, similar to the situation of the British Isles. Hail is a major peril in Russia, affecting around 5000 km^2 of agricultural area each year (Abshaev and Malkarova, 2006). The Northern Caucasus is sometimes considered the region with the highest hail hazard in Eastern Europe (e.g., Abshaev et al., 2003). Malkarova (2011) found natural crop loss ratios of 2.5% in Crimea and 4.4% in the region of Odessa (Ukraine), but more than 6% for the provinces in the N Caucasus. According to Abshaev et al. (2009), the loss ratio in agriculture exceeds 8% over several parts of the Caucasus foothills, and 2% everywhere over the foothills. These ratios are in the range of those found for other highly exposed countries such as N Italy or NE Spain. A map of hail frequency in Russia based on station data from1958 to 2008 (Abshaev et al., 2009) shows more than two hail days per year and station over the N Caucasus but also in Western Russia and the Voronezh Oblast, and several regions in the Asian part. The pattern matches relatively well with those estimated by Williams (1973). Most of the European part of Russia south of 62° N experiences around 1-2 hail days per year. Out of 40–50 hail days per year in the Southern Federal District of Russia (418,500 km^2), 5–6 lead to widespread 'emergency' damage (Abshaev and Malkarova, 2002; Abshaev et al., 2006). Point hail day frequencies in S Russia and the N Caucasus reach from 0.5 for the dry steppe, to 0.5–1 for the plain, 1 at the Black Sea coast, and finally 1–2 in the Central N foothills of the Caucasus (Abshaev et al., 2012). The 8 stations in the NW Caucasus flat-to-hilly district of Krasnodar had hail frequencies of about 1.1 on average (Malkarova, 2011).

1.3.6 China

Based on 1961 to 2005 data (Fig. 1.14) hail occurs most frequently in the high mountainous areas and northern plains. As a result, hail frequency is generally higher in northern China than in southern China. The hail frequency is highest over the central Tibetan Plateau. Hail seasons start in late spring and end in early autumn in northern and western China; they start mainly in spring in southern and southwestern China. On the diurnal time scale, hail events occur mainly between 1500 and 2000 local time in most of China except in Guizhou and Hubei Provinces (central western China), where hail events often occur during nighttime.

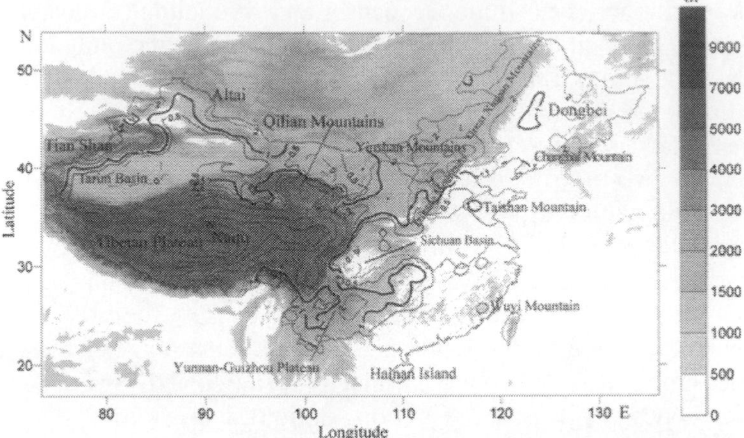

Fig. 1.14 The geographical distribution of mean annual hail frequency in China during 1961–2005. The contour interval is 0 (dashed line), 0.5, 1 (thick line), 2, 3. . . . Terrain is shaded, with scale bar on the right (m).

1.3.7 Australia

Total number of hail days based on grid at 0.5° for period 1998 to 2012 are shown in Fig. 1.15.

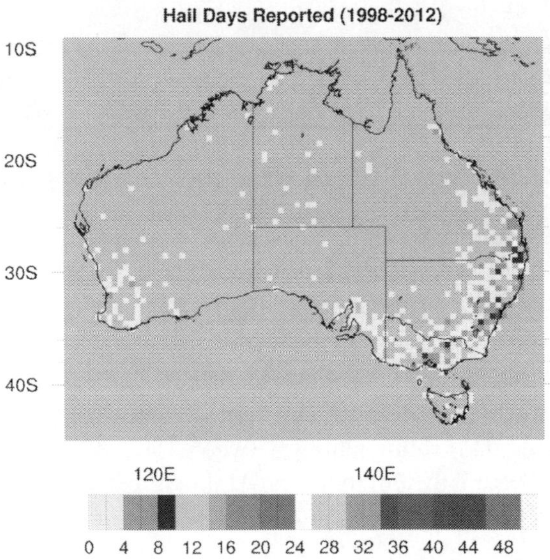

Fig. 1.15 Total number of reported hail days per grid cell (1998-2012). Australia Bureau of Meteorology: Severe Storms Archive. [http://www.bom.gov.au/Australia/stormarchive/] (Colour Plates Pg. No. 353)

The vast majority of reports occur along the southeast coast, where most of the population resides e.g south eastern South Australia, Victoria, South eastern Newsouth wales and adjoining southeastern Queens land. Major cities can be identified by local maxima in hail days. The grid cells containing Sydney and Brisbane both report about 3 days per year on average. South western region of Western Australia and Northern Territory also experiences sporadic hailstorms.

1.4 Hailstorms over India

In the Indian Meteorological Memoirs, Vol VI, Part 6, Sir John Eliot discussed the hailstorms which occurred in India during the period 1883-97. The data used in the above discussion consisted of reports of hailstorms communicated to Meteorological office by Collectors of districts, particularly during the period 1893-97, when special efforts were made by Sir John Eliot to collect a large amount of information. Most of these reports were fairly detailed as they were intended primarily for purpose of revenue assessment. They contained information of great value and interest regarding the intensity, extent and damaging effect on crops of the several hailstorms. The discussion of these data is the first attempt to study the occurrence of hailstorms over India. Based on Eliot's data Ramdas et al (1938) compiled the annual and monthly hailstorm frequency. Using the data based on 10 years report from 1957-1966, Philips and Daniel (1976) again presented monthly and annual charts showing the average number of hailstorms over India. As per their figures which exclude Pakistan, Bangladesh, Srilanka and Burma, the annual frequency of hailstorms is highest in the Assam valley, followed by Uttaranchal, Jharkhand and Vidarbha in the eastern parts of Maharashtra.

In the extreme North-East of the Assam Valley (Fig. 1.16(a)) Dibrugarh has an average of about 15 occasions. Other regions where high incidences of hailstorms are reported are the states of Tripura, the Uttaranchal and Jharkhand with about 6 occasions each. Since for each station considered the number of years of data were not same, Philips and Daniel (1976) converted their results into frequency of days with hailstorms during 100 years. Taking the year as a whole we find that the annual frequency is up to ten per year over the western Himalayan region, and up to five per year over the Eastern Himalayan region and the mountains tracts of the North-West Frontier Province; this decreases rapidly to once in two years over the adjoining plains. Over Gangetic West Bengal a hailstorm is usual once a year. The area consisting of the

Madhya Pradesh and the adjoining districts of Jharkhand, Bihar and the Utter Pradesh forms another centre of frequency of the order of once a year.

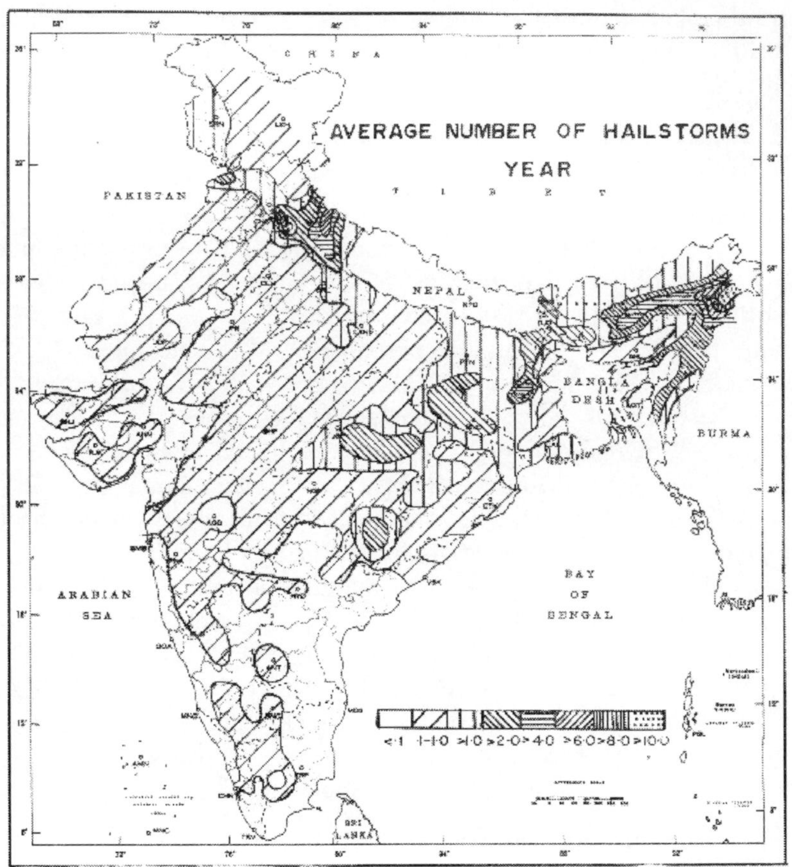

Fig. 1.16(a) Philips and Daniel (1976).

Philips and Daniel (1976) had made quite a detailed presentation of climatology of hailstorm. Due to lucidity of their presentation many recent authors (De et al, 2005; Ramesh, 2009) still refer to their climatological report. More recent statistics with respect to hailstorms based on 30 years data collected during 1961-1990 by the National Data Centre of India Meteorological Department was analysed by the author. Annual frequency of hailstorm based on these data is presented in figure 1.16(b). Frequency lines have been drawn at the interval of 0.1. But over northwestern and northeastern regions, due to strong gradient, isolines at the intervals of 0.2 or 0.4 or higher have also been drawn to avoid congestion.

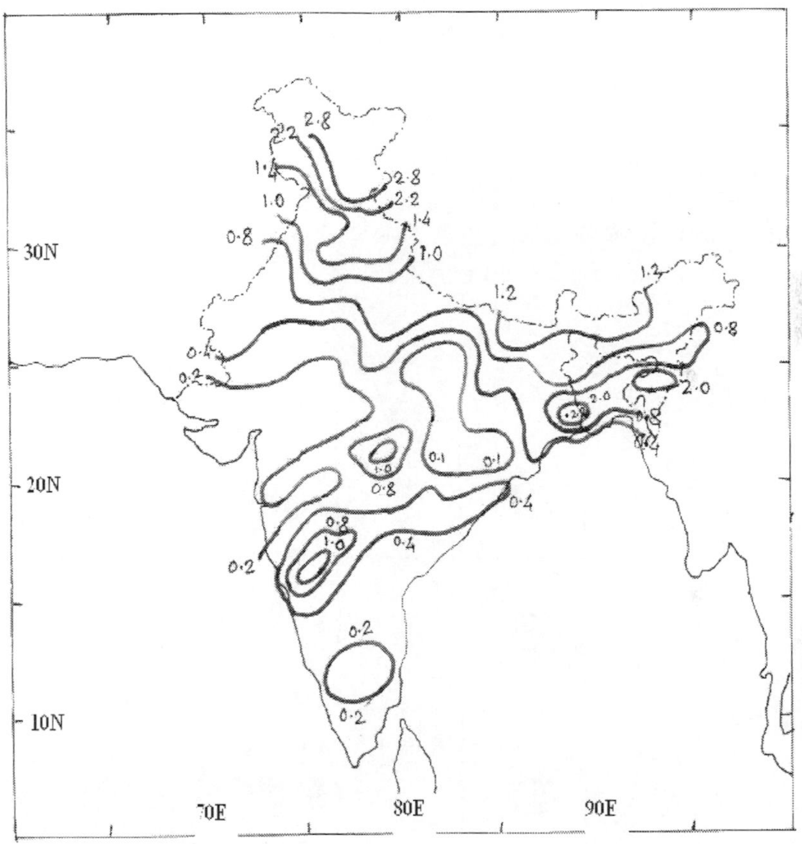

Fig. 1.16(b) Annual frequency of occurrence of hailstorm based on the data from 1961-1990 by the National Data Centre of India Meteorological Department.

Comparison of Figs. 1.16(a) and 1.16(b) shows more or less the same pattern except for the North-East India. In figure 1.16(a) High Frequency Hailstorm Zone (HFHZ) in North-East is broadly confined to the north eastern Assam and adjoining Arunanchal Pradesh region whereas in Fig. 1.16(b) the region of high frequency is shown over south Assam and adjoining Mezoram region. A similar southward shift of HFHZ could be noticed over West Bengal where in Fig. 1.16(a) it is located on the foothills of Himalaya and in Fig. 1.16(b) it is shifted to the Gangetic West Bengal.

1.5 Seasonal Variation of Hailstorm in India

Over India, Philip and Daniel (1976) followed by Mishra and Prasad (1980) reported that north of 30°N during the month of December, January and February on 25% to 40% occasions thunderstorms may be associated with hailstorms. Frequency of hailstorm days in a month per thunderstorms days in a month (i.e. N = number of days of hailstorms/ Number of days of thunderstorms) are shown in Figs.1.17(a), (b), below.

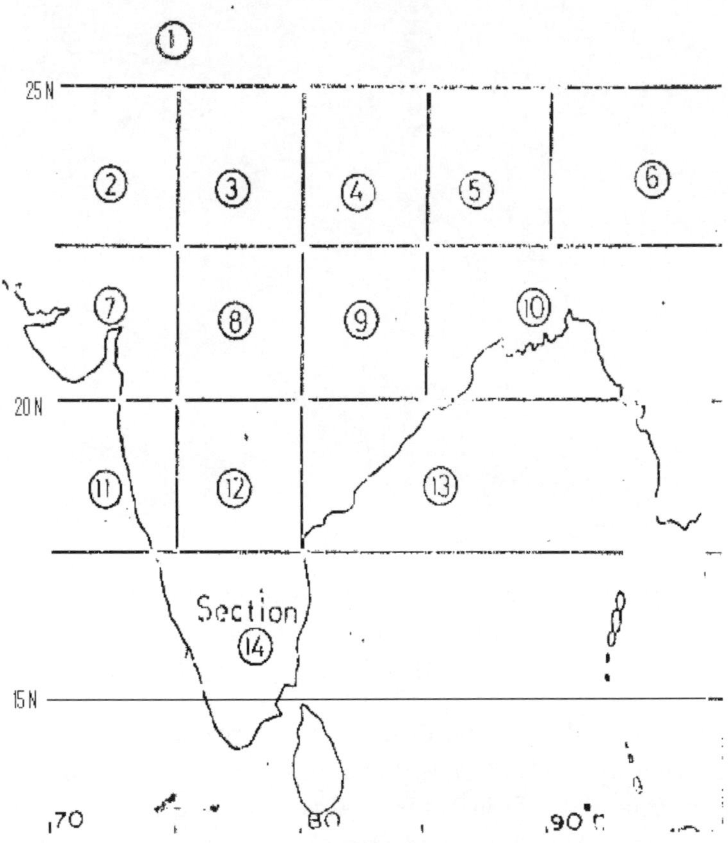

Fig. 1.17(a) Regions are marked by numbers.

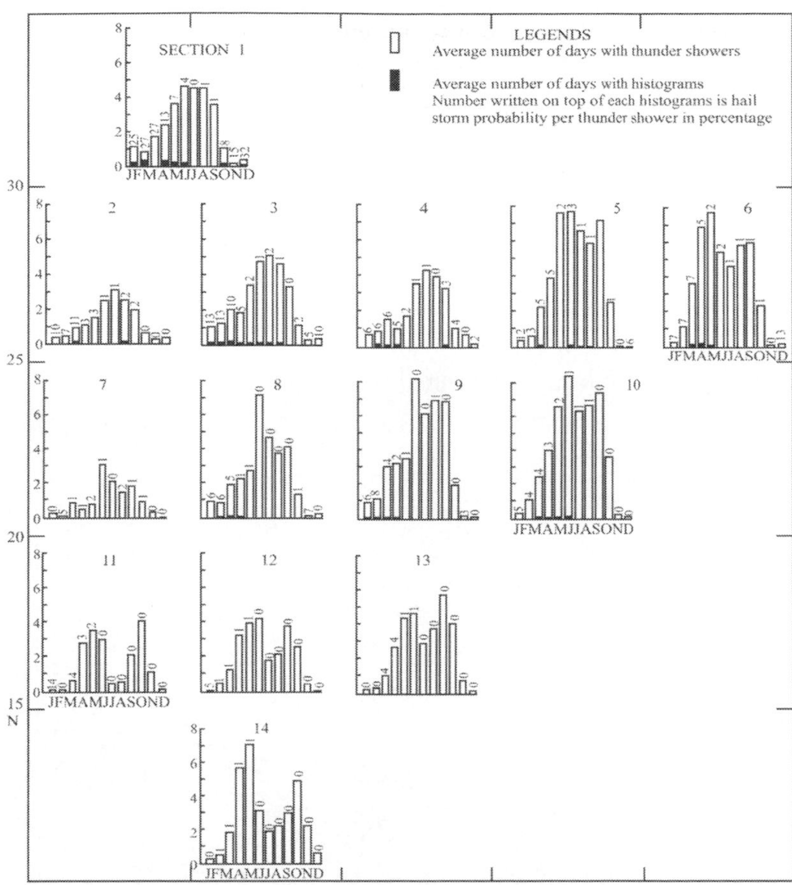

Fig.1.17(b) Histograms for each region by Mishra and Prasad (1980) showing frequency of hailstorm days in a month per thunderstorms days in a month (i.e., N= number of days of hailstorms/ Number of days of thunderstorms).

Remarkable coincidence is indicated by the seasonal and regional frequency of hailstorms with the presence of Jet stream above the thunderstorms. It may be inferred that Jet stream works as a ventilator for the latent heat of fusion to get out of the thunderstorm so as to prevent melting back of the ice-crystals already formed.

In this chapter previous monthly climatology compiled by Ramdas et al (1938) based on data collected during 1883-1897(15 years), for India, Pakistan, Bangladesh and Burma and by Philips and Daniel (1976), based on data collected during 1957-66 (10 years) have been presented on the same maps and the climatology based on the latest data collected during 1961-90 (30 years) has been put on separate maps for comparision.

Figs. 1.18(a) to 1.19(a) are maps showing the frequency of days with hailstorms based on Ramdas et al (1938), as the **isolines of frequency of days of hailstorms in 100 years** for each month of the year and also data compiled by Philips and Daniel (1976), as the hatched region for **average number of hailstorms each month** of the year. Details regarding the frequency intervals, at which Ramdas et al (1938) isolines have been drawn, are indicated in the maps. Month-wise climatology based on the recent data during 1961-1990 by the National Data Centre of India Meteorological Department and analysed by the author are presented in figures 1.18(a) to 1.19(b).

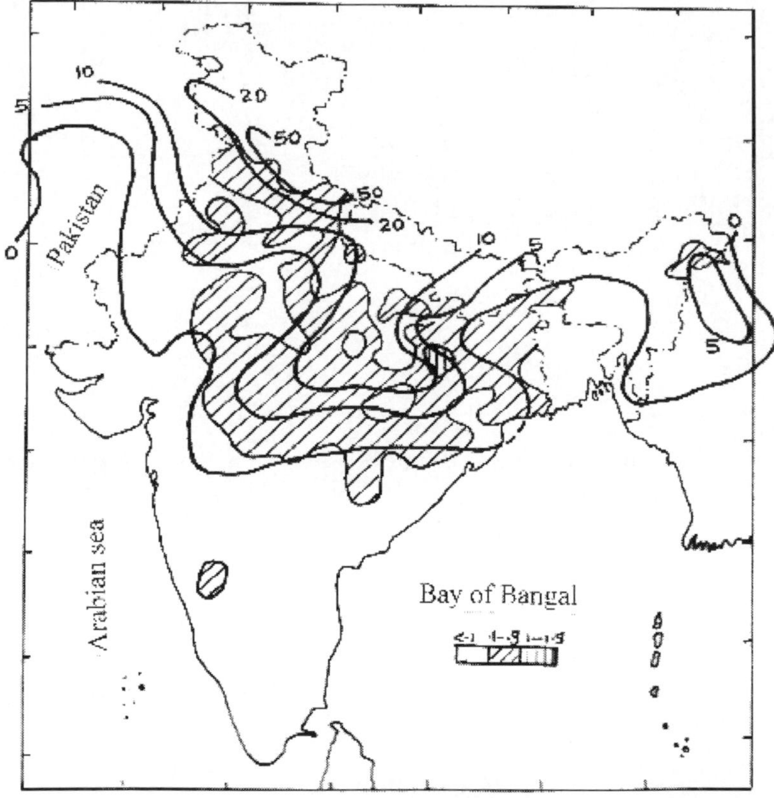

Fig. 1.18(a) Average number of hailstorms in January. Hatched scheme by Philips and Daniel (1976) and isolines of frequency of days of hailstorms in 100 years by Ramdas et al (1938).

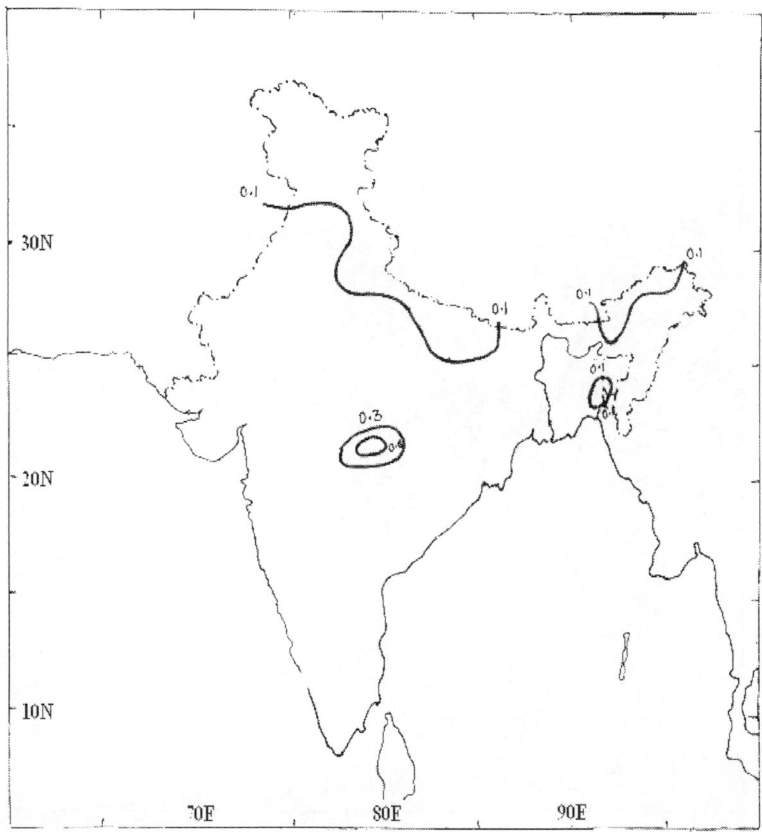

Fig. 1.18(b) Frequency of occurrence of hailstorm in January based on the data from 1961-1990 by the National Data Centre of India Meteorological Department.

Comparison of Figs. 1.18(a) and 1.18(b) also reveal gradual decrease in the hailstorm frequency in general. Also hailstorm region have shifted closure to the foothills and northern Gangetic plains. Central Madhya Pradesh emerges as the HFHZ during January month.

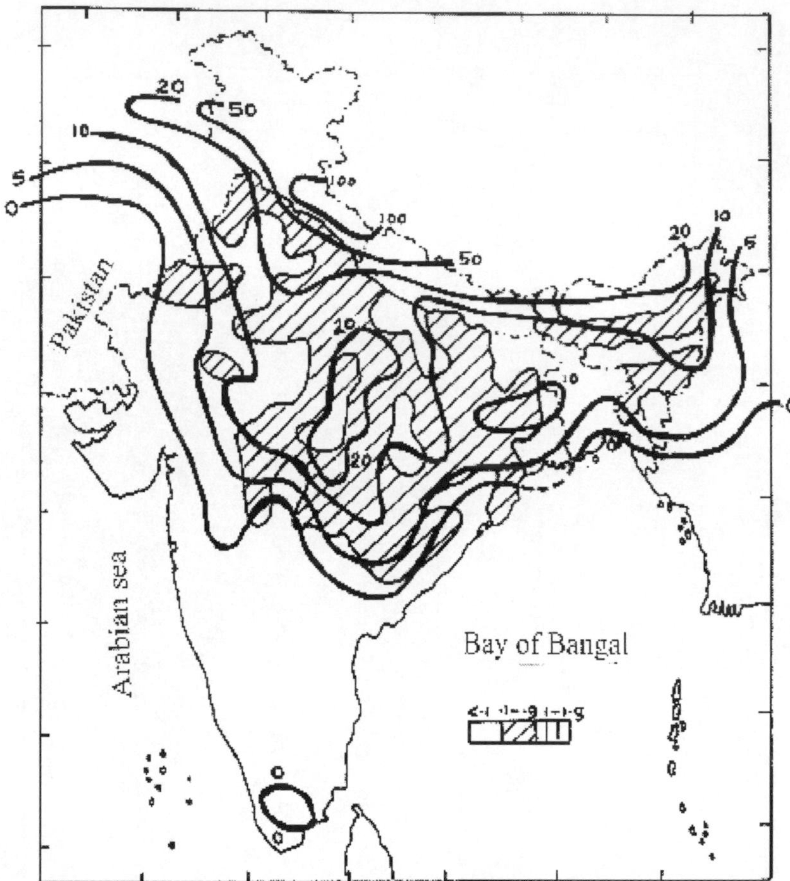

Fig. 1.19(a) Average number of hailstorms in February. Hatched scheme by Philips and Daniel (1976) and isolines of frequency of days of hailstorms in 100 years by Ramdas et al (1938).

Comparison of Figs. 1.19(a) and 1.19(b) also reveal that relatively high frequency of hail over foothills of Himanchal Pradesh, Punjab and Haryana region persists. But region of hailstorms over the central India has drastically shrunk. South-West shift of HFHZ over northeast India from north Assam and Arunanchal Pradesh to south Assam and adjoining Tripura region may also be noticed.

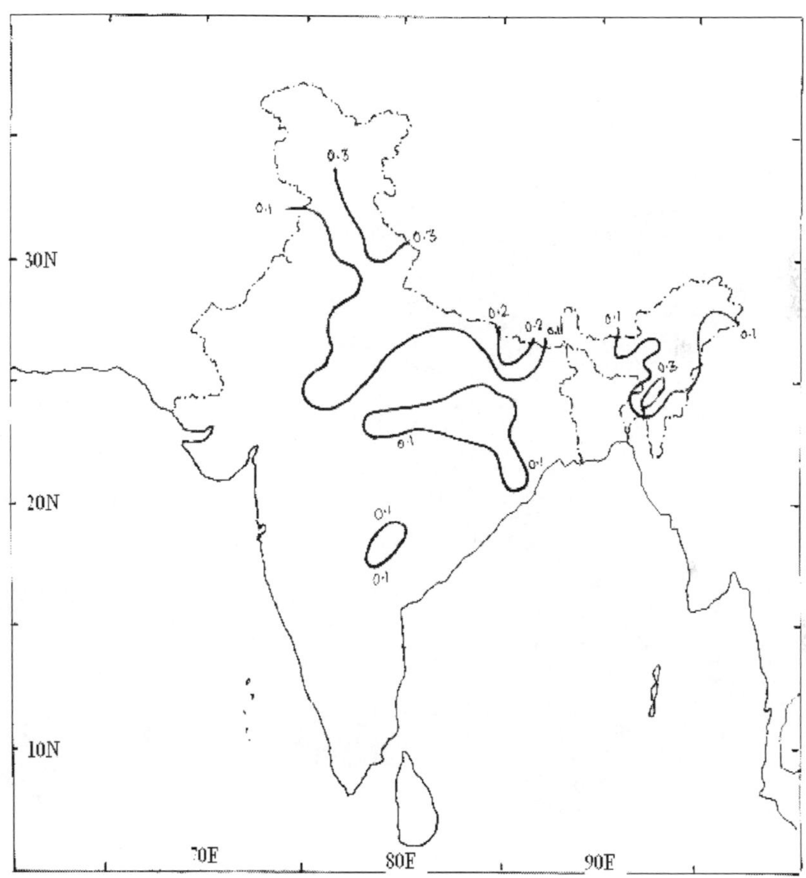

Fig. 1.19(b) Frequency of occurrence of hailstorm in February based on the data from 1961-1990 by the National Data Centre of India Meteorological Department.

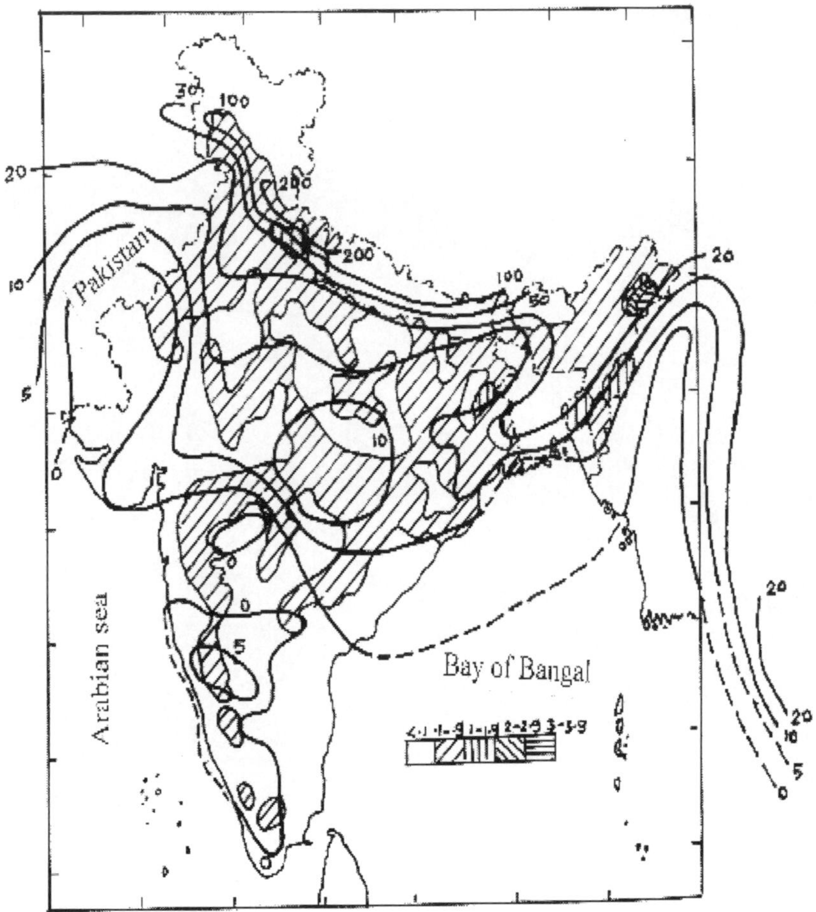

Fig. 1.20(a) Average number of hailstorms in March. Hatched scheme by Philips and Daniel (1976) and isolines of frequency of days of hailstorms in 100 years by Ramdas et al (1938).

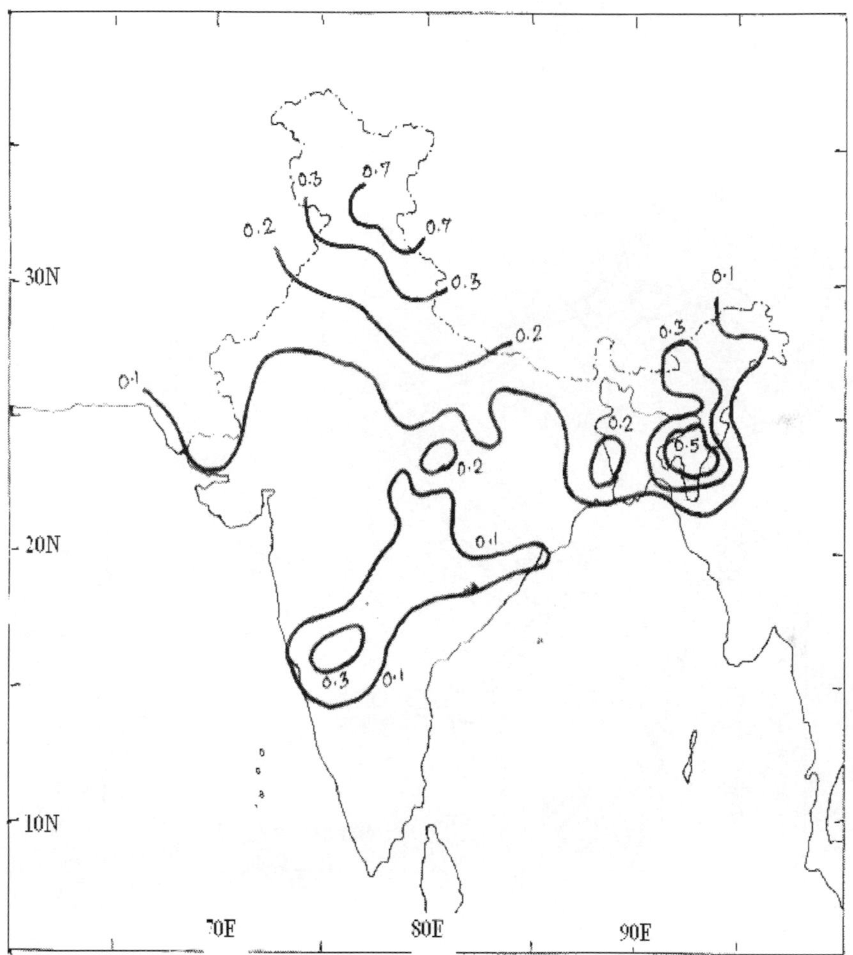

Fig. 1.20(b) Frequency of occurrence of hailstorm in March based on the data from 1961-1990 by the National Data Centre of India Meteorological Department.

Comparison of figures 1.20(a) and 1.20(b) show marked decrease in the hailstorms over the central Himalayas. High frequency of hailstorms over northeast India can be marked over the southern Assam, Mizoram and adjoining Burmese region.

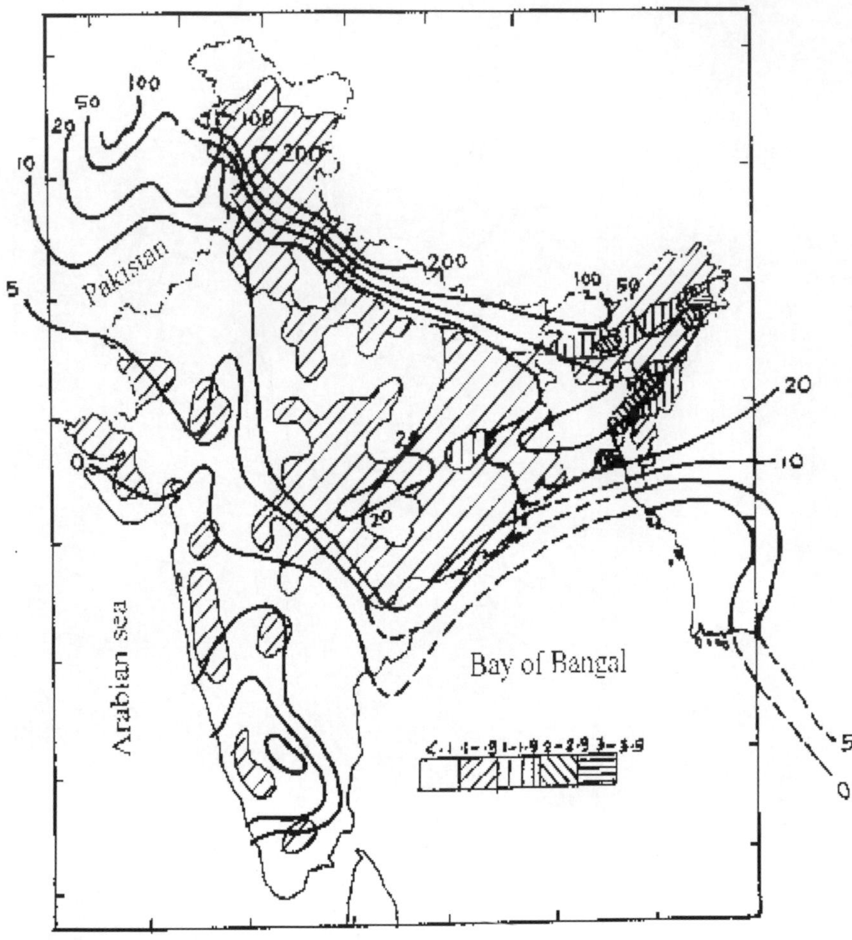

Fig. 1.21(a) Average number of hailstorms in April. Hatched scheme by Philips and Daniel (1976) and isolines of frequency of days of hailstorms in 100 years by Ramdas et al (1938).

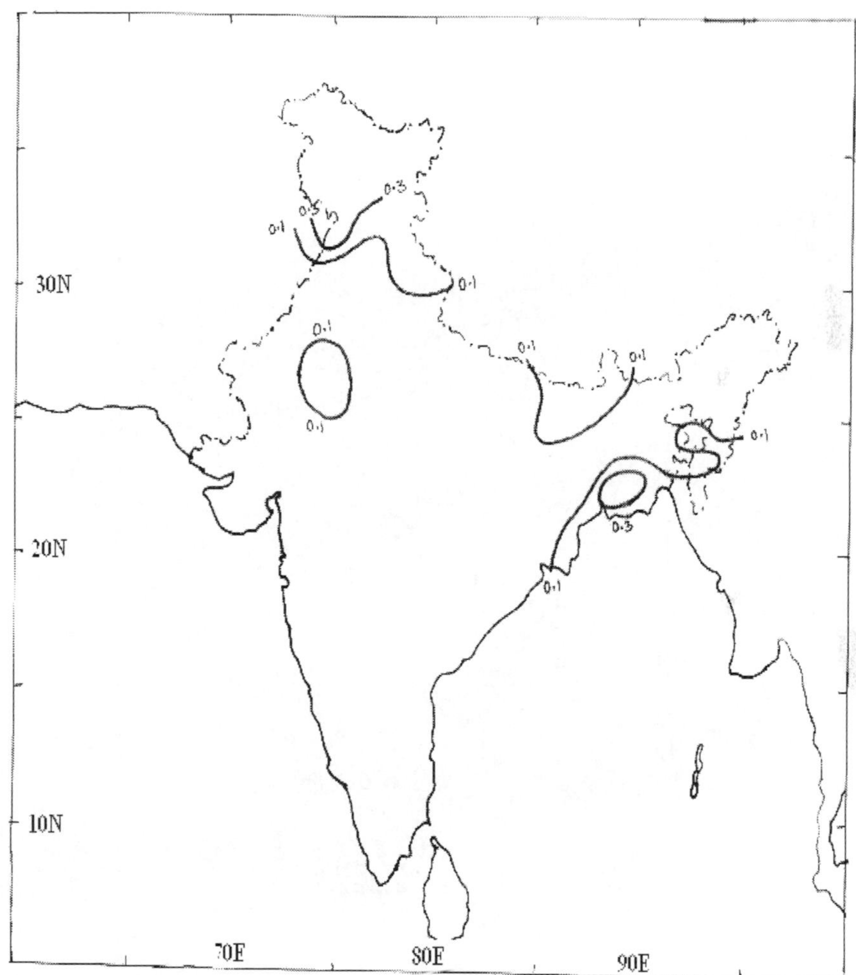

Fig. 1.21(b) Frequency of occurrence of hailstorm in April based on the data from 1961-1990 by the National Data Centre of India Meteorological Department

It may be noted that occurrence of hailstorms over peninsula has totally disappeared in Fig. 1.21(b) as compared to the previous Fig.1.21(a).

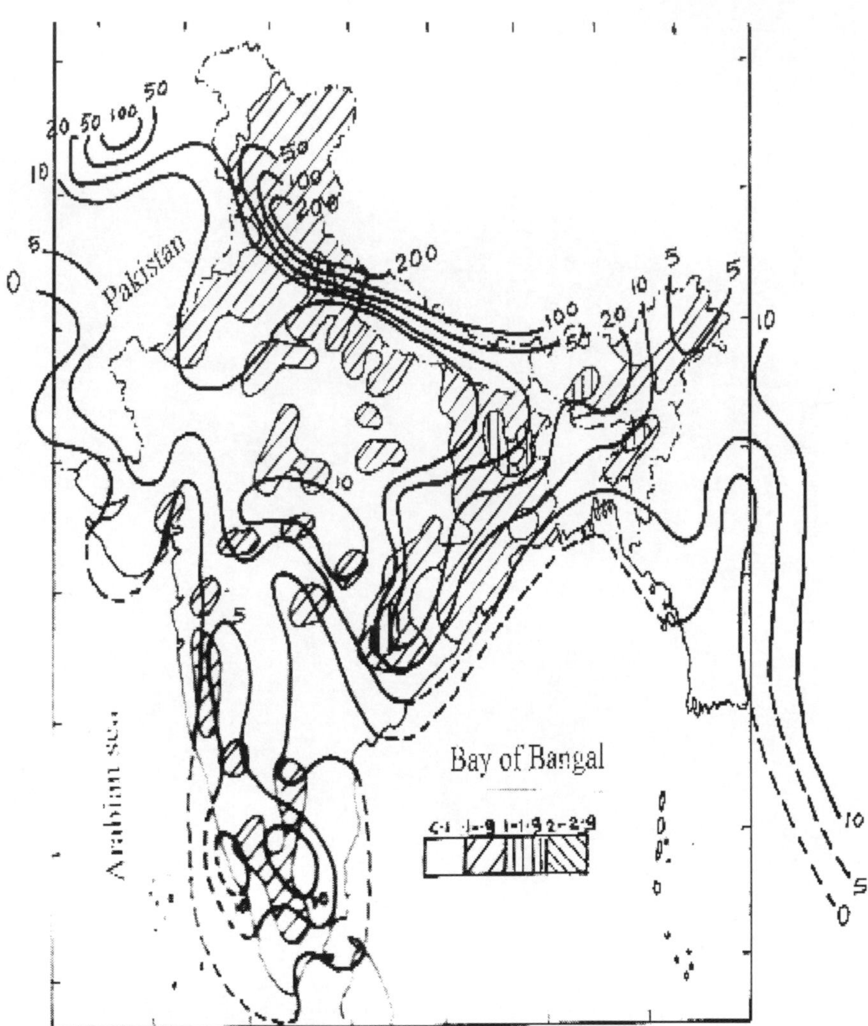

Fig. 1.22(a) Average number of hailstorms in May. Hatched scheme by Philips and Daniel (1976) and isolines of frequency of days of hailstorms in 100 years by Ramdas et al (1938).

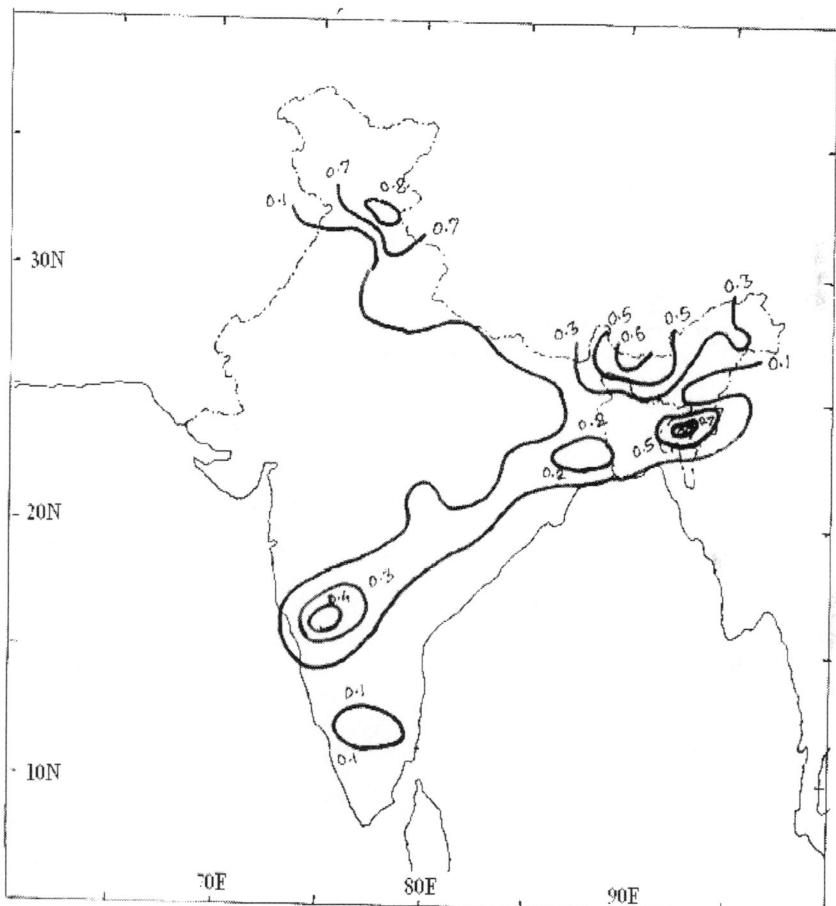

Fig. 1.22(b) Frequency of occurrence of hailstorm in May based on the data from 1961-1990 by the National Data Centre of India Meteorological Department.

Southward shift of HFHZ over northeast towards south Assam and Mizoram can be noted.

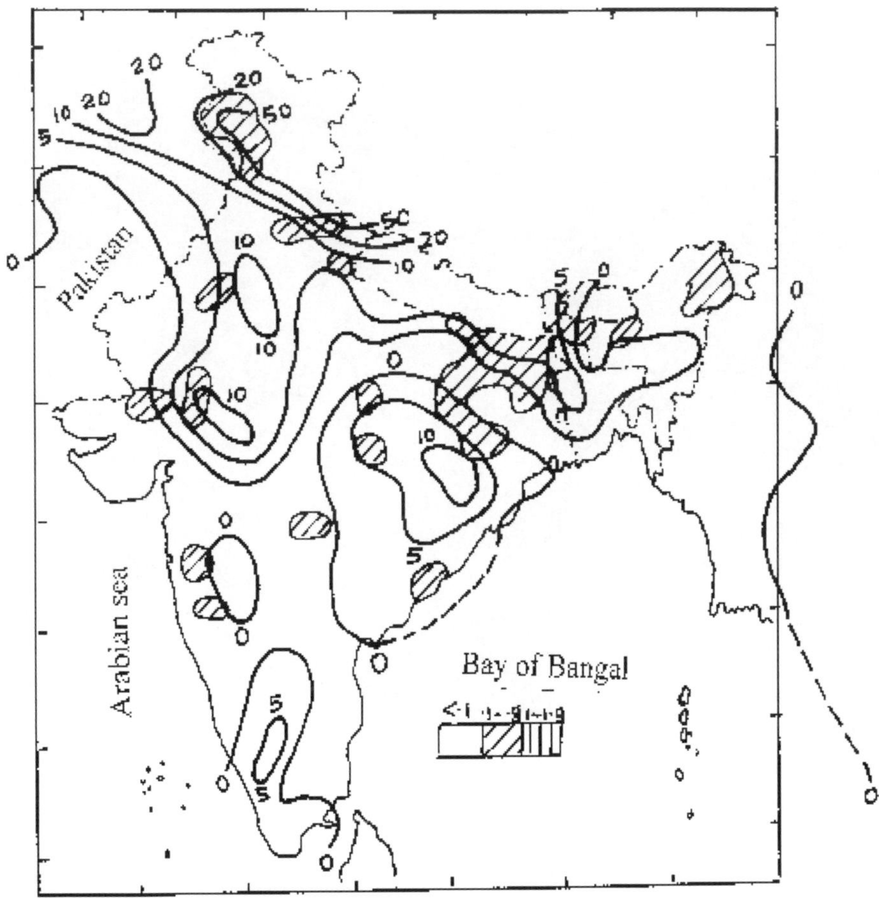

Fig. 1.23(a) Average number of hailstorms in June. Hatched scheme by Philips and Daniel (1976) and isolines of frequency of days of hailstorms in 100 years by Ramdas et al (1938).

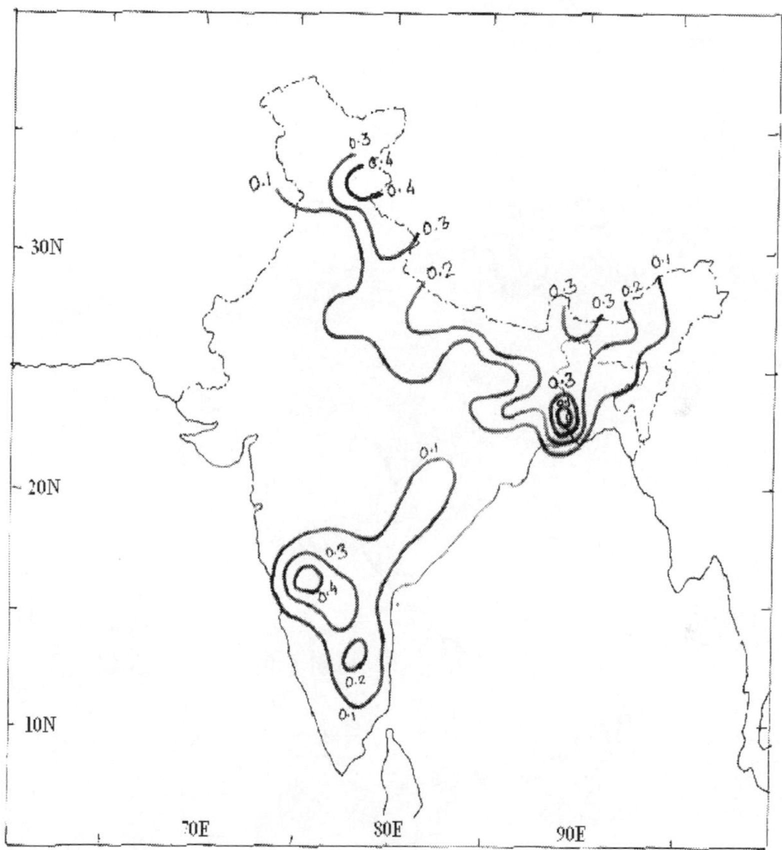

Fig. 1.23(b) Frequency of occurrence of hailstorm in June based on the data from 1961-1990 by the National Data Centre of India Meteorological Department.

HFHZ may be noticed over the Gangetic West Bengal region in Fig. 1.23(b) in contrast to Fig 1.23(a).

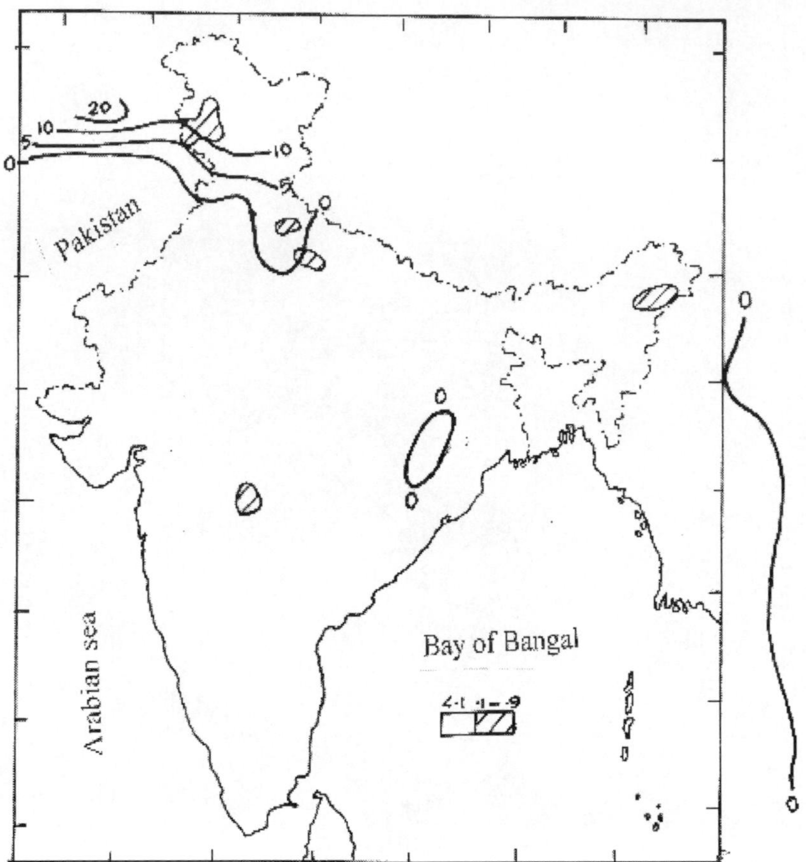

Fig. 1.24(a) Average number of hailstorms in July. Hatched scheme by Philips and Daniel (1976) and isolines of frequency of days of hailstorms in 100 years by Ramdas et al (1938).

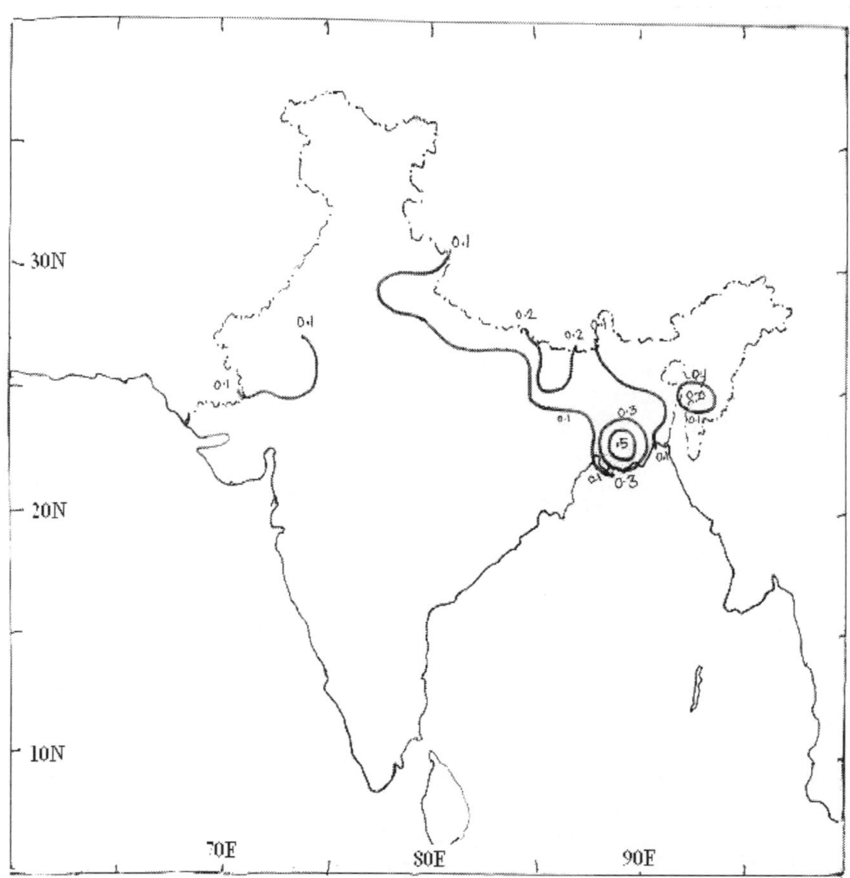

Fig. 1.24(b) Frequency of occurrence of hailstorm in July based on the data from 1961-1990 by the National Data Centre of India Meteorological Department.

Generally July is the month of relatively low hailstorm activity. However, marked shift of high frequency hailstorm activity from western Himalayas Fig. 1.24(a) to the Indo-Gangetic plains in Fig. 1.24(b) and emergence of HFHZ over Gangetic West Bengal may be noted.

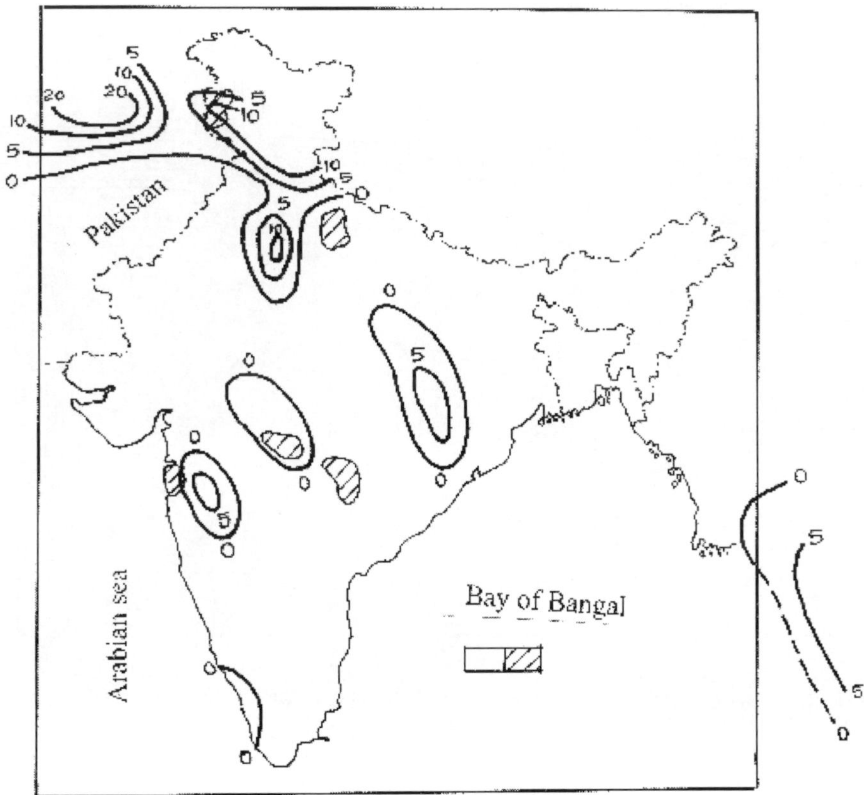

Fig. 1.25(a) Average number of hailstorms in August. Hatched scheme by Philips and Daniel (1976) and isolines of frequency of days of hailstorms in 100 years by Ramdas et al (1938).

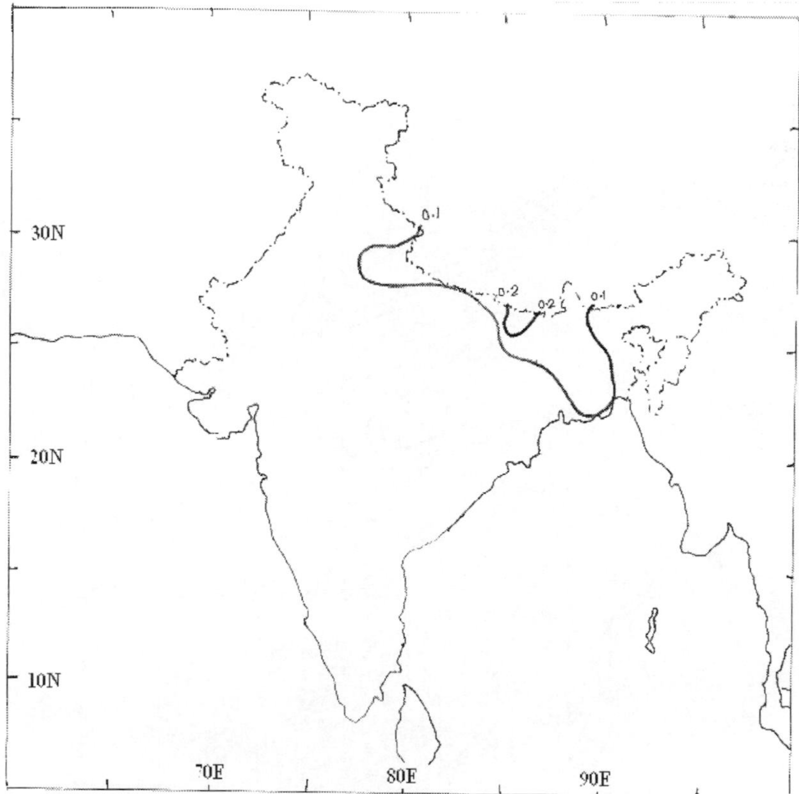

Fig. 1.25(b) Frequency of occurrence of hailstorm in August based on the data from 1961-1990 by the National Data Centre of India Meteorological Department.

During August month hailstorm activity has greatly reduced over the central and peninsular India. Over North Indian region high frequency zone shows marked eastward shift from western Himalayas and Kashmir region to Bihar and West Bengal region.

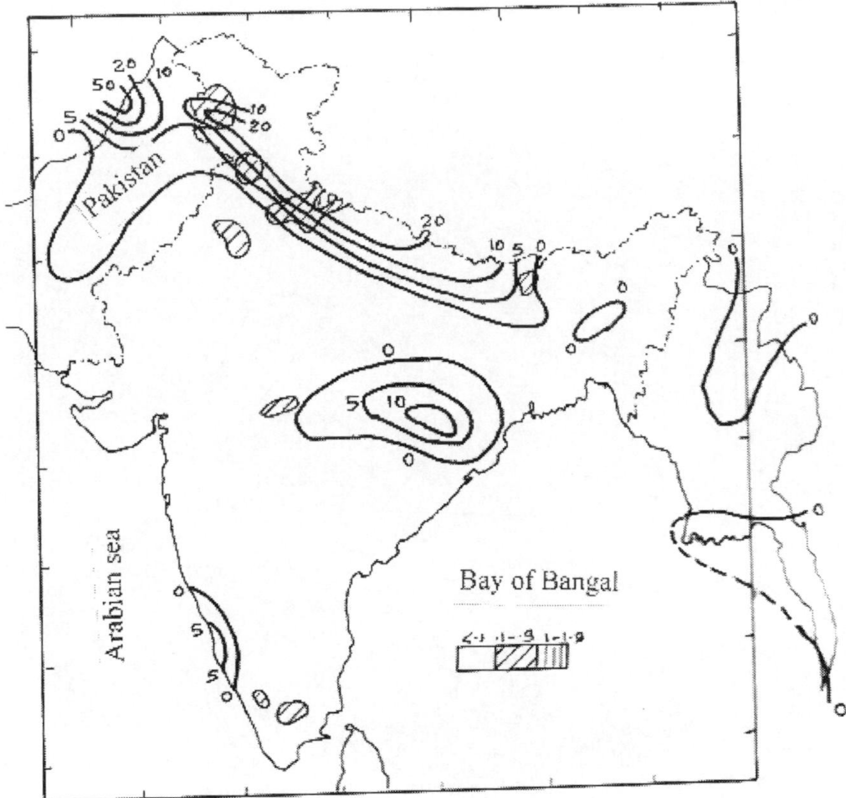

Fig. 1.26(a) Average number of hailstorms in September. Hatched scheme by Philips and Daniel (1976) and isolines of frequency of days of hailstorms in 100 years by Ramdas et al (1938).

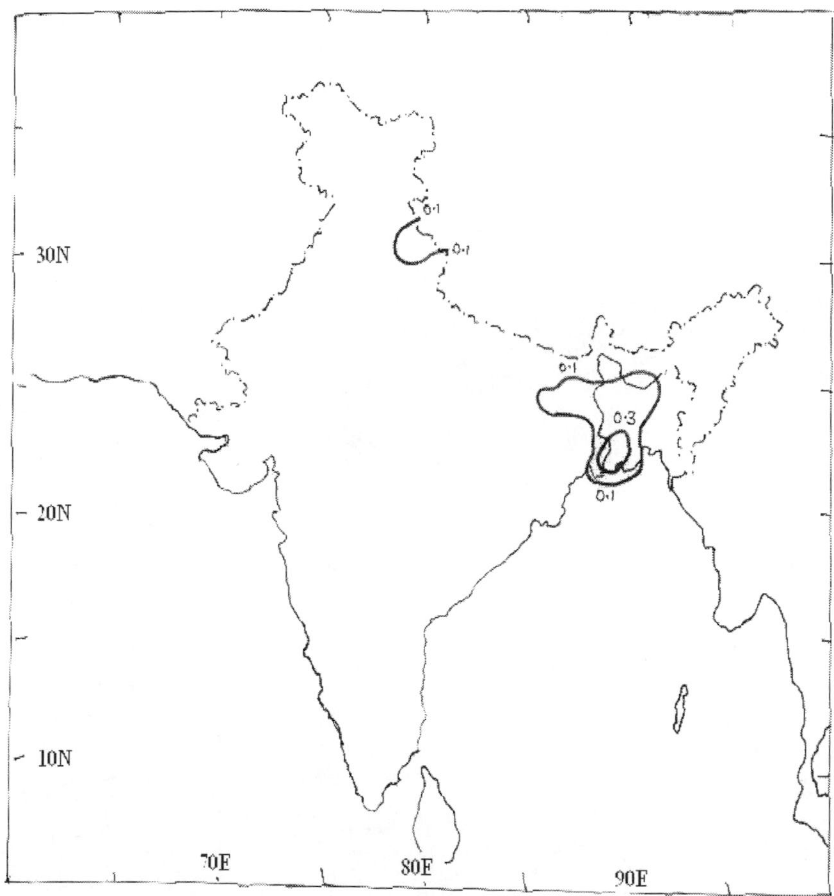

Fig. 1.26(b) Frequency of occurrence of hailstorm in September based on the data from 1961-1990 by the National Data Centre of India Meteorological Department

A comparision of Figs. 1.26(a) and (b) readily reveals significant shrinking of hailstorm activity over North-West India and northeastward shift of HFHZ from Orissa and adjoining Madhya Pradesh region to Gangetic West Bengal.

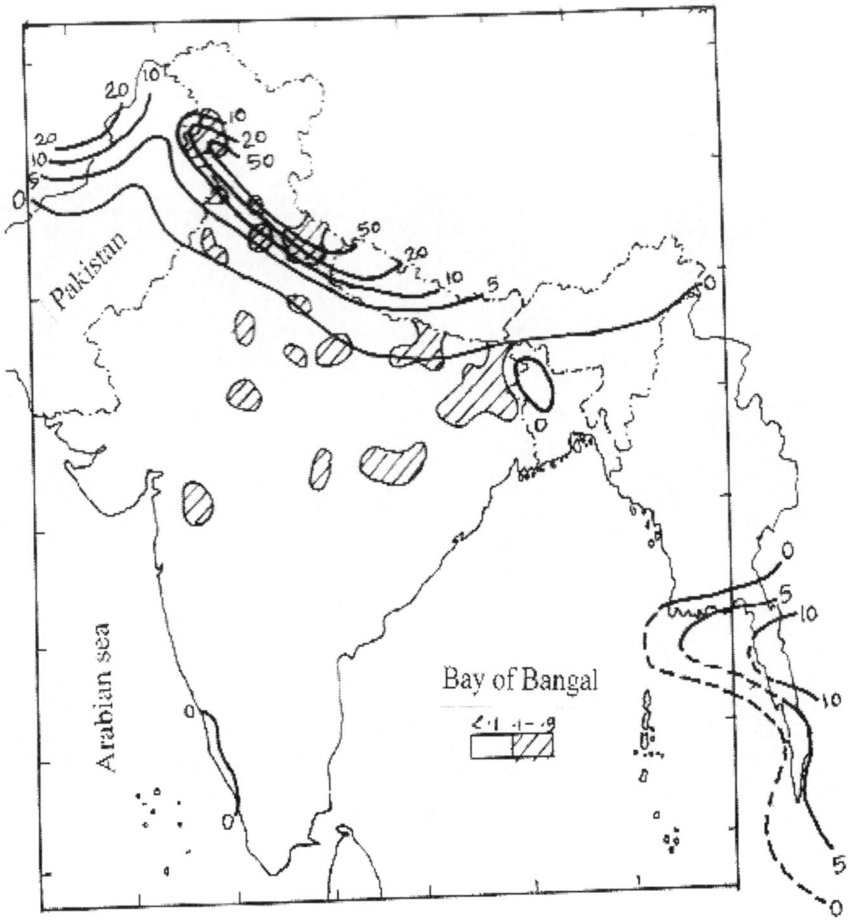

Fig. 1.27(a) Average number of hailstorms in October. Hatched scheme by Philips and Daniel (1976) and isolines of frequency of days of hailstorms in 100 years by Ramdas et al (1938).

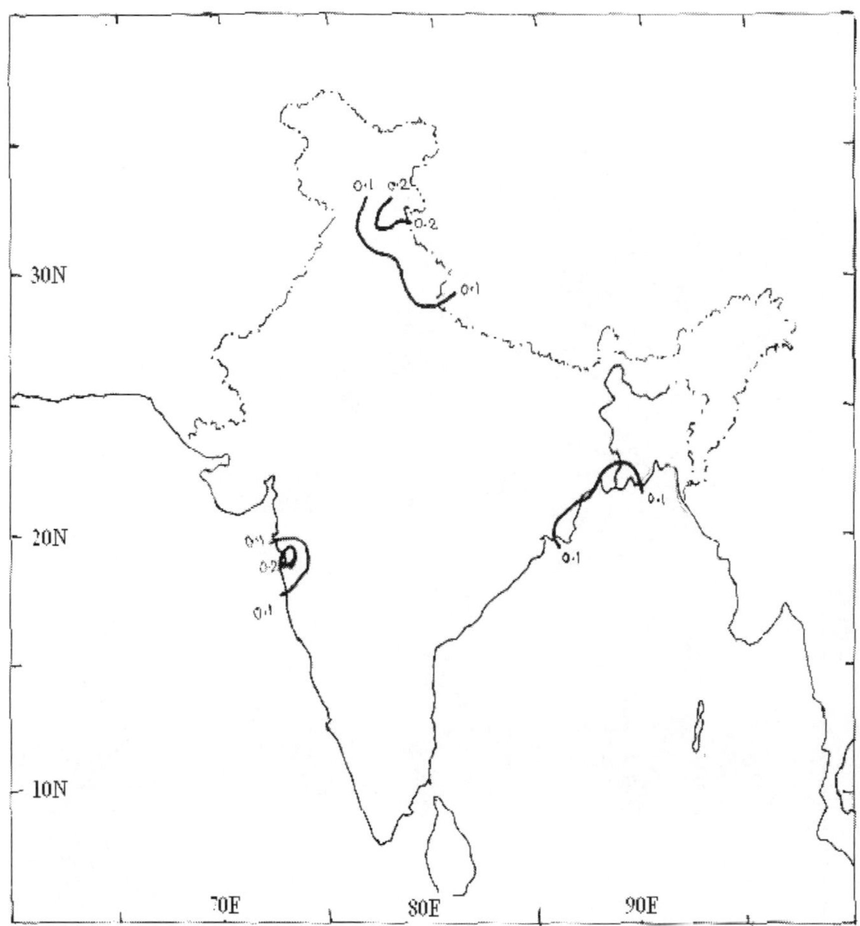

Fig. 1.27(b) Frequency of occurrence of hailstorm in October based on the data from 1961-1990 by the National Data Centre of India Meteorological Department.

A significant shrink in the hailstorm region over the foothills of Punjab, Haryana and Himanchal Pradesh may be noted during October.

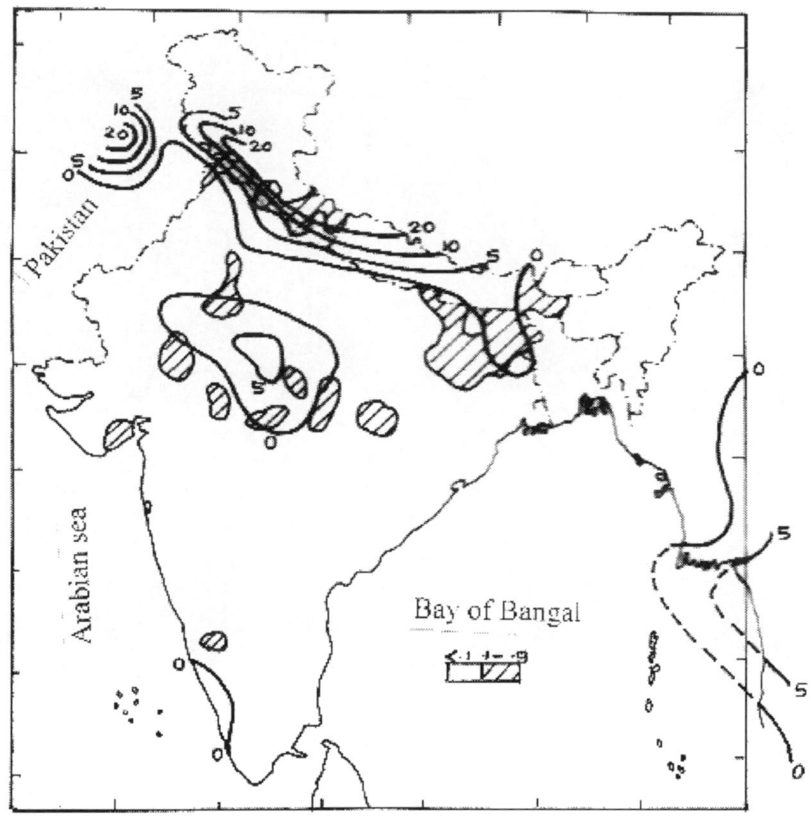

Fig. 1.28(a) Average number of hailstorms in November. Hatched scheme by Philips and Daniel (1976) and isolines of frequency of days of hailstorms in 100 years by Ramdas et al (1938).

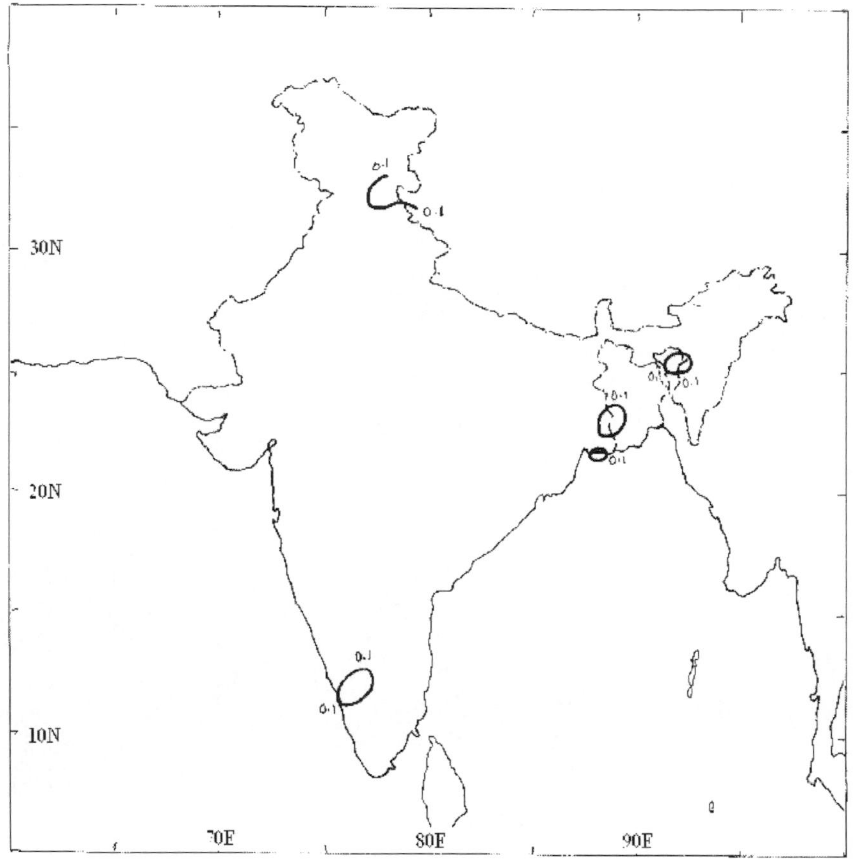

Fig. 1.28(b) Frequency of occurrence of hailstorm in November based on the data from 1961-1990 by the National Data Centre of India Meteorological Department.

Complete disappearance of HFHZ from central India in figure 1.13(b) as compared to previous data (Figure 1.28(a)) can be noted during November month.

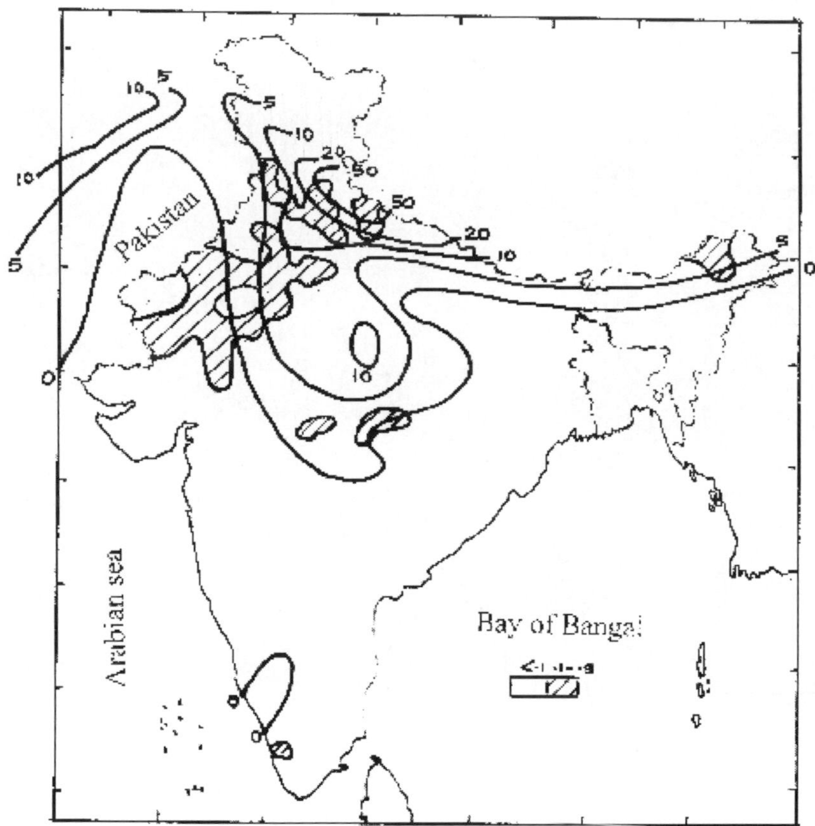

Fig. 1.29(a) Average number of hailstorms in December. Hatched scheme by Philips and Daniel (1976) and isolines of frequency of days of hailstorms in 100 years by Ramdas et al (1938).

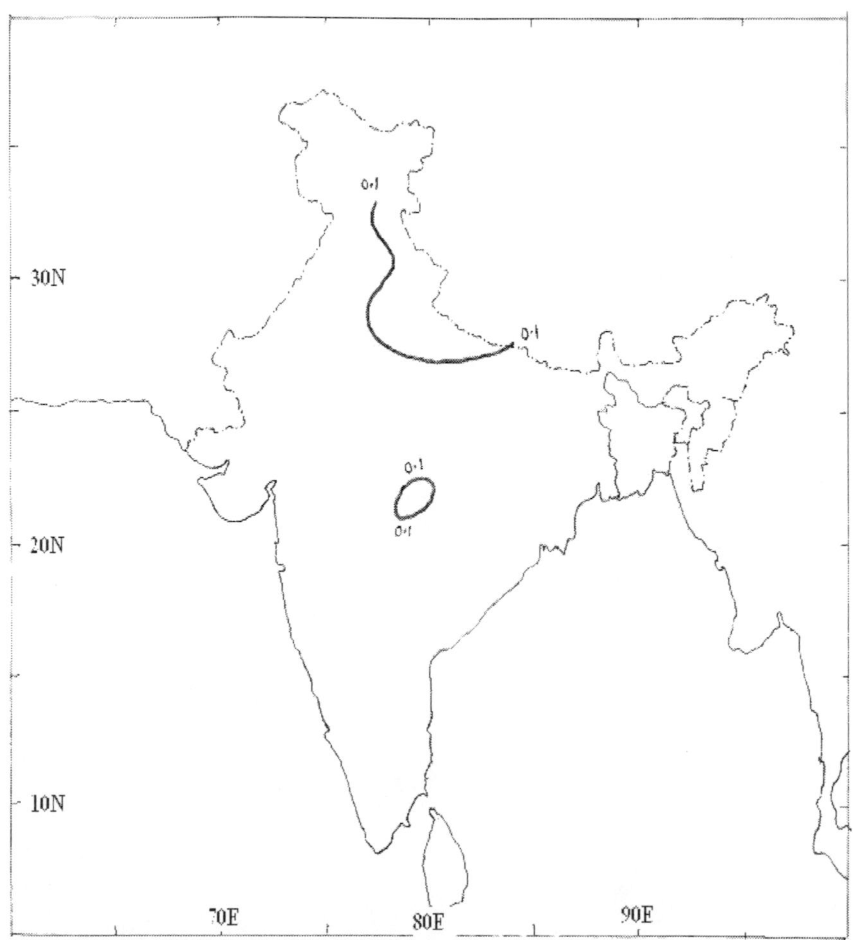

Fig. 1.29(b) Frequency of occurrence of hailstorm in December based on the data from 1961-1990 by the National Data Centre of India Meteorological Department.

Marked northward shrink of hailstorm zone from west central India towards northwest Uttar Pradesh and Uttaranchal may be noted during December.

Figures 1.30 and 1.31 by Philips and Daniel (1976) show for the various parts of the country the month and season normally having most hailstorms. It may be noted in figure 1.30 that in general from February to May most of the regions of central and north India experience most-hailstorm. Over peninsular India in general, most hailstorm occurring months are not quite specific and they are isolated regionwise and sporadic monthwise. As the latest statistics based on the 1961-90 data is not available for the months or seasons of most hailstorm days, based on the monthwise observations made in this section (figures 1.3 to 1.14) , reduction in monthwise and seasonwise activity may be inferred by the readers, while interpreting the figures 1.30 and 1.31 by Philips and Daniel (1976).

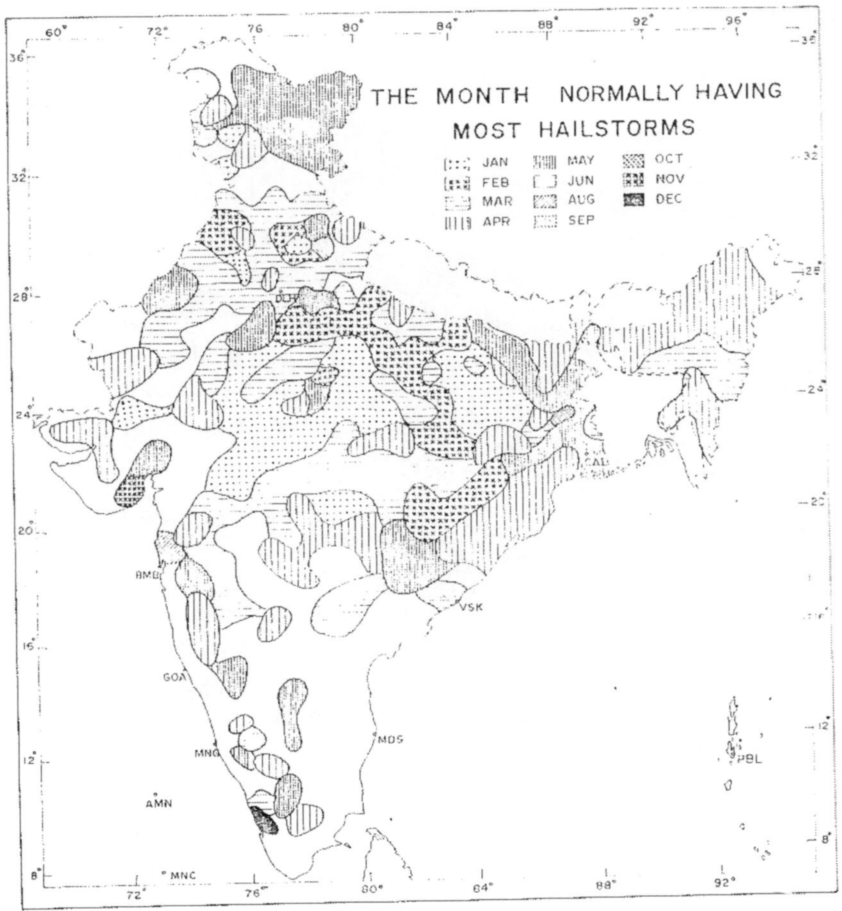

Fig. 1.30

Fig. 1.31 shows that season-wise, west central and North-West India, excluding Jammu and Kashmir region, experience most hailstorms. Over Jammu and Kashmir region, Assam and east central India April to May is the most hailstorm season, in general.

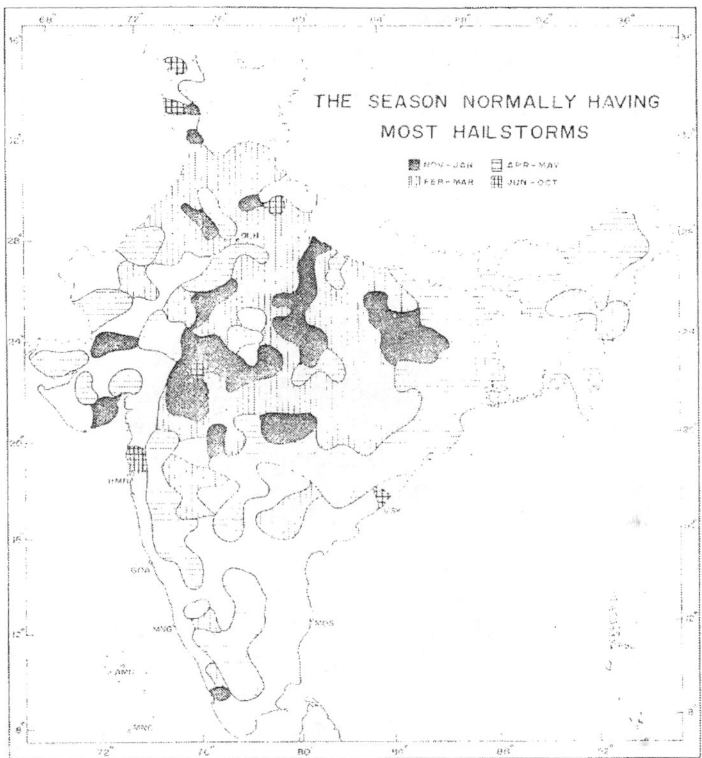

Fig. 1.31

1.6 Hail Path, Hail Streak and Hailswath

Damage of hailstones may be better understood by knowing the characteristics of the path made by the fallen hails on the earth surface. It has been investigated by many people. Prohaska (1900) gave data of hail paths in Austria (1888 – 1900), Gigeishvili (1960) in eastern Georgia (1938 – 1948), Chepovskaya (1966) in North Caucasus and Kozminski (1963 – 4) in Poland, Lemons (1942) and Schaefer (1948) in U.S.A. Gigineishvili (1960) noted that in number of cases the direction of the hail motion may change by 90^0. By studying different hail paths, he named the places where two of them intersect as "Hail – Nodes". It has been observed that when a cloud grows during a frontal process and acquires a mean velocity of about 15 –20 km/hr, then the mean length of hail path is observed to be around 15 –22 km.

Hailpath was the term used to define the road formed by hailstones fallen from the hailstorms. During mid 20th centuries observations revealed that areas with nearly continuous damage were the result of several individual streaks of hail occurring within the same general area during 1-12 hour period (Chagnon and Stout, 1966). This led to the concept of hailstreak. A "Hailstreak" is defined as areas of hail continuous in space with temporal coherence (Chagnon et al., 1967). The average hailstreak represents a fast-moving, short lived, and relatively small phenomenon. 80% of the hailstreaks in U.S.A. had area less than 40 Km^2 and median affected area was $20.5km^2$. Extremes may range from 2.3 – 2040 Km^2. In central Europe the area affected by hailstreaks is usually less than 500 km^2 (Punge and Kunz, 2016). Prefered location of occurrence of hailstreak is along the major axis of horizontal cloud base projected on the surface and during the mature stage of the convective cloud. Normally a convective cloud produces one hailstreak but 20% of them could produce four or more (Chagnon, 1970). Series of hailstreaks, often partially overlapping with each other, are termed as "Hailswath". Hence hailswath is defined as an enveloped area of hail comprising two or more hailstreaks separated by no more than 20 minutes during 12 hours or less. A hail producing system is a convective precipitation system that contains one or more hailstorms during its period of precipitation production. A hail producing system in an area of 4000 Km^2 may normally produce five hailstreaks with average separation distance of 24 km.

The results of the variou investigations with respect to the dimensions of hailswaths are given in Table. 1.1.

Table 1.1 Dimension of Hailswaths according to the data of different authors

		Length (Km)		Width(Km)	
Authors	**Location**	**Max**	**Mean**	**Max**	**Mean**
Prohaska*	Austria	Several hundred	10 – 20	10 – 20	8 – 9
Schaefer*	Grand Island	48	-----	8	----
Gigineishvili*	Eastern Georgia	90 – 100	20 – 30	10	5 – 7
Lemons*	U.S.A.	-----	----	----	1 – 6
Chagnon**	Illinois, U.S.A.	240	-	48	-
Chepovskaya*	Northern Caucasus	400	15 – 20	8 – 10	1 – 6
	Mean	----	15 – 20	----	4 -

*Sulakvelidge(1969) ** Chagnon (1970)

The edges of the hailstreaks are usually parallel, but the width varies in the direction of motion. Some times it becomes wider as the hail fall intensity rises. The most intense hail fall is observed in the central part of the hailstreak, while within hailstreak small hail free regions may be encountered. The majority of hailstreak have a length of about 4-7 km. The width of severe hail damage caused by hailstreak, though may vary from a few meters to several kms but in majority of cases it is to 1.5 km. Length of most of the hailswaths are 15-20 km and width ~ 4 km.

According to the few available studies, the mean hailstreak length is around 50 km with an exponential decrease of the number with length. In very rare cases, hailstreaks may persist over several hundreds of kilometre. Puskeiler (2013) who reconstructed hailstreaks between 2005 and 2011 from 3D radar data and appropriate post-processing. Accordingly, most of the 2632 identified hailstreaks had a length of 20 km or less, whereas only 13 events reached a length of 300 km or more.

The mean including standard deviation and median were 48.0, ± 46.7 km and 40.0 km, respectively. Specifically for the region of León(Spain), Fraile et al. (1992) found an average hailstreak area of 44 km^2. For Moldova in Eastern Europe, Potapov et al. (2007) estimated typical hailstreak length of 20-25 km with a typicalwidth of 0.2–4 km.

In view of large literature on the dimension of hailstreak, over different regions of the world a typical Hailstreak may be said to have the width of 1-2.5 km and lenghth of about 10 km (Chagnon, 1970) is shown in Fig. 1.32.

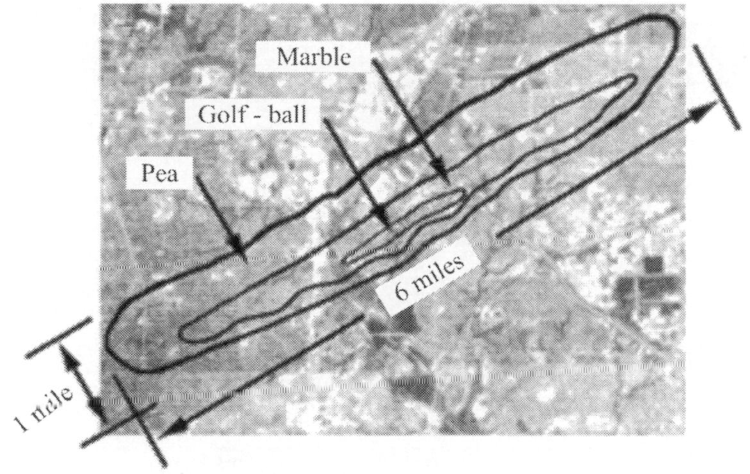

Fig. 1.32 A typical Hailstreak. (1 mile = 1.609 km). For nomenclature refer table 2.2.

1.7 Duration of Hail Fall

Usually the time during which hail inflicts damage at fixed place is taken as the duration of Hail fall. Duration of hail fall varies from a few seconds to sometimes even more then 60 minutes. Mean duration of hail fall however is normally 6 –7 minutes. Maximum duration has reached to even 90 minutes in France (Jeneve – 1961). Average point duration of hail in U.S.A is reported to be 3 minutes (Chagnon, 1970). Hail fall is in short spells followed by a short interval before the next spell. A documented report for the hail fall duration over Bamrauli (Allahabad) in Uttar Pradesh, India is mentioned below.

"On 07th Feb. 1961 hailstone of the size of 0.5 to 1 cm diameter lasted for two minutes. After a brief interval the subsequent spell was less dense and was comprised of larger size hailstone of diameter of 1 to 2.5 cm. This spell also lasted for two minutes. Again after a brief interval very large size (4 to 5 cm diameter) again fell, though, they were fewer and further apart (1 to 2.5 cm). This spell lasted for ½ minute. Last spell comprised of quite large hailstone a few among them were weighing even 35 gm".

Duration of hail fall lasted as much as 1 hour in one of the hailstorm over between Bhind in Madhya Pradesh, India on 30th Oct. 1961. But on an average hail fall duration lasts only 05 to 20 minutes over India. Duration of hail in other parts of the world is reported in Table. 1.2

Table 1.2 Duration (minutes) of Hail Fall According to Different Authors

Author	Location	Minimum	Maximum	Mean
Prohaska*	Austria	0.6	50	8 – 10
Gigineishvili*	Eastern Georgia	5 – 10	----	----
Bechwith*	Mountain regions USA	0.6	45	5
Defur*	Belgium	5	20	----
Teverskoi*	Rostov Region	5	20	10
Pastikh Sokhrina*	European path of the CIS	2	20 – 30	15
Chepovskaya*	Northern Caucasus	3	30	5 – 10
Jeneve*	France	1	90	5 – 10
Weickmann*	USA	----	85	5 – 10
Chagnon**	USA	-	-	3
Mean				6 – 7

* Sulakvelidge (1969) ** Chagnon (1970)

1.8 Duration of Hail Damage

When the cloud moves, for the time that hail falls, along the whole hailswath, from its beginning to its end, is known as the 'Duration of Hail Damage'. Duration of hail damage is identical to Duration of Hail Fall if the cloud is stationary. Duration of hail damage on the average does not exceed 80 minutes. But a maximum of 360 minutes was also observed over northern Caucasus on 10 January 1966.

Insurance sector, however, frequently use a term 'Damaging Event'. Damaging event is defined as extent of all settled claims in the reference area over a time span of 72 hours. But the time span may depend on business sector and regional practices.

References

1. Abshaev, M.T. and Chepovskaya (1966) of funktsii raspredeleniya grada(Hail distribution function) Turdy VGI No.5.
2. Abshaev, M.T., Malkarova, A.M., 2002. Effectiveness of missile artillery systems for hail protection 92 pp. 145–172.
3. Abshaev, M.T., Malkarova, A.M., 2006. Evaluating the Effectiveness of Hail Prevention. Gidrometeoizdat, St. Petersburg (in Russian).
4. Abshaev, M.T., Abshaev, A.M., Malkarova, A.M., 2012. Estimation of antihail projects efficiency considering the tendency of hail climatology change. Proceedings of the 10th WMO Scientific Conference on Weather Modification.
5. World Meteorological Organization, pp. 1–4 URL http://www.wmo.int/pages/prog/arep/wwrp/new/documents/WWRP_2012_2_Proceedings_19_June.pdf.
6. Australia Bureau of Meteorology: Severe Storms Archive. [http://www.bom.gov.au/Australia/stormarchive/]
7. Barnes, G., 2001: Severe local storms in the tropics. *Severe Convective Storms, Meteor. Monogr.,* No. 50, Amer. Meteor. Soc., 359–432.
8. Beck, R.E. 1959, Hail aloft summary, MATs Flyer, 6 No. 2, pp13.
9. Brooks, H. E., J. W. Lee, and J. P. Craven, 2003: The spatial distribution of severe thunderstorm and tornado environments from global reanalysis data. *Atmos. Res.*, 67 & 68, 73–94.
10. Cecil Daniel J., and Blankenship Clay B., 2012, Toward a Global Climatology of Severe Hailstorms as Estimated by Satellite Passive Microwave Imagers, BAMS, Am. Met. Soc, 687-703.

11. Chagnon, S.A., 1976, Scale of Hail, J. Appl. Meteor., 12, 38pp.
12. Chowdhury, A., and A. K. Banerjee, 1983: A study of hailstorms over northeast India. *Vayu Mandal*, **13**, 91–95.
13. Chepovskaya O.I. 1966, Preliminary results on studying hail distribution on earth surfaces, Turdy VGI No. 3(5).
14. De U. S., Dube R. K. and Prakash Rao G. S., Extreme weather events over India in the last 100 years, Journal of the Indian Geophysical Union, 9, 173-187 (2005).
15. Flora, S.D. 1956, Hailstorm of United states, University of Oklahama on, Norrmann, Oklahamam, 201 pp.
16. Fraile, R., Sánchez, J.L., de la Madrid, J.L., Castro, A., Marcos, J.L., 1999. Some results from the hailpad network in León (Spain): noteworthy correlations among hailfall parameters. Theor. Appl. Climatol. 64 (1-2), 105–117.
17. Friedman E.G. 1976, Hail suppression impact on property insurances, Illinois state water survey, TASH working paper 11, Urbana, 69 pp.
18. Frisby, E. M., and H. W. Sansom, 1967: Hail incidence in the tropics. *J. Appl. Meteor.*, **6**, 339–354.
19. Gigineishvili V.M. 1960, Hail Damage in East Georgia, Gidomeleoizdat, Leningrad.
20. Handa, Will H. and Cappellutib, G., 2011, A global hail climatology using the UK Met Office convection diagnosis procedure (CDP) and model analyses, METEOROLOGICAL APPLICATIONS *Meteorol. Appl.* 18: 446–458 (2011), Published online 4 January 2011 in Wiley Online Library (wileyonlinelibrary.com) DOI: 10.1002/met.236
21. Harrison H.T. 1956, The display of weather echoes on 5.5 cm airborne radar, United Airlines met circular, No. 39, Denever, Colorado (Abridged version: Aeronautical Engg. Review 15, pp 102 – 109).
22. Houze, R. A., Jr., D. C. Wilton, and B. F. Smull, 2007: Monsoon convection in the Himalayan region as seen by the TRMM precipitation radar. *Quart. J. Roy. Meteor. Soc.*, **133**, 1389–1411.
23. Jeneve, P. Lagrele (Publie Sous la direction de M.A. Viant), Paris, 1961.
24. Kozminski, Cz, 1963, Westepne Bedaria nad motodyka statyslycznych opracowan opadov Gradu w palsce, leszyty naukow wyzczy szkoly Rolniegy w czezecinie, Vol. 11.

25. Kozminsh, Cz, 1964, Westpne Geographiczne nozine szezenie wie Kszych lurz gradowych lanolowanych na obszarze palski w latach 1946 – 50.
26. Le Roux and Oliver, 1996, Hail day frequency in south Africa, Internet: *dimtecrisk.ufs.ac.za/atlas/atlas_252t.htm*
27. Lemons H. 1942 Hail in High and Low Latitudes, Bull. Am met Soc, Vol. 23, p 61.
28. Malkarova, A., 2011. Estimation of physical efficiency of hail protection accounting for changes in hail climatology. Russ. Meteorol. Hydrol. 36 (6), 392–398 (Jun.).
29. Mason B.J. 1957, The Physics of Clouds – oxford University press.
30. Matsudo, C. M., and P. V. Salio, 2010: Severe weather reports and proximity to deep convection over northern Argentina. *Atmos. Res.*, **100**, 523–537, doi:10.1016/j.atmosres.2010.11.004
31. Mishra P. K. and Prasad S.K. 1980, Mausam 31, 3, pp 385 – 396.
32. Philip N.M. and Daniel C.E. 1976 Met Monograrph, Climatology, No. 10 IMD.
33. Potapov, E., Burundukov, G., Garaba, I., Petrov, V., 2007. Hail modification in the Republic of Moldova. Russ. Meteorol. Hydrol. 32 (6), 360–365.
34. Postukh, V. P. and Sokhtina, R.F. 1957, Hail over USSR Turdy GGO No. 74.
35. Prohaska K. 1900, Die Johrliche und laglicke Periode der Gewitter und Hagel falle in Sleiermarks and Karnter, Meteorology Zs.
36. Punge, H., Bedka, K., Kunz, M., Werner, A., 2014. A new physically based stochastic event catalog for hail in Europe. Nat. Hazards 73 (3), 1625–1645.
37. Puskeiler, M., 2013. Radarbasierte Analyse der Hagelgefährdung in Deutschland (Ph.D. thesis) Institute of Meteorology and Climate Research, Karlsruhe Institute of Technology (KIT) [in German]. URL

    https://www.imk-tro.kit.edu/download/Dissertation_Puskeiler_Marc.pdf.
38. Ramdas, L.A., Satakopan, V. and Gopal Raom S. 1938. The Indian Journal of Agricultural Sciences, Vol. VIII, Part VI, December.
39. Ramesh Kumar M.R., Disasters associated with Extreme Weather Events over India, Disaster Advances., Volume 2, No. (4), October (2009).

40. Romatschke, U., S. Medina, and R. A. Houze, 2010: Regional, seasonal, and diurnal variations of extreme convection in the South Asian region. *J. Climate*, **23**, 419–439
41. Sansom H. W. 1971, Hailstorm in Kericho, area, Tech. Memo No. 22 East Africa Meteor Deptt. Nairobi. 6 pp.
42. Schaeffer V. J. (1948) The Natural and artificial Formation of snow in atmosphere, Trans Amer Geophysics, Vol. 29. No. 4.
43. Schuster, S. S., R. J. Blong, and M. S. Speer, 2005: A hail climatology of the greater Sydney area and New South Wales, Australia. *Int. J. Climatol.*, **25**, 1633–1650
44. Stuart Piketh and Roelof Burger, 2014, Hail in South Africa, North-West University, Noordwes-Universiteit, Guest Lecture. Contact: roelof.burger@nwu.ac.za.
45. Sulakvelidze, G. K. 1969, Rainstorm and Hail, IPST Press, Jerusalam.
46. Tuovinen, J.-P., A.-J. Punkka, J. Rauhala, and H. Hohti, 2009: Climatology of severe hail in Finland: 1930–2006. *Mon. Wea. Rev.*, **137**, 2238–2249.
47. Webb, J., D. M. Elsom, and D. J. Reynolds, 2001: Climatology of severe hailstorms in Great Britain. *Atmos. Res.*, **56**, 291–308
48. Williams, L., 1973: Hail and its distribution. Studies of the Army Aviation (V/STOL Environment), Army Engineer Topographic Laboratories Rep. 8, ETL-SR-73-3, 27 pp.
49. Xie, B., Q. Zhang, and Y. Wang, 2010: Observed characteristics of hail size in four regions in China during 1980–2005. *J. Climate*, **23**, 4973–4982.
50. Yamane, Y., T. Hayashi, A. M. Dewan, and F. Akter, 2010: Severe local convective storms in Bangladesh: Part I. Climatology. *Atmos. Res.*, **95**, 400–406
51. Zhang C, Zhang Q. and Wang Y. 2008, Climatology of hail in China: 1961-2005, J. Appl. Meteor. Climatology, 47, 795-804.

CHAPTER 2

Hailstones

The term hail or hailstone is applied to a variety of solid hydrometeors that fall from super cooled clouds. Simple hailstones are smooth and sphereoroidal, conical or generally irregular shapes in shapes. Their spheroidicity was explained aerodynamically quite some times back by Reynolds (1879), Shvedev (1881) and Barkow (1908). A hailstone has density comparable to that of solid ice. Several efforts have been made in experiment and theories to study natural hailstorms and various methods have been adopted which permit comprehensive description of their physical characteristics. It may be mentioned here that chief difficulty involved in the various experiments made on the fallen hailstone, comes from the change due to transportation delay that occurs between the appearance of the hailstones on the ground and the time of investigation. It is due to this fact that as a rule hailstone has to be transported and stored below 0^0C temperature prior to examination. But when this is done, phase transition is likely to take place which, as shown by List (1959), changes the liquid water often contained in the system of capillaries within the hailstone. Also it is impossible at present to relate the size, shape, structure and density of such solid precipitation elements to the physical characteristics of the parent cloud. However, incidence of hail is usually associated with thunderstorm. But in a few cases, although rarely, small hails may fall from clouds, which do not reach the thunderstorm proportion.

2.1 Classification of Hail

Before going into detailed physical study of the hailstones we shall distinguish between the various forms of ice particles which are sometimes called hail, but which range from large stones of true hail to the smaller graupel or soft hail. The following definitions are in accordance with the recommendations made by the International Commission for ice and snow.

2.2 Hail

Ice balls as stones, ranging in diameter from that of medium size raindrop (≈5 mm) to many times more. As per WMO recommendations at least 5 mm or more diameter stone can be referred as hail. European Severe Weather Database (ESWD) defined large hail for diameter > 20 mm. They may fall, singly or frozen together, into irregular masses. They are composed either of clear ice of alternate clear and opaque layers.

The successive concentric layers of alternatively clear ice and opaque ice are of successive accretion of water drops around a small kernel of ice falling through a thick cloud, as each drop is frozen into the nucleus, it may form a new shell giving it onion like cross-section. They are having 5-150 mm or more diameters. Thickness of successive layers has been found increasing and thickest layer is the outermost one. 3-5 of such layers are commonly seen. Maximum of 30 layers have also been reported. The formation of opaque and transparent ice layers in hailstone growth in natural condition is probably due to successive passage of growing hailstone through layers of high and low liquid water content in super cooled part of the cloud which results in moist and dry hail growth condition. Thermodynamic hailgrowth condition, even in the region of constant liquid water contents, may also cause such stratification. Table 2.1 (a) gives the thickness of a typical hailstone at Gauhati (India) given by Rakshit (1962). 2.1 (b) shows the structure of another hailstone collected over Jodhpur (India) on 18th Mar 1957 with 7 layers.

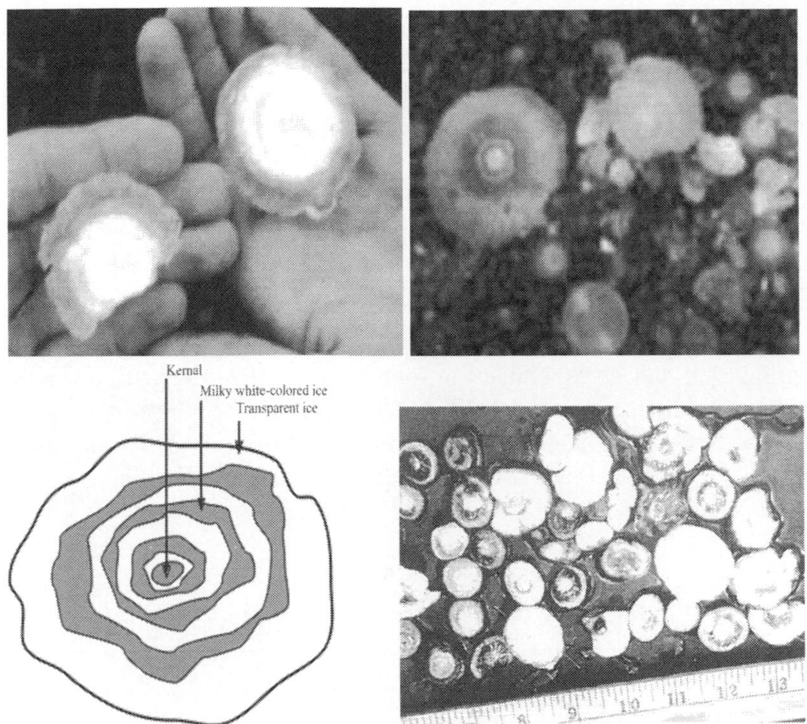

Fig. 2.1 Hailstone layers (Left); Assorted Hailstones of various sizes (Right) (Credit: NOAA Photo Library).

Table 2.1(a) Layers in Hailstone and their Typical Dimension

Elements	Appearance	Averages Diameter (mm)
Core	Opaque	6
First Layer	Transparent	12
Second Layer	Opaque	18
Third Layer	Transparent	24 – 27

Table 2.1(b) Structure of Multi Layered Elementary Hailstone

Element	Appearance	Diameter (mm)
Core	White	4
1st Layer	Transparent	6
2nd Layer	White	7
3rd Layer	Transparent	8
4th Layer	White	9
5th Layer	Transparent	11
6th Layer	White	12
7th Layer	Transparent	13

Among the reports archived by the ESWD, the fraction of hail greater than 40 mm (large hail) is 0.223. Estimating entire hail events rather than considering individual reports and assuming an exponential distribution of maximum hail size per storm, Punge et al. (2014) found that 23.8% of all reported hail events include hailstones greater than 20 mm, and 3.5% greater than 40 mm. These estimates are probably biased towards central Europe, where a large fraction of all ESWD entries were reported. Furthermore, they are biased towards larger hailstones, which are more attractive for both the media and storm chasers.

A hail stone shape is spheroidal at smaller sizes and becomes more irregular at larger sizes. Since the shape is close to spherical, a guide for reporting hail size by comparing it to spherical objects of the same diameter is used for simplicity in reporting. Normally hailstone are ≥ 5 mm and < 5 mm size are soft hail or grauple or may be referred as shot. Other nomenclatures are mentioned in Table 2.2.

Table 2.2 Nomenclature for hailstone size

Pea	6 mm
Dime or grape	19 mm
Quarter or walnut	25 mm
Golf Ball	44 mm
Tennis Ball	63 mm
Base Ball	69 mm
Orange	80 mm
Grapefruit	100 mm
Giant	> 100 mm

Giant hail stones (refer Fig.2.2) are usually more irregular in shape. This is because irregularities of a smaller size hail stone are exacerbated as the hail stone gets bigger. Also, smaller hailstones can merge onto a bigger hailstone. When this happens the hail stone will have bulges and will have a larger diameter in certain directions. In addition, giant hail stones impact the earth's surface with a great force. This force can distort the shape of the hail stone.

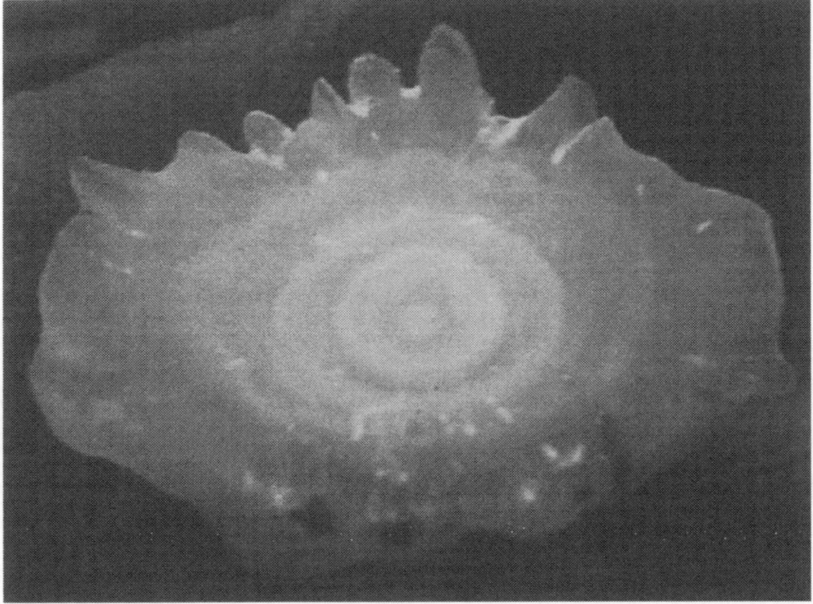

Fig. 2.2 Image of a giant hailstone. This is the Coffeyville, Kansas, U.S.A hailstone that fell in 1970. The hail stone was 1.67 pounds and over 5.5 inches in diameter.

For sizes 5-10 mm they are generally spherical or conical in shape. 10-20 mm size hailstones tend to be ellipsoidal or conical. 20-50 mm hailstones take on ellipsoidal shapes with lobes; while still larger stones between 40-100 mm appear irregular including disc shapes with protuberances (Field et al, 2009).

2.2.1 Graupel or Soft Hail or Snow Pellets

These are white, opaque, round or conical pellets of diameter up to about 6 mm. They are consisting of a central crystal covered with frozen cloud droplets (rime). They are generally of low density (0.3 – 0.6g/cm^3), easily compressible and may shatter on striking a hard surface. Observations made on top of the Hohenpussnberg during 1946-49 revealed that hail and graupel rarely fell from clouds with base temperature below – 10 ^{0}C (Weickmann - 1950), the precipitation then consists mainly of rime aggregates of snow crystals. Clouds with base temperature near 0 ^{0}C produced graupel, soft hail. They are crisp, easily compressible and do not exceed a few millimeters in diameter (≤.6 mm). True hail is more frequent from the cloud base which is warmer than 0^{0}C. In Alberta

(Canada) during 1965-66 44% of the hail fall reported soft hail. The greatest percentage frequency of soft hail was reportedly found during the beginning and end of the hail season, which mainly extends between May to September there and reached a minimum in July. In 97% of cases it was observed that the largest or one size less than the largest falling was the soft hail.

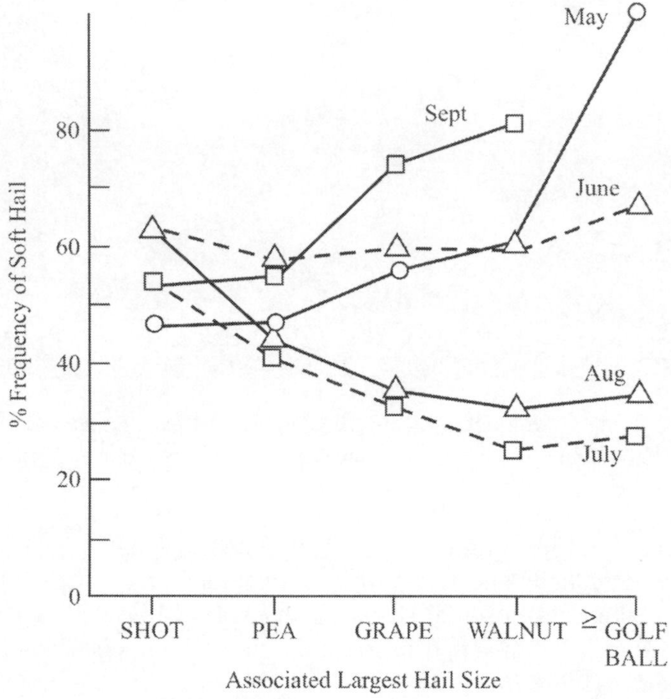

Fig. 2.3 Relation between percentage of hail and maximum hail size by months.

In Fig.2.3 systematic seasonal changes is evident when the percentage frequency of soft hail is related to the maximum size of hail. In May it was found that there was strong tendency of soft hail to occur in association with large hails. In June, and August most of the hail tend to be hard. September reverts to regime to that similar in May. It may be noted here that in India hail season mainly extends between February to May whereas in Alberta it extends between May to September. For Indian region however, no such statistics is available which relates types of hail with the months.

2.2.2 Small Hail

In consonance with WMO definition of hail (≥ 5 mm) and large hail for >20 mm size diameter, 5-10 mm diameter stones may be referred as small halstone and 10-20 mm size may be referred as moderate size hailstones. It consists of semi-transparent, round or conical grains of frozen water a few millimeters across. There are pellets with soft hail nucleus and thin outer coating which gives it glazed appearance. They are little larger than graupel but still a few millimeter in diameter. The grains are wet when they fall in a temperature above freezing point and often mix with raindrops.

2.2.3 Ice Pellets (American Sleet)

It is also known as frozen rain as they are transparent, more or less globular grains ice and about the size of rain drops. In British islands sleet is known as mixture of rain and snow.

2.3 The Shapes of Hailstone

Hailstones include large variety of aerodynamically explicable shapes which are formed by accretion of drop moisture or by melting at positive air temperature. Complex, aggregates of simple hailstone frozen together by large drops of supercooled water also result in aerodynamically explicable shapes. They are of many shapes and forms, although most of the true hailstones are either roughly spherical, conical, sphericoncal (pear shaped), ellipsoidal or discoidal in shape. On the basis of the thermodynamic growth condition hailstones of diameters upto 1.5 cm should have mainly conical shape. Larger hailstones may subsequently take spherical or ellipsoidal form etc. According to Macklin (1963), 60-70 percent of the stones from a severe storm in SE England were oblate spheroids and only a few percent were spherical. Browning and Beimers (1967) reported that the oblateness of large hailstones tends to increase during growth. Preferential melting at the two ends of the minor during descent to the ground may add to oblateness. This may, rarely sometimes lead to apple shapes, too. The spherical form, however, is most common especially when hail is small. Sometimes jagged irregular shapes with many external irregularities and protuberances are found. Weickmann (1962) point out that conical hailstones fall most often in warm showers during spring and hailstones of irregular form are more likely in summer. According to him large hailstones of diameters 5 cm and more often fall as shapeless piece of ice. These observations, however, are to be tested for other regions of the world.

Many peculiarly irregular shapes of the hail have been found. Fig. 2.4 shows a few over Australia in January.

UNUSAL HAILSTONES

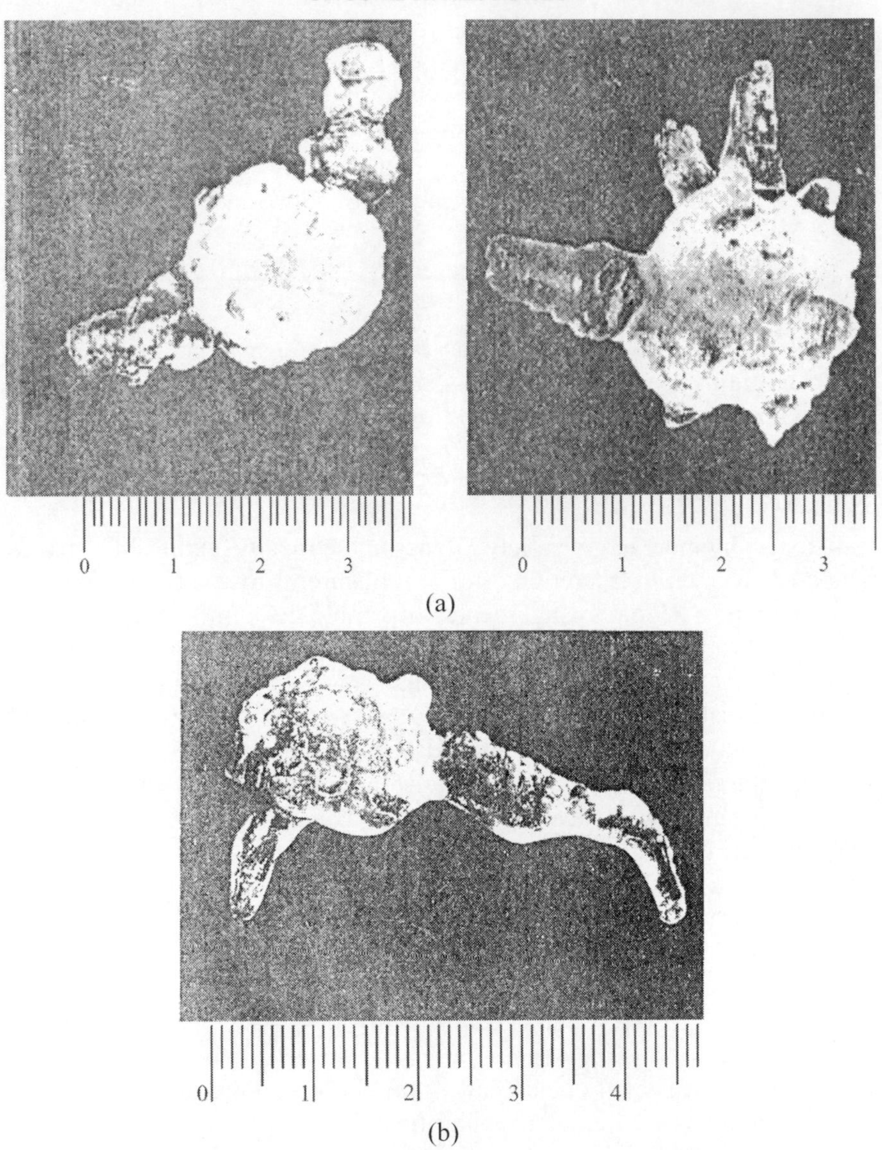

Fig. 2.4 Peculiarly irregular shapes of the hail have been found over Australia in January.

Medford Pear Shippers Association and Oregon state University conducted a hail survey in Rogue River Valley around Medford, Oregon,

USA from 1 May to 10 October 1959 (Decker et al – 1961). The predominant features were conical shape, even when it was apparently embedded in ice, which spilled around the skirt of the cone towards the tip of the cone. Only 10% of the stones resembled tooth-shape (see Fig. 2.5 (3-C)).

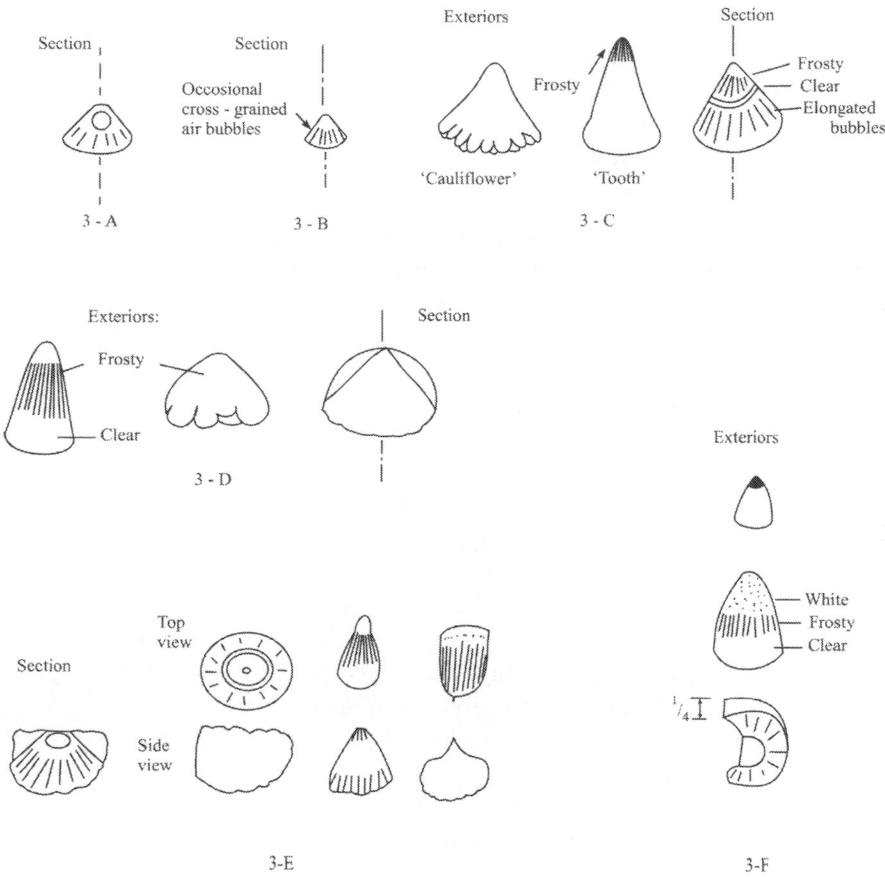

Fig. 2.5 Hailstones collected north of Medford, Oregon on 10 September 1960.

Fig. 2.5 (3-A) shows the fracture cross section of the stone. This stone had been split along the longitudinal axis shown and was originally symmetrical about the axis. Thus the widest part of the section shown, formed a circular skirt around the axis. The diameter of these stones was about 1.3 cm. Hail shaped like cauliflower and tooth types structure were also collected in another hailstone sample. Rounded and blunt cone

(Fig. 2.5(3-D)) is also not less common. The cross section of these stones helps to reveal the cause for the rounding (see Fig. 2.5(3-A)). Evidently the straight side of the cone which may have been present originally becomes rounded by accretion of additional ice in the last stages of the formation of stone. Structure of another sample of stone (larger one) is shown in Figure 2.5 (3-C). It showed quite a different structure which dominated the other samples of smaller stones. Sizes in the larger samples ranged from 2.0-3 cm in diameter. The cross-section shows a flattened surface apparently resulting from accretion of additional ice to continue the rounded cauliflower skirt out beyond the straight conical side. Fig.2.5 (3-F) shows typical samples ranging in size from 1.6 to 2 cm diameter.

A hailstone in northern suburbs of Sydney, at 1515 local on 3 Jan 1971 produced hailstone of most unusual shape, (Mossop -1971). The fall of these jagged stones (Fig. 2.3(a)) lasted for about a minute only and was followed by a light fall of stones of more unusual shapes and then rain. These hailstones were almost entirely of clear ice only. Small region of milky ice of about 5 mm in diameter occurred at the center of some stone. List (1959) and Macklin (1961) had shown that clear ice indicates that the main hailstones growth took places in 'wet condition'i.e. the stones were sweeping up the cloud drops too fast for all the water to freeze immediately. This water is retained throughout on the surface of the stone with a network of ice growing throughout. Only when the rate of collection becomes very high will water be shed from the stone. The more exaggerated finger-like growth of the stone seems to indicate that water was being shed, probably from favoured lob-type protuberance which would then continue to grow more rapidly than the neighbours. Hailstone which fell at Sydney (Fig. 2.3(b)) shows slightly flattened knobby sphere about 2 cm in diameter with four fingers 1-2 cm long inclined at about 30^0 to the principal plane of the centre of stones. Instead of tumbling randomly in the cloud this stone probably becomes settled in an attitude such that the principal plane was horizontal and water was being shed from backward pointing finger.

A number of stones when carefully split and illuminated from behind reveal the interior structure of the core when their exterior appeared to be more or less spherical (Fig. 2.6(a)). In (Fig. 2.6 (b)) a tooth stone split longitudinally shows transverse zones of bubbles which are still visible despite the internal whitening of clear ice which occurred during storage for several weeks before the picture was taken. The tip reveals the suggestion of possible kernel stone, on which the large mass of ice was collected. Careful examination indicated that these stones were not fractured from larger hail balls. Fig. 2.6(c) shows another stones clearly conical in shape, with a single transverse layer of bubbles near the blunt end. This stones appeared to have grown by addition of ice on the blunt end to the broadest portion after starting as a more slender cone similar to that in Fig. 2.6 (b).

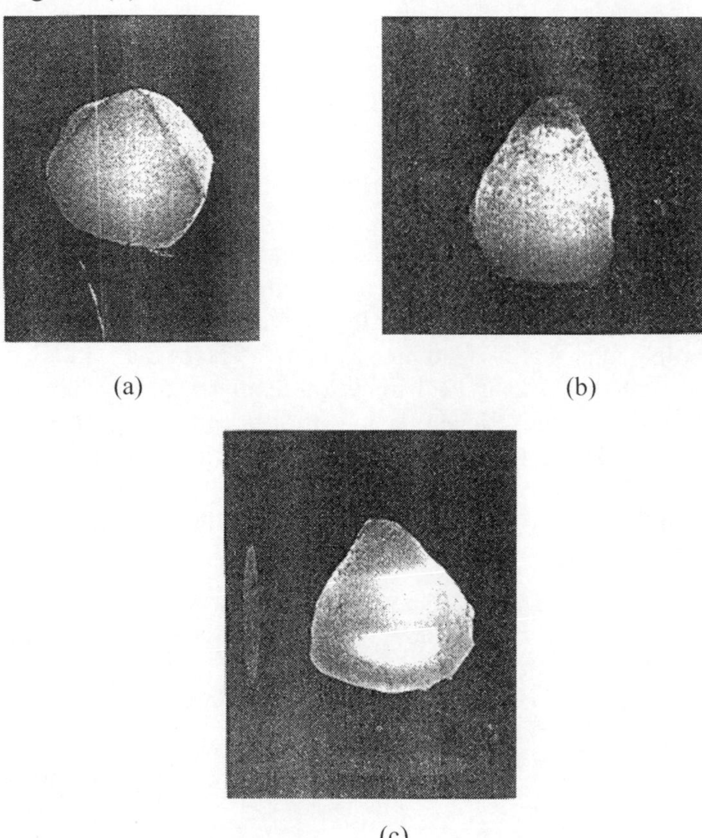

(a) (b)

(c)

Fig. 2.6 (a) External spherical shape concealed the internal cone revealed in this split section; (b) Slender cone with transverse layer of bubble and a small conical stone near the tip; (c) This widened cone possibly resulted from the addition of ice to a slender initial cone, as suggested by the shoulder and layer of bubbles outlining an interior.

Without information on the diameters in which the stone were falling it cannot be, however, definitely stated that what the top of the stone was. However, the Cauliflower appearances of many of the stones strongly suggests that they grew by accretion of ice when the rapidly falling stones encountered super cooled water drops which froze on to the falling stone while smearing over the surface. This gave peculiar petal appearances to the exterior surfaces of the blunt end of the stone near the skirt (below) with the smear extending towards the more sharply pointed end (top) of the stone and suggesting wet growth i.e. sweeping of cloud droplets too fast for all the water to freeze immediately (Ludlam – 1958).

2.4 Size of the Hailstone

Most hails are a quarter of an inch or less in diameter. But some grow much bigger and do damage. There have a few systematic observations on hailstones size; observers tend to report the maximum and mean sizes of hailstones but not the smaller size, so that we have very little reliable information on size distribution of hailstones measured on the ground in various regions of world have been quoted by Souter et al (1952) and Sulakvelidze (1969), (Table 2.3). Almost everyone who investigated this aspect points out that for no more than 3-5% of hail falls, had the diameter exceeded 3 cm. Hailstones of diameter 5 cm and more are relatively less frequent. Leaving asides the exceptions, most of the large hail diameters fall between the range 6 to 8 cm. This is because the collection efficiency of hail of diameter larger than 5.0 cm decreases sharply, although they are not rare. Pea or golf ball-sized hailstones are not uncommon in severe storms.

Table 2.3 Sizes of fallen hailstones

Author	Region of hail fall	Hail Maximum Diameter, cm			Maximum Weight (gm).
		Maximum	Minimum	Mean	
Prohaska	Austria	3.0	0.3	-	-
Gigineshvili	Eastern Georgia	4.0-5.0	-	-	400
Blair	USA (Nebraska)	13.7	-	-	153
Schaefer	USA (Grand Island)	6.2	-	-	-
Mason	USA (Grand Island)	13.8	-	-	-
Sauter	India	-	-	0.6-3.0	3400
Emerson	France	-	-	0.5-2.0	-
Emerson	Central Europe	-	-	2.0-3.0	-
Emerson	USA	-	-	1.2-1.8	-
Bevit	USA	7.5	0.3	1.0-1.8	-
Jeneve	France	3.0	0.3	0.5-1.5	-
Voronov	Eastern Georgia	2.4	0.3	1.4-1.8	-
Chepovskaya	Northern Caucasus	8.0	0.3	1.4-1.6	350
Battishvili	Eastern Georgia, Samsati Region Mean	3.0	0.2	1.2-1.5	-

The literature dealing with the study of hail only describes individual cases of hail fall and contains hardly any data on microstructure of hail clouds and the size distribution function of hailstones. Abshaev and Chepovskaya (1966) studied size spectrum of fallen hail by taking 20 instances of hail fall (38 spectra). 4 hail spectra were taken from the studies of Doughlas and Hitschfeld (1958), 4 from a paper by Atlas et al (1961) and the remaining 30 spectra were collected in spring-summer months of 1962-1964 on the Mushta and Samsari Mountains. In each case 300 to 2000 hailstones (on an average 600) were measured. To reduce the changes in the size spectrum to the minimum, due to any transformation in the presence of a cross wind etc the hail samples were collected from the middle of hail path. Since the samples were collected in high mountains with surface air temperature below 10^{0}C, transformation in the size spectrum because of melting in the fall path, was reduced. It may be understood that only small numbers of the smallest hailstones might have been affected due to melting. Of the 20 cases considered, 15 displayed a unimodel asymmetric distribution. In five cases unimodal and bimodal distribution were encountered along hail path. Refer Fig. 2.7.

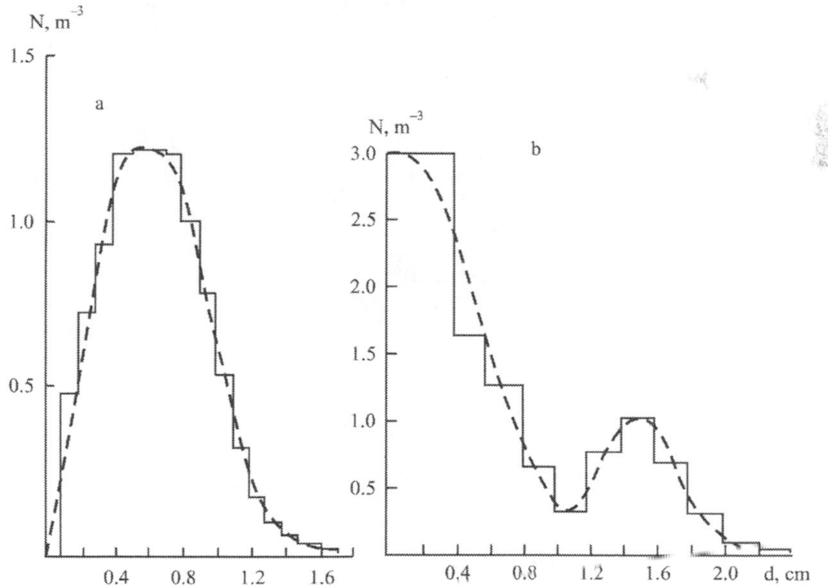

Fig. 2.7 Unimodal and bimodal hailstone-size distributions; a- Alberta (Canada) and b- Samsari (Georgia).

Size distribution is most satisfactorily described with the aid of a gamma function in the form,

$$\eta_{\mu,\beta}(d) = \frac{N}{\Gamma(\mu+1)\,\beta^{\mu+1}} d^{\mu} e^{(-d/\beta)}$$

Where $\eta_{\mu,\beta}(d)$ is the density of the size distribution of particle, N is the number of hailstones per unit volume, $\Gamma(\mu + 1)$ is the gamma function and μ and β are the distribution parameters. It was estimated that for various hail spectra μ lies between 0 -10 and β between 0.043 – 0.76 cm and standard deviation σ between 0.12 – 0.67 cm.

These hail spectra are also satisfactorily described by the one sided truncated normal distribution (Fig. 2.8) as

$$\eta_{\varphi,\xi,\sigma}(d) = \frac{N_0}{\sqrt{2\pi\sigma[1+\Phi(\varphi)]}} e^{-((d-\xi)/\sqrt{2\sigma})^2}$$

Where N_0 is the total number of hailstones in the sample and φ is the normalized truncated point, determined by φ = - (ξ/ σ) where ξ and σ are the mean and standard deviation; $\Phi(\phi)$ is the degree of truncation (the portion of the sample eliminated from the initial set). For truncation at d = 0 degree of truncation may be 30% (sulakvelidize 1969).

If γ_μ^{-1} is the reciprocal of the incomplete gamma function of argument μ and has the form

$$\gamma_\mu(z) = \frac{1}{\Gamma(\mu+1)} \int_0^z z^{\mu} e^{-z} dz$$

and if Φ^{-1} is reciprocal of the normalized normal distribution

$$\Phi(x) = \frac{1}{\sqrt{2\pi}} \int_0^x e^{-(t^2/2)} dt$$

In Fig. 2.8 abscissa is showing hailstone diameter d and the ordinates represent $y_1 = \gamma_\mu^{-1}[H(d)]$ and in Fig. 2.9 $y_2 = \Phi^{-1}[H(d)]$. H(d) is the distribution of the sample i.e. relative number of hailstones in the sample with diameters smaller than d.

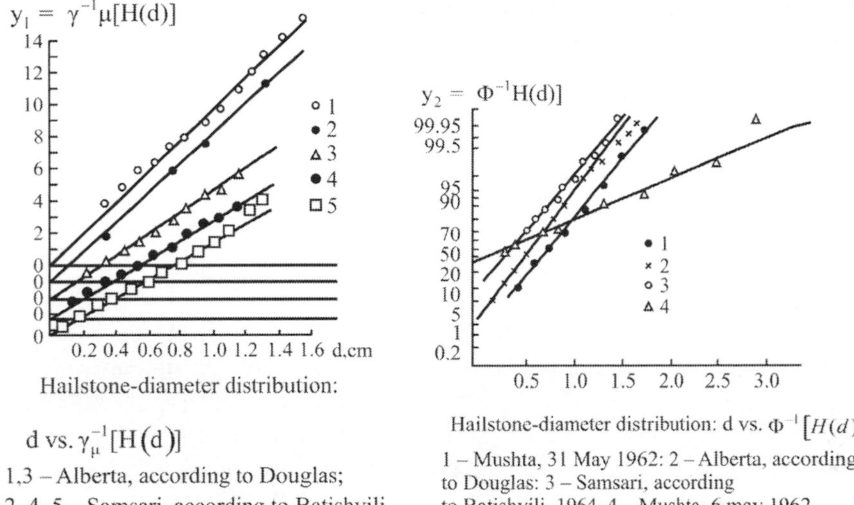

Hailstone-diameter distribution:

d vs. $\gamma_{\mu}^{-1}[H(d)]$

1,3 – Alberta, according to Douglas;
2, 4, 5 – Samsari, according to Batishvili

Fig. 2.8

Hailstone-diameter distribution: d vs. $\Phi^{-1}[H(d)]$

1 – Mushta, 31 May 1962: 2 – Alberta, according to Douglas: 3 – Samsari, according to Batishvili, 1964, 4 – Mushta, 6 may 1962

Fig. 2.9

Abshaev (1966) observed that when unimodal and bimodal distribution of hailstone sizes are observed in the same hail fall, a bimodal distribution of hailstones sizes are encountered in the central part of the hail path. Normally unimodal maximum is considerably large than the bimodal ones. In east European countries the average ratio of maxima had been reported as 3.6 and ratio of modal sizes of the hailstones spectrum for unimodal and bimodal respectively as 3.1. It may be mentioned; however, that ratio of maxima varies from place to place. But there is smooth transition from the unimodal to the bimodal distribution. A bimodal spectral distribution of hailstones can be described by the sum of two gamma distributions namely $\eta(d) = \eta_{\mu 1,\ \beta 1}(d) + \eta_{\mu 2,\ \beta 2}(d)$, where the parameter $_{\mu 1}$ takes values from -1 to 1and the parameter $_{\mu 2}$ taken values from 4 to 8. Hail spectra are independent of the climatic and geographic conditions i.e. unimodal spectral hail distribution is described as independent of the sampling location by a gamma distribution and a one sided truncated normal distribution.

As for the size distribution within a hailstorm, most authors assume an exponential or power law probability distribution function (Ludlam and Macklin, 1959; Fraile et al., 1992), even if literature on the spatial variation of this distribution obtaining statistically significant results is scarce (see, e.g., Sánchez et al., 2009). According to the study of Simeonov (1996) hailstones hitting hailpads within a sample of several hail days exhibit an inverse Rayleigh distribution. Fraile et al. (1999) propose a gamma distribution to consider also melting processes eliminating part of the smaller stones.

2.4.1 Large Stones

In India Eliot (1900) while discussing the hailstorm distribution in India has cited cases of hailstones of enormous sizes and weights from central India. He reports hailstones of even 0.4 to 1.8 kg weight. But biggest sizes of stones (more than 1.8 kg) were reported by him in NWFP (Pakistan). His statistics is, however, not very reliable due to lack of any scientific method those days, for hail measurement. Hariharan (1950) reported from Andhra Pradesh, Hyderabad (India), hailstone weighting 3.4 kg, on the ground (diameter ≈ 21 cm). On 25 of March 1996 a severe hailstorm over Imphal, Manipur State of north east India is reported to have lashed with large hailstones of sizes varying from 10 to 25 centimeters in diameter. Orography plays the most important role for the location of zones of large hailstones (Punge and Kunz, 2016). Bilhem and Relf (1937) quote a report of large stone which fell over potter western Nebraska in 1928 which had a diameter of 13.7 cm and weighed 681 gm. Its density was about 0.5 gm/cc. In eastern Colorado, USA, a favourite hunting ground of hailstorm, about one of every seven storm contains a few stones exceeding 2.5 cm. in diameter. Maximum sizes of hailstones in various parts of the world are shown in the Table 2.4.

Table 2.4 Maximum sizes of hailstones in various parts of the world

Region	Maximum diameter in mm
India	250
Germany	250
U.S.A.	203
Bangladesh	180
Italy	170
France	150
U.K.	150
China	40

Shown in Fig. 2.10 is the largest hailstone that fell near Vivian, South Dakota, U.S. during summer (2010) on July 23rd. It measured 20.3 cm in diameter, 47cm in circumference, and weighed in at 878.8 gm.

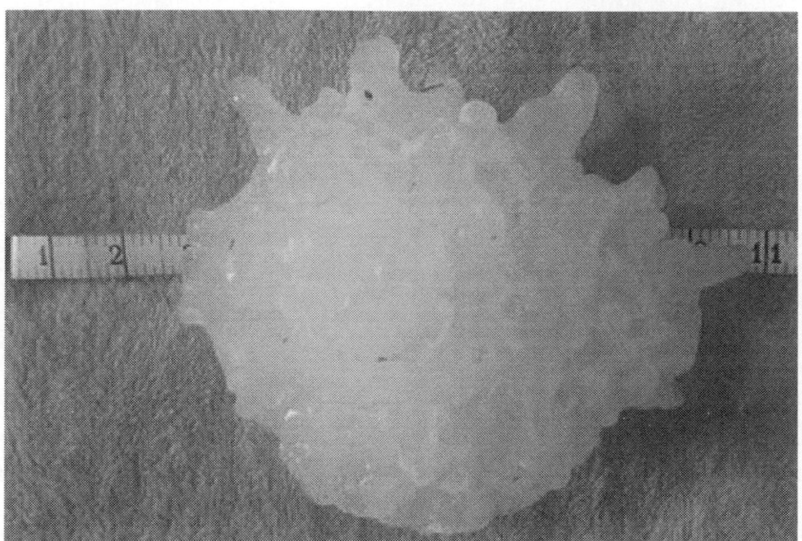

Fig. 2.10 The largest official hailstone ever collected in the U.S. An eight-inch monster that fell at Vivian, South Dakota on July 23, 2010. (Colour Plates Pg. No. 353)

On 10 May 1964 in northern Caucasus near the Malka River, large hailstone were observed consisting of friable snow graupel; their dimension reached 6-8 cm on 30 August 1965. Also in north Caucasus near the village Kichmalka, shapless pieces of ice fell whose average size was up to 8 cm. The June 1966 hail damage near Krasnodar and Gnozny in the mountainous regions of the Northern Caucasus was attributed to individual hailstones weighing 300 gm. Almost every year in this region there are one or two instances of hail damage in which the hailstones reach diameter of 5 cm although common size of hailstone there is 1-2 cm. Large hailstones usually fall in the Caucasus during July and August. In Voronezh on 14 August 1961, the maximum diameter of the hailstone was 8 cm and the weight of individual stone exceeded 300 g (Antonov 1962). These were not the conglomerates consisting of individually developed stones.

Hailstones size at higher levels is difficult to determine but some concept of the sizes of the stones aloft may be obtained from the size of the dent and holes made in the body of the aeroplane. Documented evidence exists that severe damage has been caused to aircrafts while flying between 5 to 9 km amsl. Refer section 11.5 for details. During one flight between Karanchi (Pakistan) to Delhi (India) at 5.7 km the largest size of the hole in the aircraft was 5" across which suggested the size of the hailstones to be in the proximity of 5". The size of the largest dents being 15" x 10" would make one tentatively conclude that diameter of the largest hailstones within cloud at that height where the viscount aircraft was hit with it, might had been 8" or so, if not more. There is however, possibility that the dent was repeatedly made by several hits made by different stones in close proximity on the airframe. Enlarged hole may also be attributed to this possibility. Since in its journey from super cooled water droplets region in a cumulonimbus cloud above 0^0 isotherm to the ground hailstones is likely to undergo considerable change through several processes, it is difficult to conclude about their shape and size aloft. While in some cases the size is likely to decrease due to evaporation or melting in other it is likely that because of collision with rain and other cloud particles at the subfreezing temperature it may increase by regelation and freezing or by conglomeration of more than one stone.

The rate of melting of a hailstone while falling through clear air will be determined by the rate of transfer of heat from its surroundings by conduction and convection. If its surface temperature is below the dew point of the air then melting process would be guided by the rate at which latent heat is liberated by the condensation of water vapour upon its surface. Hence melting occurs when net positive heat transfer takes place towards the stones from the air; which exceeds the heat transfer by it into air. The temperature of the air through which the hail falls below 0^0 isotherm level (i.e. in more than zero temperature regions) varies from layer to layer till it reaches ground. Melting condition will be determined by this surrounding air only. If 0° isotherm lies above 4 km of the earth surface then most hailstones with $d < 1.6$ or 1.8 cm normally melt while falling through the warm part of the atmosphere (i.e. above 0^0C) and reach the ground as liquid precipitation. Thus a convective cloud may give hail during winter but with the same height of 0^0 isotherms it may give liquid precipitation during summers. Also same cloud in which hail of diameter $d < 1.8$ cm forms, may give different types of precipitation at different orographic heights. In mountain regions this may give hail due to shorter path between the 0^0 isotherms and surface of earth and over plains it may give only shower. Hail frequency over the hilly region of India (Fig. 1.2(b)) is, therefore very high. Another point to be noted is that melting without water breakaway may be expected only for hailstones of diameter $d < 0.5$ cm since threshold radius of freely falling water drop in air ≈ 0.25 cm. For such spherical stones falling through still, clear air conditions Ludlam (1958) showed that following equation gives the change of radius,

$$R_0^{7/4} - R_g^{7/4} = 0.58;$$

which gives values of radius on ground (R_g) and initial radius of hail when it begins to fall in atmosphere from 0^0 isotherm level (R_0), as in Table 2.5.

Table 2.5

R_g (cm)	0	1	2	3	5
R_0 (cm)	0.73	1.3	2.2	3.2	5.1

These figures show that a smooth, spherical hailstone must have a diameter rather larger than 1 cm at 0^0C level if it is to reach the ground in hot weather. On the other hand if the size at the level is appreciably greater than this minimum value the shrinking during the fall to the ground is not very great. A modification to Ludlam equation was made by Baily and Macklin (1968) for lobed stones with radius more than 3 cm. This is because heat transfer from such stones is three time to that smaller stones. For these stones radius changes as

$$R_0^{7/4} - R_g^{7/4} = 1.53.$$

Changes for diameters 3, 4 and 6 cm on ground are shown in Table 2.6.

Table 2.6

Rg (cm)	3	4	6
Ro (cm)	3.37	4.30	6.23
% Change in R	12	7.5	3.8
% Change in Mass	41	24	12

Table 2.5 indicates that even very larges stones can undergo appreciable melting during their fall to the ground if they are lobed. Rate of melting of small hailstones, spheres of solid or rime ice, in falling from 0^0 level to the ground was computed by Mason (1956). The calculations showed that, in a saturated atmospheres with a temperature gradient of 6.5 ^{0}C/km, solid ice sphere of R = 3 mm would melt completely after falling 2.5 km and soft hail pellets of the same size, but density 0.3 g/cm^3, would melt completely after falling only 1 km from the 0^0C level. Laboratory measurements by Drake and Mason (1966) on ice particles

suspended in air stream of controlled temperature humidity and velocity give melting times that agreed fairely well with calculated values especially if allowances was made in the latter in the heat contents of the water retained by the particles during melting. The melting times of small ice cones of various shapes did not differ from those of spheres of equal volumes by more than 10 percent.

Rassmussen et al (1982, 1984a, b) showed that Mason's theory leads to an underestimate of the rate of melting hence it computes longer time for melting. The reason for the discrepancy were traced to the following: 1) the ice particle floats up in the liquid coat, the two forming an eccentric spherical pair, which increases the rate of heat conduction within the water coat and 2) convection within the water coat further increases the rate of heat transfer. Rassmussen et al developed equations to include these effects. They showed the results of calculations in good agreement with their wind tunnel experiments.

References

1. Abshaev, M.T. And O.I Chepovskaya (1966), Hail distribution function (Russian) Turdy VGI NO. 5.
2. Antonov V.S. (1962), Unprecedented Hail in Voronegh (Russian) Meteorologiya I Gidrologiya, NO. 3.
3. Atlas D, Harper w. and Ludlam F. (1961), Multiwavelength Radar Reflectivity of Hailstorm – QJRMS, Vol. 89.
4. Baily I. H. and Macklin W.C. (1968), Heat transfer from artificial hailstorm QJRMS 94, 93 (287, 346, 357 – 358, 368).
5. Barkow E. (1908), The formation of Groupel, Meteorolgiche Zeitschrift, 25 (10) 456 – 458. Gidromeleoizdat, Lengingrad.
6. Beckwith W.B. (1956), Hail observation in Denner Area, United Airlines Met Cir, No. 49 (257 – 8, 259).
7. Bilohem E.G. and Relf E.F. (1937), Dynamics of large hailstones QJRMS. 149 (349).
8. Browning K.G. and Beimers JGD (1967), The oblateness of Large Hailstones J. App. Met. 6. 1075 (333).
9. Carte (1963), Hailstones in Pretoria – witwalersrand area (CSIR) South Africa, News latter NO. 145 (334).
10. Decker W. Fredand Lyle D, Calvin (1961), Vol 42 No. 7 PP 457 – 480.
11. Douglas R.H. and Hitschfeld (1958), studies of Alberta hailstorms sci Rep. MW – 27 Mc Gill Univ., Montreal.

12. Drake J.C. and Mason B.J. (1966), Melting of Small ice spheres and cones. QJRMS 92, 500 (368, 529).
13. Eliot, J. (1900), India Met. Department Memoirs, 6, 6, pp 237.
14. Field P.R., Hand W., Cappelluti G., McMallian A., Foreman A., Stubbs D. and Standarisation – Final report for EASA, 2008. OP. 25, Date 14 November 2009.
15. Fraile, R., Castro, A., Sánchez, J.L., 1992. Analysis of hailstone size distributions from a hailpad network. Atmos. Res. 28 (3), 311–326.
16. Fraile, R., Sánchez, J.L., de la Madrid, J.L., Castro, A., Marcos, J.L., 1999. Some results from the hailpad network in León (Spain): noteworthy correlations among hailfall parameters. Theor. Appl. Climatol. 64 (1-2), 105–117.
17. Hariharan, P.S., (1950), IJMG, 1, 1, pp73.
18. Rasmussen, R. M. and H.R. Pruppacher, 1982: A wind tunnel and theoretical study of melting behavior of atmospheric ice particles. I: A wind tunnel study of frozen drops of radius < 500µm. J.Atm. Sc39, 152-158.
19. Rasmussen, R. M. V. Levizzani and H.R. Pruppacher, 1982: A numerical study of heat transfer through a fluid layer with recirculating flow between concentric and eccentric sheres. Pure Appl. Geophys., 120,702-720.
20. Rasmussen, R. M, V. Levizzani and H.R. Pruppacher1984a: A wind tunnel and theoretical study of the melting behavior of atmospheric ice particles. II: A theoretical study of drops of radius < 500 µm. J. Atmos. Sci., 41, 374-380.
21. Rasmussen, R. M, V. Levizzani and H.R. Pruppacher, 1984b: A wind tunnel and theoretical study of the melting behavior of atmospheric ice particles. III: Experiment and theory of spherical ice particles of radius> 5000µm, J.Atmos. Sci., 41, 381-388.
22. Levin L.M. (1961), Studies in the Physics of coarse dispersed Aerosols (Russian). Izdatel stro ANSSR, Moskva.
23. List R. (1959), Wachstum Von. Eis – Wasseryeminischem in Hagelwrsuchskanal Helv. Phys. Acta. 32, pp 293 – 296.
24. List R. (1960), zur Thermodynamik Teilweiso Wassriger Haglkorner Z angew Math. Phy. 11, 273 – 306.
25. Ludlam, F.H. (1958), (a) The Hail Problem Nubila 1 59.
26. Ludlam F.H. (1958), (b) The Hail Problem Nubila 1, 12 (351 – 7, 366 – 8).

27. Ludlam, F.H., Macklin, W.C., 1959. Some aspects of a severe storm in SE England. Nubila. 2, pp. 38–50.
28. Macklin, W.C., (1961), Accretion in mixed clouds, QJRMS, 87, pp 413-424.
29. Macklin, W.C., (1963), Heat transfer from hailstone, QJRMS, 89, 360, pp 351, 357, 366-68.
30. Mason B.J. On the Meeting of Hailstones QJRMS 82, 209 (368).
31. Massop, S.C., (1971), Weather, Vol 26, No. 5, pp 222.
32. Parson J. (1959), Report of Hail Survey Operation and observation (Unpib file copy). Medford organ Medford pear shipper Assocition.
33. Punge, H., Bedka, K., Kunz, M., Werner, A., 2014. A new physically based stochastic event catalog for hail in Europe. Nat. Hazards 73 (3), 1625–1645.
34. Punge H.J, Kunz, M., 2016, Hail observations and hailstorm characteristics in Europe: A review, Atmospheric Research 176–177 (2016) 159–184
35. Rakstit D.K. (1962), IJMG, Vol. 13, 77 – 80 pp.
36. Reynolds O. (1879), On the manner in which raindrop and hailstones are formed, Library of Philosophical Soc. of Manchester, Memoirs, 3: 48.
37. Sánchez, J.L., Gil-Robles, B., Dessens, J., Martin, E., López, L., Marcos, J., Berthet, C., Fernández, J.T., García-Ortega, E., 2009. Characterization of hailstone size spectra in hailpad networks in France, Spain, and Argentina. Atmos. Res. 93 (1-3), 641–654.
38. Shvedev F. (1881), Chrorakoe grad, Russkoe Fizko-Khimicheskoe obshchestuo, Otdel 1, 1, Chast Khimiia, No. 1 p 71; No. 3 p 92.
39. Simeonov, P., 1996. An overview of crop hail damage and evaluation of hail suppression efficiency in Bulgaria. J. Appl. Meteorol. 35, 1574–1581 (Sep.).
40. Souter R.K. and Emerson J. B. (1952), Summary of available Hail literature and the effect of hail on aircraft inflight, NACA, Washington, Tech Note No. 2734 (259).
41. Sulakvelidze, G. K., (1969), Rainstorm and Hail, Israel Program for Scientific Translation.
42. Weichmann, H., (1950), The Problem of the formation of precipitation in cumulonimbus clouds, proc. Chicago. Conf on third Elec. P 49 (259).
43. Weickmannn, H. (1962), The Language of Hailstones and Hail – Nubila Anno. Vol.5.

CHAPTER 3

Hail Sampling and Analysis

Collection of hailstones and their examination for shape, size, water, content, structure, density, temperature, roughness etc. give various clues to their origin and growth. Basic difficulty in hail collection are their breaking after impact on the ground and delay in their collection which results in their melting and deshaping from their original size and shape. Storing hails at subzero temperature to avoid melting may cause freezing of the water content inside the capillaries of the hail's inner structure. Much early hail research was motivated by the economics of crop and aircraft damage. General weather research and storm forecasting techniques and the availability of meteorological radar and automatic data processing gave fillip to the hail research. Crop damage promoted further research on damage assessment techniques; and various projects on hail suppression in the 1960s.

Initially, at great majority of stations in India, the meteorological observation were made by part time observers by non scientific means which used to cause their melting and deshaping. Observers often used to fail to follow instructions regarding the reporting of weather remark correctly. During 1960's with the expanded interest in hail, special purpose network of trained volunteer observers was set up in many parts of the world to record detailed information about hail occurrences and associated meteorological phenomena. As widely dispersed, uniform and objective measurement of hail parameters are desired, various instruments were developed to provide the measurements of hail and related parameters. Development in hail measurement systems have taken the form of refinements of observation reporting methods, by stratification of estimates and expansion of instrument capabilities by addition of technological improvements for observer dependent data systems including hail chase operations, manual sampling and also remote data collection instruments. Detailed review of surface hail measurements has been provided by Nicholas (1977). In the present chapter we describe the various aspects of sampling, measurements and analyses of hail.

3.1 Hailstone Sampling

Accurate and timely sampling of hail is the starting point for any type of study in hail and hailstorm. Various methods have been suggested both manually and through instruments for proper sampling. However, even the most reliable instruments and methods suffer from inability to adequately sample hail at a point. Due to extreme horizontal variability of hail, even on small scales and relative scarcity of hail occurrences, it is very difficult to get an adequate sample of hail from a single storm. Hence network and statistical analysis must be employed to fill in the information gaps.

Appropriate sampling is preceded by appropriate hailstone collection.

3.1.1 Hailstone Collection

There are two approaches of collecting hailstones i.e. either without the presence of any person by instruments or manual collection immediately after fall.

(a) Static Instrument Collection

(i) ***Fly-screen Funnel**:* Douglas and Litchfield (1958) suggested one fly-screen funnel acts as shock absorber with aperture angle 135^0 and filled on the base, with dry ice to prevent any melting. List (1961) suggested that smaller aperture of 60^0, as large angle of 135^0 may cause large number of incident particles to bounce back and get broken. The disadvantage of this type of hail sampler is that it has to be checked every two days and refilled with dry ice. Although it has advantage of remote placement, and can be deployed in an array but suffers from sampling problems such as limited size and lack of time resolution i.e. it does not convey the temporal order, in which hail might have fallen. Data reduction and analysis also requires considerable time and resources.

(ii) ***Recording Hail Samplers:*** The Swiss mobile hail sampling system by Federar (1976) is shown in Fig. 3.1. It collects time resolved hail samples and preserves them in chilled hexane.

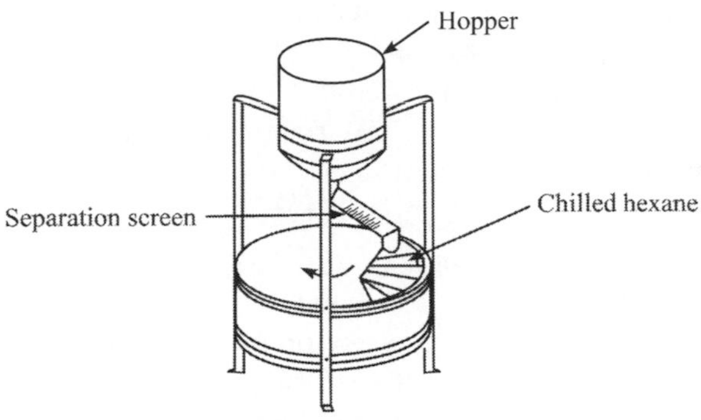

TIME RESOLVED HAIL COLLECTOR

Fig. 3.1 Schematics of Swiss recording sampler (Federer, 1976).

Fig. 3.2 A photograph of one of the instrumental mobiles used for sampling. The precipitation collectors are on the boom extending towards top right.

(b) **Mobile Hail Chase and Collection:** Hail collectors mounted on a vehicle could be used to collect hails all along the path of the hailstorm. Time and location of the surface collection may be chosen from the most intense hail or rain echoes on radar screen emanating from the storm and collection vehicle may be moved there. Aircraft reports of hailstorm will also assist in positioning the vehicle at suitable place for sampling. Such mobile hail chase provides most detailed data on hail falls. One of such vehicles is shown in Fig. 3.2. Hail samples on such vehicles are collected in polythene bags in plastic buckets being supported by metal rings atop a mast. The mast could be lowered for insertion and removal of liners and hoisted into the air for the duration of collection. Collectors are approximately 4m above the ground level and positioned so as to be away from any obstacle from which water could splash into the collectors. After collection the liners are carefully removed from holders, and placed on dry ice to freeze the sample.

It may be mentioned here that fluid or solid base, cooled below 0^0C, into which the hailstone may be allowed to fall directly, is only to be recommended where no measurement of temperature or water content are to be made. An exception only lies in the case where the collection container has itself been filled up as a calorimeter, so that one of these values can be measured directly.

(c) **Manual Hail Collection:** Volunteers may be selected for picking up the hail stones from the ground. While picking up hails it should be kept in mind that size, spectrum and depth of fallen hail vary considerably along and across a path. Hail can best be picked up from lawns or grassy surfaces, where the impact is deadened and the insulation is adequate, which is an important factor up to the moment of collection. Smaller ice particle, such as all forms of graupels, must definitely be collected with container cooled to a temperature below 0 °C. In case of graupels the speed of fall is so considerably reduced that chances of disintegration of these particles are rare.

3.2 Measurement of Shape and Size of Hail

The characterization of the shape and size of a hailstone provides an important source of information. It gives values for the condition determining the aerodynamic behaviors of the ice particle as it falls and in some cases indications of the conditions in which it originated. Size and shape vary considerably along the length and width of hailpath.

(a) **Physical Examination:** Examining hail in cold laboratory (Fig. 3.3) and their photograph serves as an important piece of evidence and may even be taken stereoscopically. Manually collected hail can also be measured in cold laboratories. The advantage of this method is that in case of excrescences, length and thickness of spines or protuberances can be measured separately.

Fig. 3.3 Physical examination of hailstones.

(b) **Hail pad:** For determination of size Parson (1959) used one foot square of styrofoam, one inch thick covered with aluminium and mounted about one meter above ground on an inverted fruit box (Fig 3.4). Schleusener and Jennings (1960) and Decker and Calvin (1960) also used similar hailpads (Fig. 3.5) for size measurements. They measured the major and minor ones of elliptic dents left in the aluminium foil and related them to stone size by calibration using steel sphere. These were dropped in the pad from a height sufficient to match the kinetic energy of a theoretical hailstone falling at its terminal velocity. It was assumed that the stones were hard, smooth spheres of known homogeneous density falling through undisturbed air of known density and viscosity. Many variations in the composition of the styrofoam foil and in the instrument calibration and data reduction procedures have occurred since the appearance of the same. Strong (1974) has discussed many aspects of hailpad measurements and operations features. Because of its simplicity low cost, and rugged construction it is ideal for widespread distribution in the field. Hailpads are effective indications of the presence of hail. When properly calibrated, they provide an estimate of the hail size distribution at a point, integrated over the duration of hailfall. Because these can be correlated with crop damage the hailpad has been considered a useful tool for objective assessment of crop damage, although there are many uncertainties in the correlation, (Changnon 1971). The hailpad, which does not record time, may suffer from overloading by heavy hail resulting in overlapping of hailstone dents. It is also important to note that the limited size of the hail pad sampling area (typically 30cm square) may insufficiently record the size distribution of a light hail. Finally there is no way to distinguish between two successive hailfalls occurring between service events of the hailpad. Thus its value is limited to studies of individual storm. In normal operation, the hailpad is said to provide hail size distribution to an average accuracy of ± 20 %.

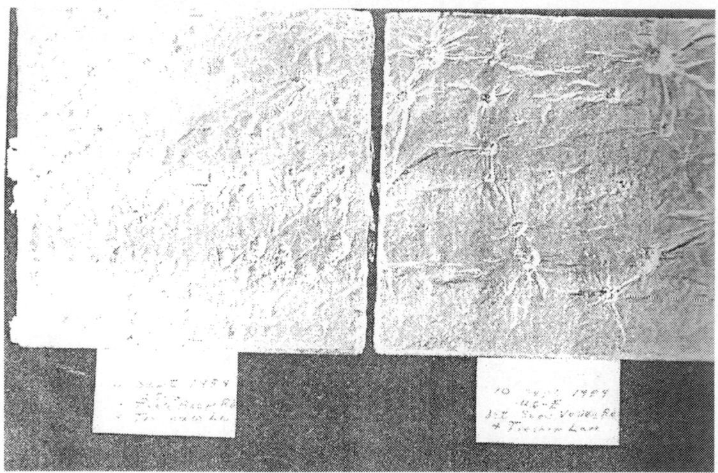

Fig. 3.4 Hail recorders; each one foot square, dented by hailstones.

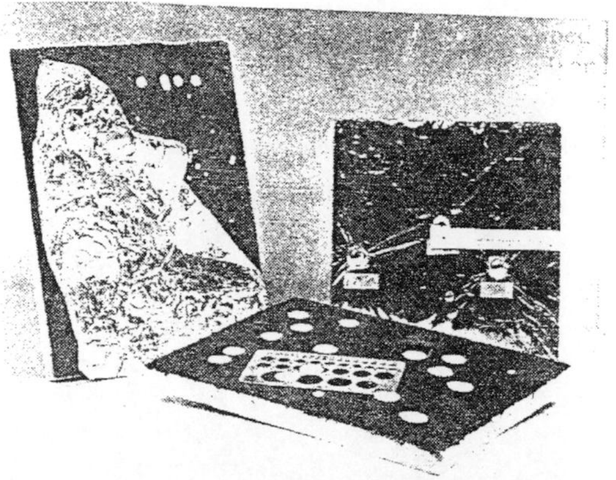

Fig. 3.5 Hailpads used by National Hail Research Experiment (NHRE), U.S.A. along with data reduction tools. Dents on pads were highlighted by the roller painting foam before measuring.

(c) **Hail Stool:** Wind-blown hail is measured on hail-stool. The instrument consists of a round, horizontal surface supported by a foil-covered styrofoam cylinder (Fig.3.6). Wind blown stones impacting at an angle less than 14^0 from vertical appear as dents only on the top surface. For greater vertical angles, stones impact the cylinder and an estimate of mean wind speed and direction can be made by measuring the distance from the top surface to the uppermost dent on the cylinder.

Fig. 3.6 Illinoise State Water Survey hail stool and hailpad (Towery and Chagnon, 1974).

(d) **Hail Cube:** A more successful method for estimating wind accompanying hail was developed in italy (vento, 1972). the device shown in Fig. 3.7 consists of a cubical frame fitted with one horizontal and vertical hailpads.

Fig. 3.7 Illinois State Water Survey hail cube mounted on fence post (Towery and Chagnon, 1974).

The relating numbers of stones on the vertical and horizontal surfaces yield estimates of the mean horizontal velocity of the stones and the direction of impact.

3.3 Instruments Related to Hailfall

Swiss hail sampler in Fig. 3.1 although supplied time-resolved hail sampling, had limitation of capacity and duration. Duration of hail fall at any place, its time, and estimate of hail-rate is necessary for verifying radar measurements of hail and for many other indepth studies. However due to significant cost and complexity of these instruments they are not useful for widespread distribution and data collection. Some are described below:

(a) **The Geophone:** Meant for recording the impacts of hailstones on the surface, this instrument has a momentum sensor consisting of a seismic geophone transducer and associated circuitry. it is designed to produce signals generated by hail impact so as to be distinguishable from other phenomena such as rain and records the information on cassette magnetic tape. the major disadvantages of the present model are its inability to record clock time and its small (15 cm) diameter sampling area and its inability to respond to wind-induced horizontal moment.

(b) **NOAA Momentum Sensors:** This instrument (Fig. 3.8) consists of aluminium impact surface of 36 cm diameter, mounted on spring. It deflects in response to the momentum of impact of a hail stone. Once activated by the impact of a hailstone the chart driveruns until 90 seconds after the impacts. The instruments are battery powered and event actuated. It records time of hail to the accuracy of ± 1 min and has momentum measurement accuracy of ± 2.5%. It should be noted, however, that the instrument does not respond well to hail stones less than 1 cm diameter. Instrument's sensitivity to wind gusts in the field environment is one of the handicaps.

Fig. 3.8 The NOAA hail momentum sensor showing expanded aluminium surface and strip chart recording system (Towary and Chagnon, 1974).

(c) Ballistic Pendulum Momentum Sensor: The instrument records hail momentum based on the principle of ballistic pendulum. It is self powered, simple, relatively inexpensive for a recoding instrument and is capable of measuring the momentum of 6 cm to 7 cm diameter hailstone occurring every 1.5 seconds for as many as three to 10 minutes period. Deflection in the 30 cm diameter platform are recorded and related, through calibration, to the momentum of a falling hailstone. Maintenance problems and sensitivity of the instrument to wind gusts are disadvantages of this instrument.

(d) Recording Rain – Hail separators:

(i) ***Swiss Recording Rain hail separator spectrometer*:** This instrument is normally mounted in an mobile chase van. The foam rubber-lined rotating bottom plate is photographed every 30 seconds while hail is falling.

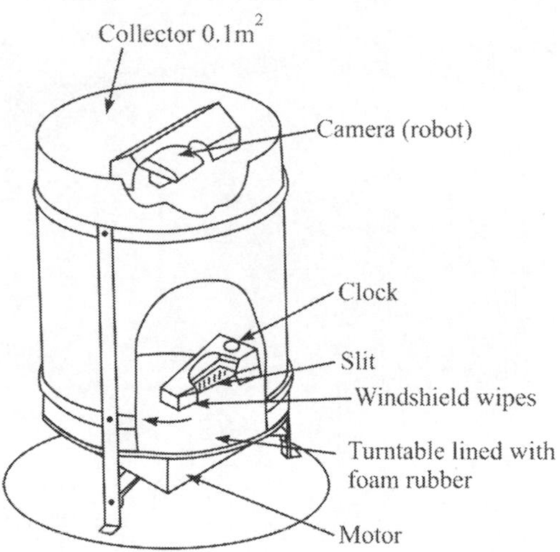

HAIL SPECTROMETER

Fig. 3.9 Swiss photographic spectrometer showing turntable which rotates every 30 seconds revealing new hail distribution for photography (Federer, 1976).

At the end of 30 seconds period the stones on the plate are removed by wiper blades. The photogrtaphs give a time-dependant hail size spectrum as sampled by the 0.1 m^2 opening. Fig. 3.9 shows the instrument diagram.

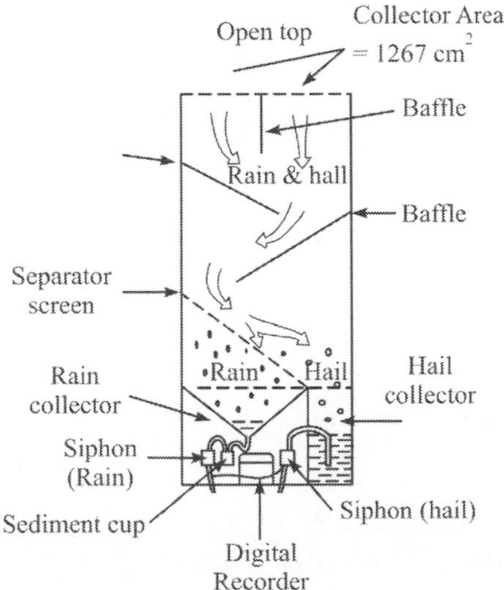

Fig. 3.10 Schematic of the National Hail Research Experiment (NHRE) hail/rain separator showing recording system.

(ii) ***NHRE Rain Hail Separator:*** In 1973 the NHRE (USA) designed rain hail separators with flow rate transducers and a digital recorder was designed to measure one minute mass amounts of rain and hail. Fig. 3.10 shows a cross section of the separator and the recording system. Hail falling into a water-filled reservoir displaces its mass in water and the overflow is monitored by another flow digitizing device which feeds into a second channel in the recorder.

(e) Moving Foil Hail Recorder: This instrument was used during Alberta, Hail Project (Koren 1969). This is battery operated moving foil hail recorder as shown in Fig. 3.11. In this, hail falling through the 46 cm square opening is separated from rain by an inclined screen. Hail closes a switch and then falls onto the foil from a fixed height. The switch closure produces a time mark on the recorder and starts the foil drive motor. The instrument produces a continuous record of hailstones impressions.

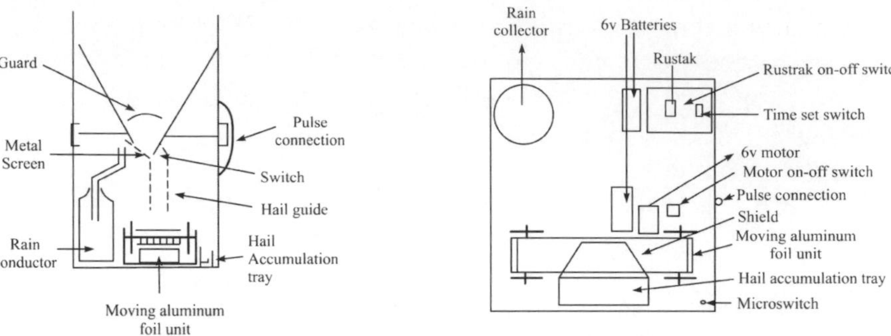

Fig. 3.11 Side and plan views of moving foil hail recorder from Alberta (Koren, 1969).

(f) **Optical Disdrometer:** A recording hail spectrometer (Fig. 3.12(a)) was developed by NHRE (USA). A light source shines on a parabolic mirror which creates parallel rays which travel across the sampled region and are intercepted by a horizontal array of 100 phototransistors.

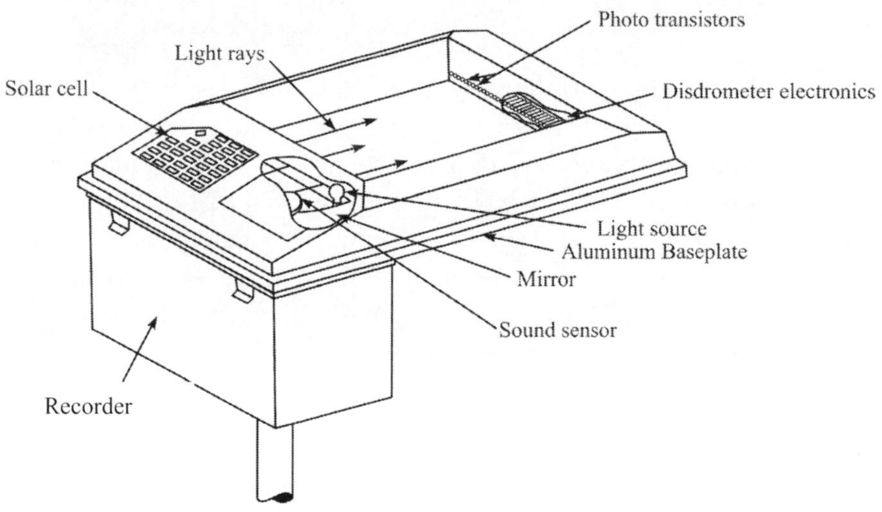

Fig. 3.12(a) Schematic of NHRE optical hail Disdrometre (Nicholas, 1977).

A hailstone falling through the 25 cm X 37 cm sampling area intercepts a beam of light as wide as its maximum horizontal dimension. Electronic scanning of the photocell at 10 kHz permits an estimate of the fall speed of nearly spherical stone. The instrument provides a time resolved vertical kinetic energy spectrum as well as size distribution. One difficulty with the instrument is its inability to distinguish small (5-6 mm diameter) hailstone from the larger water drops. It is activated upon impact of a hailstone on the case and records up to 10 hailstones per second on 35 mm film.

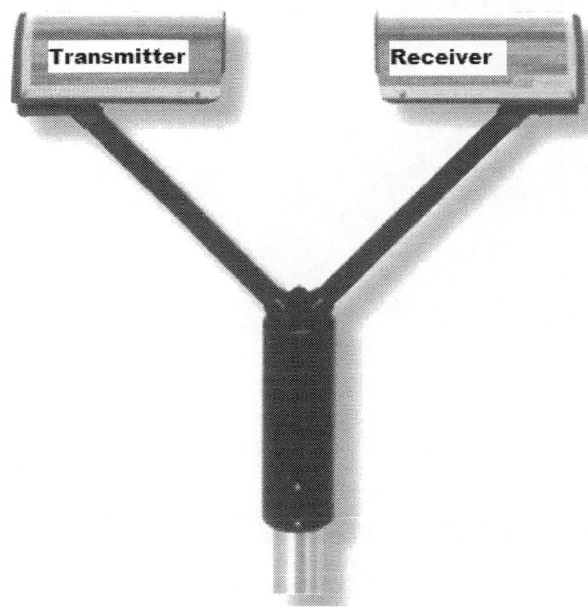

Fig. 3.12(b) Laser Optical disdrometer for capturing particle size and velocity of liquid and solid precipitation. (Courtesy OTT Parsivel; www.OTT.com).

Recently multifunctional Laser Optical disdrometer is developed to measure all types of liquid and solid precipitations. Refer fig. 3.12(b). The transmitter unit of the sensor generates a flat, horizontal beam of light that the receiver unit converts to an electrical signal. This signal changes whenever a hydrometeor falls through the beam anywhere within the measuring area (54 cm^2). The degree of dimming is measure of the size of the hydrometeor and the fall velocity is derived from the extinction signal duration. Accuracy of solid precipitation measurement by this instrument is ±20%.

(g) **Hail Wind Detector:** It consists of a curved, gridded impact surface mounted on a wind vane which orients itself to the on coming hail and records the average angle of incidence of the hail. Two problems with the instrument are it's small sampling area and slow response characteristics of vane (Fig. 3.13).

Fig. 3.13 Hail-wind detector. The instrument orients itself into the falling hail and provides an estimate of the average angle of impact of wind driven stones (Towery and Chagnon, 1974).

3.4 Hailstone Examination

Various physical properties of a hailstone provide useful information with respect to its origin, formation and profile before impacting the ground surface. Different methods used to determine these properties are described below:-

(a) **Specific Gravity:** Since hailstone may be composed entirely of clear ice or of opaque ice or a mixture of both, their mean specific gravity may lie anywhere between 0.3 and 0.9. The first reliable information concerning the actual specific gravity of hailstone was made available by List (1958), Vittori (1960). Vittori and Di Caporicco (1959) carried out very precise measurement of hailstone density to within ± 0.0005 g/cm^3. Three hailstones out of 40, in their sample, had the minimum density of 0.80g/cm^3 and the remaining 37 densities varied from 0.873 to 0.915g/cm^3. Hail density varies from 0.5 to 0.9 g/cm^3. The density of dry (opaque) hail is at present usually taken as 0.6 g.cm^3 and that of moist (transparent) hail as 0.9 g/cm^3. Table 3.1 shows the density range computed in different experiments.

Table 3.1 Experimentally determined densities of grauple and hail (≥ 5mm) and their size range*

Size Range in mm	Density Range in kg m^{-3}
0.5 – 3.0	50 – 450
0.5 – 1.0	450 – 700
1.0 - 2.0	250 – 450
0.4 - 3.0	80 – 350
0.5 – 3.0	850 – 890
0.5 – 6.0	500 – 700
0.8 – 3.0	130 – 130
8.0 – 19.0	310 – 610
1.0 – 7.0	200 – 700
26.0 – 36.0	834 – 856
11.0 – 31.0	810 - 900
9.0 – 39.0	870 - 915

*.(Courtesy Visvarajah Somasundaram, School of Aerospace, Mech and Manu. Engg., RMIT Univ., Australia)

Fig. 3.14 Apparatus for measuring the buoyancy of hailstones.

While measuring the density of hail, assessment of weight and volume must be undertaken at a temperature below 0^0C. Volume measurement of hailstone can be in rare cases, undertaken geometrically i.e. when the structure is exactly spherical, rectangular or some kind of shapes measurable geometrically. Otherwise it is possible to know the buoyancy of hailstone by means of a wire device immersed to a fixed point which presses the hailstone inside the fluid. The extra weight created in this way can be weighed and corrected for the buoyancy of the wire device, which enable the desired buoyancy of the hailstones to be calculated. The most advantageous sealing fluid is tetralin, though other fluid are possible, the only conditions are that the fluid should not attack ice and that it is saturated with water, so that it does not absorb H_2O. Refer Fig. 3.14.

Measuring the densities of porous particles and in particular of soft hail is difficult because the air capillaries suck in the liquid designed to give buoyancy. This obstacle can be overcome, however, by having recourse to phthalic acid diethylester, a liquid which was originally used be Quervain (1954) for thin section of snow without damage to the structure. This substance has the characteristic that it freezes at approximately -7^0C and can thus easily be injected by means of hypodermic syringe at a higher temperature into the open capillary systems of the hailstone under investigation. If accurate doses are given, the surface tension of this special fluid brings about an evenly curved surface. Furthermore, if the temperature is now reduced below -7^0C, then the phthalic acid freezes and a compact particle results. Since we know the net weight of the ice and the total weight of the compound particle, we can again use the technique described above for computing buoyancy in order to measure the density. But tetralin must be replaced by oil or another sealing fluid, which does not dissolve phthalic acid.

The method described above yields the true density of loose ice structure. Earlier Nakaya (1954) had given constant density 0.125 g / cm^3 for all soft graupels. With the aforesaid method it was found that this figure was not true for all the cases - List (1958). In the method described above measurement of entirely enclosed air spaces and their separation from the capillary systems connected with the outside air has not been done. To obtain such a result, the procedure of Kusunoki (1958) could be used, which measures the quantity of air enclosed in the ice. However, since the air pressure in the individual pores is not known, reliable conclusions could be drawn with regard to density by the procedure of Kusunoki (1958), either.

In case, sufficiently accurate balance is not available and when no separate measurement of weight and volume is wanted then extremely simple method for determining the density was devised by Macklin (1961). He used a number of glass cylinders each filled with liquid of different density. By establishing the fluid in which the hailstone under observation just begins to sink, a first rough indication is given to its density, while measurement of fall in the cylinder gives a more exact figure.

(b) **Surface Temperature of Hailstone:** It is ideal to use radiation pyrometer to determine the surface temperature instantly and with sufficient accuracy. Thermometer may also be applied for temperature measurements. But its sensitivity will be limited. The temperature of fallen hail was measured at the Geophysics Institute of the Academy of Sciences of Georgia. According to its report the temperature of hail after falling on the ground is close to 0^0C but sometimes as high as 5^0C or as low as -6^0C is also noticed.

(c) **Surface Roughness of Hailstone:** It can best be carried out by the replica method of Schaefer (1956) whereby the dried polyvinyl formal skin is removed and examined under microscope for average variation in roughness. The differences in height which then show up can be directly measured by focusing the microscope on them.

In order to ascertain the influence of different liquids on the outer surface of hailstone, Vittori (1959) had used this method for determining the roughness.

(d) **Determining the liquid water content**

(i) **Method–I:** The total amount of water in hailstone can be determined with calorimetric method, but it is to be assumed that the hailstone exhibit a homogeneous temperature of 0^0C. This condition is although not entirely satisfied under all conditions since within the hailstone itself the most varied temperature gradient can occur. However, the extent of this gradient may be considered limited due to relatively good thermal conductivity of ice, Measuring the amount of water in a hailstone is then simple, provided one knows the total aggregate weight of both ice and water.

The hailstone is put into Dewar flask, which is partly filled with liquid of known amount, temperature and specific heat (water can well be used for this). One has then only to measure the decrease in temperature of the liquid after the hailstone has melted in order to deduce the original quantity of ice, which is particularly active as a result of fusion. Hence if the weight of the liquid water in hailstone is $G^{w}{}_{H}$ then,

$$G_H^W = G_H - \frac{G_L C_L (t_1 - t_2) - G_H C_W t_2}{S}$$

where, G_H^w = The weight of the liquid water in gm.

G_H = The weight of the hailstone (gm).

G_L = The weight of the liquid in calorimeter (gm).

C_L = Specific heat of the liquid in calorimeter cal/gm^0.

C_W = The specific heat of water (Cal/gm^0).

t_1 = The original temperature of liquid in calorimeter (^{0}C).

t_2 = The final temperature of liquid in calorimeter (^{0}C).

S = The heat of fusion (Cal/gm) and $t_1, t_2 > 0^0$C.

The measurement of amount of water must be done, however, with great care by means of calibrated calorimeter; List (1960). The disadvantage in the aforesaid method lies in the destruction of the hailstone, but this may be felt to matter less as such measurements have to be carried out immediately after the fall of hail when plenty of hail stones are available any way.

(ii) Method–II: This method is based on calorimetric determination on how much cold is needed to freeze liquid water in the hail and reduce the entire hailstone to a temperature below 0^0C. In this case water is no longer required as auxiliary fluid. The liquid water content in hailstone G_H^w can be measured with the following formula.

$$G_H^w = \frac{G_L C_L (t_2 - t_1) + G_H C_E t_2}{S}$$

Where t_1, $t_2 < 0^0$C and C_E = Specific heat of ice (cal/gm^0). This method of calorimetric measurement has a disadvantage since it must be carried out at a temperature of less than 0^0C and also requires much more time until the temperature equalizes through the middle of the hailstone. But unlike the first method it does not bring about the destruction of hailstone.

(e) **Removal of Liquid water by Centrifuge Force:** When the liquid water in a fallen hailstone is contained in the external zone or connected with the surface in someway, then it can be obtained by method through the use of centrifuge (Fig. 3.15). But with this method water contained in spaces sealed off from the surface will not be obtained. The portion of water in a hailstone can easily be measured by means of a centrifuge, if the hailstone is weighed both before and after being spun dried. The decrease in weight gives the weight of the water that has been removed. In using the method care should be taken that centrifuge shows a temperature near 0^0C and that the inner chamber, where the hailstones, placed on cotton wool, is brought to exactly the melting temperature of the ice by means of a mixture of ice and water. If this is not done, additional phase changes could take place and influence the measurement result.

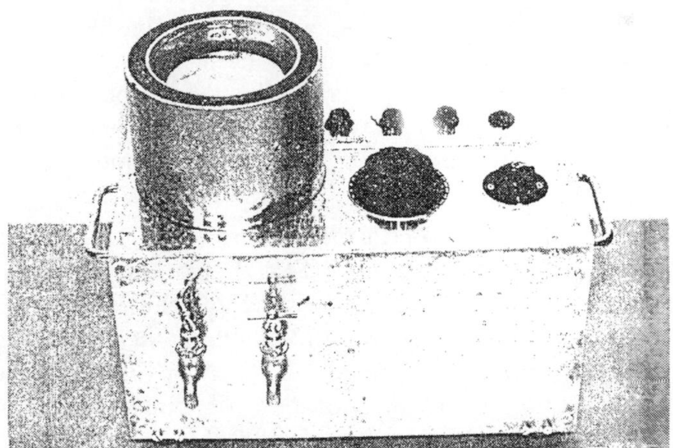

Fig. 3.15 A centrifuge cooled with a mixture of ice and water for spin-drying watery hailstones or artificial ice particles.

3.5 Storing of Hail Stones

Storing should be such that apart from freezing of original liquid water, contained within the hailstone no other changes in structure should occur. In any case, it is advisable to examine hailstone within a few months of their fall. The ideal temperature at which hailstone may be stored is between – 10^0C and -20^0C in closed glass or plastic bowls which are shut up in insulated boxes. These should possess inner or outer surface having good conductivity so that internal temperature gradient disappears. In this way any vaporization of hailstone onto the coldest part of the bowl can be prevented and thus the exterior of the hailstone can be well preserved for years. In this method hailstone does not show transformation of structure for over a period of even more than two years.

3.6 Aerodynamic Forces, Stability and Position during Free Fall

To understand the build-up and origin of a hailstone, it is of great importance to have some knowledge of the characteristic of fall shown by bodies of different shapes. This means that the aerodynamic forces acting on different types of body must be measured. By determining the coefficient of air resistance, we obtain figures for the speed of free fall in a given direction. Since, not every fall position is stable, we still have to distinguish which positions can actually occur in nature. For this tests can best be carried out with models falling through water in an experiment tank, in which case Reynold's law of similitude have to be observed. For measuring the drag coefficient of hailstone wind tunnel is the best equipment. It also gives better understanding of the physical processes controlling the hail formation. Before we go about understanding the method of measurement of drag coefficient we must understand first the design and method of operation of Hail Wind Tunnel.

(a) **Hail Wind Tunnel:** The principal feature of the Hail Wind Tunnel can be seen from the diagram (Fig. 3.16). Fan I forces the air through the temperature regulation and filtering section 2 into an angle piece which is followed by humidying and equalizing section 3 measuring about two meters in length. This leads into the measuring section, 4 in which various experiments are carried out. From here the air stream passes through a diffuser into the return duct 5 and back to the fan, thus closing the circuit. The required humidity conditions are obtained by injecting steam or water at the beginning of equalizing section 3; the ice forming seeds which determine the ice characteristics of the moisture directed on to the test objects are also added at this point. The installation is controlled from desk 6, where all important values are indicated and recorded. Other parts of the Hail Wind Tunnel are shown in the diagramme in Fig. 3.16.

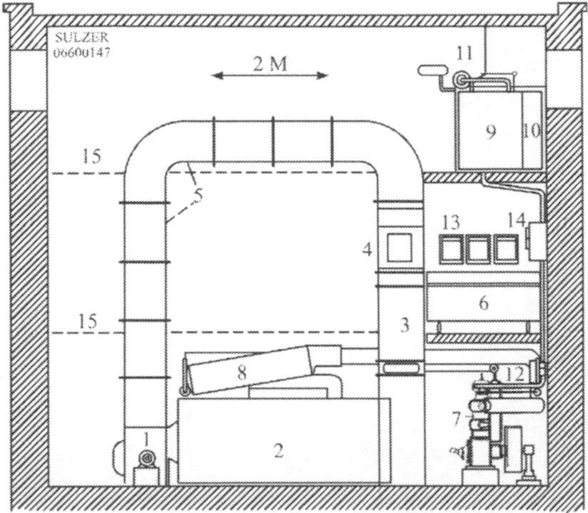

Fig. 3.16 Diagram of Hail Test Tunnel, installed at the Federal Institute for Snow and Avalanche Research, Weissfluhjoch.

1	Fan	9	Ammonia Condenser
2	Temperature Regulator	10	Compressor
3	Humidifying and equalizing section	11	Oil separator
4	Measuring Section	12	Motorized valve
5	Return Duct	13	Compensating recorder
6	Control duct	14	Temperature control system
7	Refrigerating compressor	15	Floor grating
8	Liquid Separator		

(b) Measurement of drag coefficient of hailstone: The drag coefficient of hail is a function of Reynold's number (Re), which generally exceeds 10^2. Fig. 3.17 shows drag characteristics with increasing Reynolds number for a smooth and rough sphere. A C_d range from 0.5 and 1.0 can be expected for hail impacts to vehicle body panels.

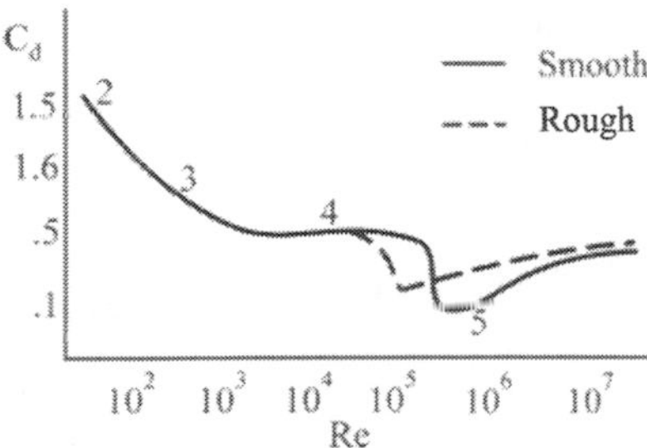

Fig. 3.17 Drag Coefficient (C_d) of a sphere against Re {http://www.slideshare.net/visvarajah/the-characterisation-of-hail-and-fraudulent-impacts-to-vehicle-body-panels}.

Note thate C_d for rough and smooth surfaces rapidly falls with increase of Reynold's number from $10^2 - 10^3$. It then stabilizes till 10^4. During $10^4 < Re < 10^5$ C_d for rough surface rapidly falls as compared to smooth surface. For $R_e > 10^5$ both curve rise after fall.

The best way of measuring the drag coefficient of hailstone is in a hail wind tunnel in which the air can be cooled or which is itself set up in cold laboratory. The ideal arrangement of the experiment is to have the hailstone frozen onto the end of a pendulum so that it hangs straight into the axis of the tunnel. The air resistance is measured by bringing the pendulum into the vertical position by means of a sensitive spring balance, in such a way that the aerodynamic forces acting on the hailstone are compensated by the spring balance. With the cross section of the hailstone determined photographically, the picture being taken along the axis of the wind tunnel, all the values are known, including the air speed and density, for calculating the drag coefficient. It is then possible to calculate the speed of free fall which corresponds to this position. Care must be taken here that the air speed at which the coefficient is measured roughly corresponds to the speed of free fall. This is the only way to eliminate the element of uncertainty which could arise from change in character of air flow. It is advisable to see that the pendulum is frozen into hailstone along the line of one of its main axes. Various different positions can also be measured by twisting the pendulum round.

The drag coefficient obtained this way sometimes differ considerably as regards speeds of fall the values for smooth spheres which are occasionally adopted in theoretical calculations. The fall speed of hail can be highly variable and can be affected by environmental factors such as updrafts and downdrafts. Therefore a simplification is made by assuming the hail to be smooth spherical object and falling at a terminal velocity (V_T) theoretically expressed as:

$$V_T = \left\{ \frac{4g\, \rho_{Hail} D_{Hail}}{3C_{dsphere} \rho_{Air}} \right\}^{1/2}$$

Where V_T is the terminal velocity (m/s), g is acceleration due to gravity(9.81 m/s^2), ρ_{Hour} is the density of hail(kg/m^3), D_{Hail} is the diameter of hail (m), $C_{dSphere}$ is the drag coefficient of the sphere and ρ_{Air} is the density of air (kg/m^3). The expression shows that as the size and density of hail increases, the terminal velocity increases. However as the drag coefficient and air density increases, the terminal velocity decreases.

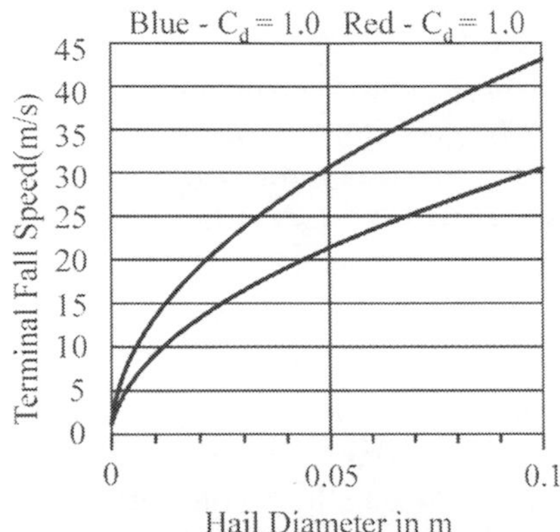

Fig. 3.18 Terminal fall speed range {http://www.slideshare.net/visvarajah/the-characterisation-of-hail-and-fraudulent-impacts-to-vehicle-body-panels}

Fig. 3.18 shows the fall speed range of different hail sizes bounded by terminal velocities for C_d of 0.5(fastest fall speed) and 1.0(slowest fall speed) calculated using hail density of 900 kg/m^3, 9.81 m/s^2 for acceleration due to gravity and air density of 1.225 kg/m^3 at 15^{0}C. Matson and Huggins (1980) measured 15–20 m s^{-1} fall speeds for 1-cm-diameter hailstones. Large hailstones, however, have much large terminal velocities (tens of m s^{-1}). Roos (1972) estimated a 47 m s^{-1} fall speed for the giant 1970 Coffeyville, Kansas, hailstone (refer Fig. 2.2).

In this connection point to be noted is that not only spherical ice particles are rare in atmosphere but natural hailstone never exhibits an entirely smooth surface. We thus obtain a flow character, naturally differing from that occurring around a smooth object.

3.7 Method of obtaining Thin Sections

Size, shape, weight, volume, coefficient of air resistance and surface temperature etc, explain only the external characteristics of hailstone. Meteorologists are interested, however, just as much in the intermediary phases in the growth of a hail stone as in the end product for our final understanding on the hail formation. Quervain (1954) had developed technique for obtaining very fine section for snow. The hailstone is cut up to produce as many thin slices of ice as are needed, having a thickness of about 0.3 mm or even less. These cross sections give us information about the hailstone the arrangement of air bubbles and of the individual ice crystals. Much depends on where the slice is made through the hailstone. Care should be taken that one slice is made through the oldest part of hailstone, the original particle which is generally a graupel. This we call the main section. The growth center, however, by no means necessarily coincide with the center of gravity or the center of symmetry in hailstone. List (1958) and List and Quervain (1953) showed that the origin of growth is generally located centrally in the hailstone and often even at the periphery.

The Thin Section Saw

It is consist of basically a circular cutting blade, adjustable in height and running horizontally. The carriage with the hailstone is mounted firmly on a turntable or slide table which travels into the revolving blade and produces an incision. The machine diagram is represented in Fig. 3.19 which is briefly described below.

The slide or mount with hailstone is engaged by suction due to a vacuum between time mounting cylinders. Excellent support is achieved by this method which cannot result in any distortion of the glass slide. This slide can be changed very quickly, particularly as the bearing surfaces cannot get dirty. Apart from the suction mechanism, the slide is also slightly fixed mechanically.

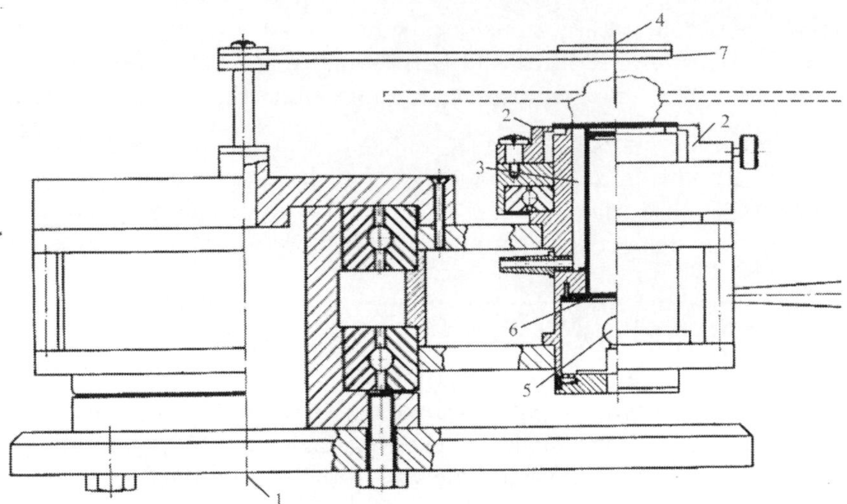

Fig. 3.19 Consecutive parts of the revolving cutting table: (1) Revolving axis of the cutting table, (2) Clamp for the slide, (3) Cylinder ring with vacuum for fastening the slide by section, (4) Axis about which the slide clamp can be turned, (5) Source of light, (6) Polarized filter- I, (7) Polarized filter- II (turned through 90^0 with regard to polaroid filter- I).

(a) The mechanical fixture of the slide enables it be turned about its own axis. This allows the hailstone to be cut from whichever side is most suitable.

(b) The circular velocity of the turntable is adjustable and automatic. The hailstone does not, therefore, need to be inserted by hand into cutting machine. A stop switch terminates cutting automatically.

(c) The thickness of the section can be measured during cutting if the analysis (a Polaroid filter) is shifted over the axis of the specimen holder. Polarized light is generated along this axis which, when analysis, gives the crystallographic picture of the ice section and the colours, which appear are suitable for thickness.

(d) The thickness of the section can, however, also be checked directly by adjusting the vertical height of cutting blade.

(e) By cooling the motor and drive mechanism of the saw, it is possible to prepare, at a temperature of even – 10 ^{0}C, a thin section from loose ice structure, too.

The Preparation of Thin Sections

To prepare thin sections from ice of high density (i.e. with few air bubbles), the following steps have to be taken;

(a) A slight thaw is initiated on one side of the specimen hailstone by placing it on a heating plate at a temperature slightly above the freezing point of water until a sufficiently large flat surface has been created. This thawed surface must be located so that it is parallel to the main section i.e. the section which we want to examine.

(b) Thin thawed surface is brought into contact with a slightly warmed glass slide. The hailstone must then be pressed against the glass so that the water remaining between them, on freezing, is not thicker than 0.02 mm. Since this intermediary layer produces no colour effects in polarized light when it is as thin as this, the additional ice produces no change in the appearance of the crystallographic structure of the thin section itself. (For the examination of normal hailstone the glass size should measure 50 × 50mm.)

(c) An intermediary stage is now necessary during which the water between the hailstone and the slide is frozen. This must be done slowly, as too rapid freezing – like a badly cleaned slide – will result in poor cohesion. It should be realized that the main pressure bears on this joint during cutting, so that it must not be weakened in any way.

(d) The slide with the hailstone on it is now fastened to the slide table of the saw, and the height of the blade set so that that cut will go through the desired zone of the hailstone. If the inner structure of the hailstone cannot be seen at the beginning, due to an opaque outer shell, then it is wisest to make the first cut at a position which will make it possible to see into the inner zones of the hailstones.

(e) Cutting is done in such a way that the mechanical strain on the part of the hailstone which is frozen to the slide is kept at the minimum. This usually means that one saws from one side only as far as the middle of the hailstone, then stops the cutter and begins again from another side. Since, depending on the quality of the ice, small pieces can be born out of the cut surface, it is advisable to make an additional cut and to set blade about0.3 mm down so that the new surface can be made smooth.

(f) This surface represents one boundary of the zone from the hailstone which we have to examine more closely. Another slide has now to be frozen onto this surface, the actual slide which will carry the section. The new slide is slightly warmer for this purpose, so that when it is placed on the cut surface sufficient ice will be melted to produce a faultless joint between the hailstone and glass. If the slide is too warm, however, superfluous water will be created which will cause interference and have to be removed artificially later.

(g) A second pause is made here to give the liquid water times to freeze to hailstone onto the slide.

(h) The actual thin section itself is now produced, by first fixing the new slide onto saw. The hailstone will now have been turned upside down to form its former position and then sawing through the hailstone at approximately only 1 mm above the thin – section slide. We then have two pieces;

    (i) The rough thin section

    (ii) The remainder of the hailstone which is still stuck to the first slide.

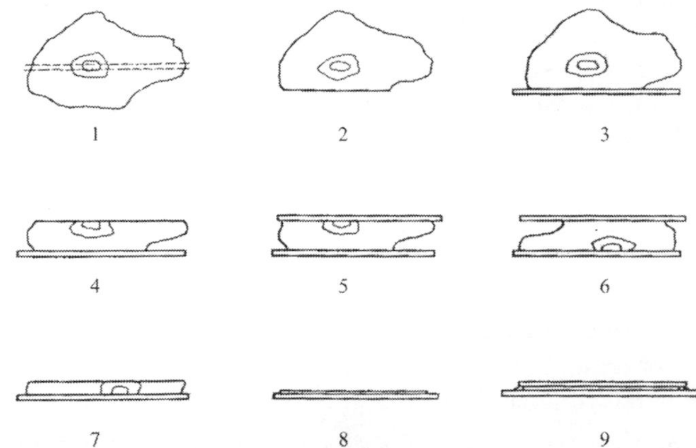

Fig. 3.20 Stages in the production of a thin section: (1) Hailstone with intended place of incision, (2) Preparing a flat surface by melting, (3) Mounting on a warmed slide and freezing, (4) Fixing the slide onto the cutting table, cutting and polishing, (5) Freezing-on the thin section glass slide, (6) Turning and laying onto a glass plate covered with liquid paraffin, filling of the intermediary space with liquid paraffin.

The former i.e. rough thin section is now reduced in height by stages between 0.3 and 0.4 mm until it has the desired thickness. For colour photographs of the arrangement of the single ice crystals, a thickness of between 0.35 and 0.45 mm is recommended, while for a black and white evaluation an ice layer of between 0.25 and .35 mm is the best. It should be said that these figures are particularly suitable for photographs of the ice structure; for observing the air bubble in the ice, and particularly for detecting their arrangement in "shells", it is far better to photograph thicker sections which can be reduced later.

(i) The thin section glass slide is now cleared and should be scrapped with an erasing knife to remove all traces of superfluous ice or water which have been caused by thawing the hailstone onto glass.

(j) To prevent vaporization of thin section, it is placed on glass plate (60 × 60 mm) covered in liquid paraffin, and the remaining space between the two pieces of glass is then filled with this liquid.

For clarity, the various stages in this process are represented in figure. 3.20. We thus have a thin section ready for observation, photography and further examination. Important point to be noted here is that aforesaid method will not be suitable for hailstone containing a large number of air bubbles in them. This is because the water which would emerge while the slide was being frozen onto the ice would be sucked by capillary action into the hollow spaces in the ice and would either entirely or partly fill them up. The fine-section picture would in this case be falsified in its most important features. Thus for obtaining thin section in such a case following steps are taken.

(a) Hailstone, graupel etc is to be dipped in a liquid, phthalic acid diethyl ester, which freezes at -7^0C, and penetrates in a still liquid condition into all the connecting air capillary system of the ice. The combined ice and phthalic acid which, in the case of smaller particles, is generally put into a container, is now frozen at a temperature of -10^0C or less. A solid body results which can be sawed without fear that lamellas of ice will break.

(b) All the steps from (a) to (i) of thin section in case of hailstone without large number of air bubbles, as described above, may now be applied, with the single difference that slide has in each case, to be frozen on not with water but with liquid phthalic acid diethylester. This intermediary layer has to be allowed for in reckoning the final thickness of the thin section.

(c) Since the sealing fluid used has the characteristic that it interferes with the double refraction of the ice, it is necessary before commencing observation, to dissolve it out chemically. This is done with tetralin. This fluid can at the same time be used as a preservative against the evaporation of the ice, in the same phase where liquid paraffin was used earlier. Unfortunately, tetralin itself evaporates relatively quickly, and this calls for frequent checks of the fine section which have been prepared and to be kept for any length of time. The result is a fine section with all its ice laminas and ice walls intact, neither broken nor disturbed in position.

The most difficult fine section to prepare, however, are those from hailstone where loose and dense ice occurs side by side. As phthalic acid diethyl ester does not produce sufficient adhesion to stick compact ice onto glass, it is necessary to go back to the old method of freezing on. This can be done very well if there are large air hollows in the ice which can be filled with the organic liquid. If not, then one has to try and freeze compact ice zone onto the slide with freezing water, while the neighboring loose-ice zone are treated and stuck with phthalic acid. To prepare a perfect slide under such conditions requires considerable delicacy of touch and much experience.

3.8 The Structure of Air Bubble and their Significance

The technique for making fine sections opens up the new world of the hailstone interior. The first knowledge to be gained is based on the arrangement of the air bubbles which are responsible for the more or less opaque zones. One can surely assume here that similar conditions of ice growth produce ice having the same constant density or which is the same thing as having on an average the same number and size of air bubble. About zones where the air bubble preserves the same character, we know that the conditions of growth did not alter greatly. This does not mean, however, that we are already in a position to judge with regard to particular ice zone what the respective conditions of origin actually were. Important for us, are the areas of instability in the arrangement of air bubble. These changes in density, where they form continuous layers either in closed shells or open segments indicate a definite moment in time when the conditions of growth changed abruptly. We are thus in a position to mark off the shape which the hailstone had at this moment. If there are lines of bubble which are not closed, these indicate that the fall position of the hailstone at that particular stage was extremely stable, so that always only the same zone of particles was iced. Fig. 3.21 shows the fine section of one hailstone. The point where growth began is indicated by the letter 0. There then developed a cone shaped graupel G. When this had reached the size of 5 to 6 mm, the conditions of growth changed and this led to a partial layer P up to stage C, then the direction of fall with regard to the shape of the hailstone remained fixed.

There then followed and intermediary zone where the original conical hailstone became rounder as it took on slowly freezing water and turned into a rotational ellipsoid with a continually changing fall position R. No transitional phase from cone to ellipsoid can however be recognized simply from the bubble structure alone. Three dimensional measurements show that the hailstone actually had a rotational ellipsoid shape at the stage of the closed circle of bubble R. The final phase can be considered as the extreme shape to which a triaxial ellipsoid approximates with, however, indentations occurring in the direction of the shortest axis and giving a doughnut appearance.

From these observations we may conclude that changes in the condition of growth result in a change in the shape of a hailstone. With this information in hand we gain a basis for determining exactly the conditions of origin, for we can artificially produce all the established intermediary shapes and ice them over until we get similar ice deposits as those which are found in the subsequent zones in the hailstone.

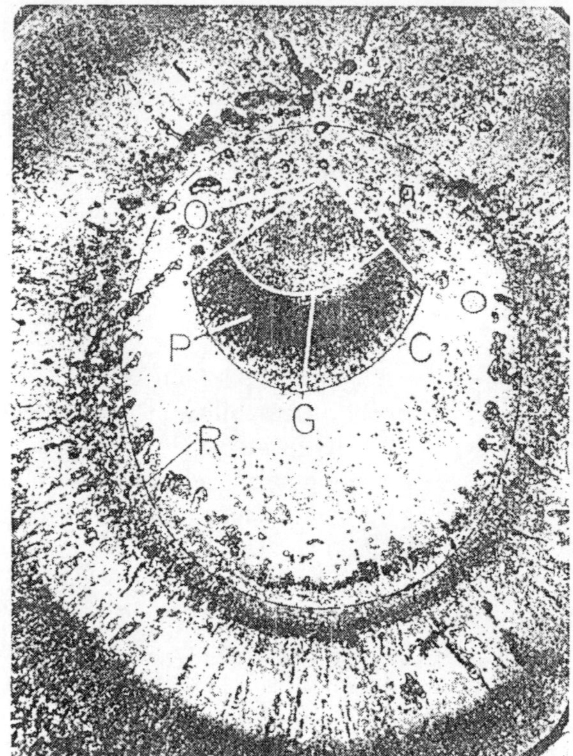

Fig. 3.21 Thin section from hailstone K 57.15 by ordinary light. O: Origin of growth, P: Partial layer, G: Maximum size of the graupel, C: Conical hailstone, R: Shell of the rotary ellipsoid (2.4:1).

3.9 The Structure of Ice Portion of Hail

The crystalline structure of ice shows us, in the same way as the structure of air bubbles, all the changes in the conditions of growth which are responsible for a particular hailstone. They also make it possible, however, to form much more finely differentiated picture in cases, for instance, where loose ice frameworks subsequently take on slowly freezing water.

Fig. 3.22 Thin section from hailstone K 57.15 by polarized light: arrangement of the ice crystals in the centre of the hailstone.

The intermediary shapes of the hailstone which are revealed by the layers of air bubbles can be in the structure; in addition, however, the individual ice crystallites also indicate, the spatial direction of growth, provide they are big enough. Fig.3.22 shows the inside of a hailstone in polarized light with the particular intermediary shapes and directions of growth clearly emphasized. The direction of growth of ice zones belonging to the same period of origin can normally be extended backwards to the point where they intersect in so-called centers of symmetry. Fig. 3.22 exhibit, for example three centers of symmetry (S_1, S_2, and S3). Each bigger change in the condition of growth generally has a dislocation of the center of symmetry, as a result. Another advantage of thin section kind is that it allows us to determine the crystal axis of single ice crystal, a factor which can sometimes be of some importance.

3.10 Painter and Schaefer Method for Internal Examination

For making a structural investigation of hailstone painter and Schaefer (1960) gave another simple method which light on the crystalline buildup of ice particles of high density. This process may well be the simplest of all more elementary examinations, though it cannot replace the more precise thin-section technique. In this technique a hailstone is stored at a temperature of – 20 ^{0}C and a cut is made through it. The new surface is then polished carefully and thereafter exposed to the cold air for 20 min. Then the open surfaces of the single ice crystal vaporize with varying rapidity according to the optical direction of their axes. If now, such a surface is painted with polvenylformal (temperature – 4 ^{0}C), then after a certain time a skin can be removed which carries an exact impression of the differences of height and therefore the structure of the hailstone.

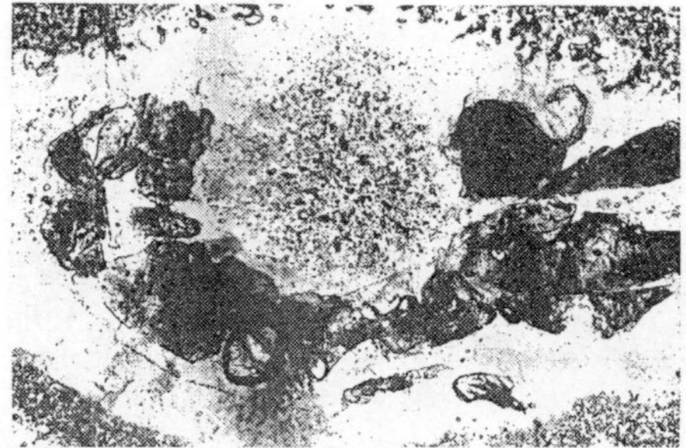

Fig. 3.23 Centre of hailstone K 59.2 with air cavities which have been filled with water at an earlier stage of growth (8.7:1).

3.11 The Structure of Water Caverns

It has been observed that during the growth of hailstone water also becomes incorporated in the ice and cannot freeze because of insufficient total heat exchange with the surroundings. Its effects further complicate the problem of hail formation. The condition which has been established experimentally under which ice- water conglomerations or pure-ice deposits can form, are presented separately by the curves shown in Fig. 3.23. It was assumed for this purpose that we were dealing with spheres of varying diameter which were freely falling and becoming iced over. One should notice that for all the water to freeze in the case of spherical hailstone 2 cm in diameter with free content of 5 gm / m^3, the temperature in the cloud must be at least –27^{0}C. In the case of hailstone of 4 cm in diameter, the limit is as low as –38^{0}C. Due importance should be attached to this factor in the study of growth of hailstone. Hence simply disregarding the water which slowly runs out of such ice – water conglomerations, or attributing it merely to melting is not correct. Thus great care must be taken in measuring the density of hailstone, because as was mentioned earlier, this is done at temperature below 0^0C.

References

1. Changnon, S.A. Jr. Donald W Staggs, 1969, Recording hail gauge evaluation, ISWS final Report, NSF Grnt GA – 1520. 47 pp.
2. Changnon S.A. 1971, Hailstone characteristics related to crop damage J Appl Meteor, 10, 270-274.
3. de Quervain M. 1954, Die Metamophose des Schneekrstalls, verhandlungen der Schweiz Naturforschenden Gesellschaft, Davps, 114-122.
4. Decker, Fred W., and Lyle D. Calvin, 1960, Hailfall of 10 September 1959 near Medford, Oregon, Bull. Am. Met. Soc., 42, 475-480.
5. Douglas R.H. and W. Hilscheld, 1958, Studies of Alberta Hailstone, Sci. Rep. MW-27, Mc Gill Univ., Montreal.
6. Federer, Bruno, 1976: Preliminary outline of a line resolved hail sampling system, Hail: Vagaries of hail science and hail suppression, Am Met Soc. Met Monograph, p 258.
7. Fremsted, P.G. 1968, Transducer for measuring momentum Final Report, NSF Grant GA-935, Inst. Atmos., Sci., South Dakota School of Mines. 47 pp.
8. Gjelsvik, Asbjorr M. 1966, A Signal processing and recording system for a hailstone momentum transducer, Final Report, NSF Grant GA-935 Int. Atmos. Sc, South Dakota School of Mine & Tech, 39 pp.
9. Koren O. 1969, Development of a moving foil hail recorder. Tech memo Tech 711, Dept of Transport Meteor. Branch, Canada 17 pp.
10. Kusunoki K. 1958, Measurements of gas bubbles content in sea iced, low temp. Sec. Ser. A 17, 123-134.
11. List R, 1958, (a) Kennzeichen atmospharischer Eispartikeln, II Teil, Z angew, Math. Phys. Pa, pp 217-234.
12. List R. 1958, (b) Kennzeichen atmospharischer Eispartikeln, I. Teil Z. angew Math. Phys. 99, 180-192.
13. List R. 1960, Zurthermodynamik Teilweise wassriger Hagelkomor. Z. angew, Math, Phys. 11, 273-306.
14. List, Ronald, 1961, Physical Methods and instruments for characterizing hailstorm, Bull. Am. Met. Soc., 42, 452-466.

15. List R 1962 Hail test tunnel, Weather pp 317-319.
16. Machline W. C. 1961, Imperial College, London.
17. Matson, R. J., and A. W. Huggins, 1980: The direct measurement of the sizes, shapes, and kinematics of falling hailstones. *J. Atmos. Sci.*, **37**, 1107–1125.
18. Moragan Griffith, Jr. and Neil G. Towery, 1974; Micro scale studies of surface hailfall, final Rep. NCAR 25-73, UCAR, 34 pp.
19. Nakaya U. 1954, Snow crystals, natural and artificial, Cambridge, Harvard University Press, 510pp.
20. Nicholas, T.R.(1977), AMS Meteorological Monographs, Vol. 16, 38, 257-267.
21. Parson, J., 1959, Report of hail survey operations and observation (unpublished) Medford, Oregon, Medford Pear Shippers Association.
22. Roos, D. S., 1972: A giant hailstone from Kansas in free fall. *J. Appl. Meteor.*, **11**, 1008–1011.
23. Schaefer. V. J. 1956, The Preparation of snow crystal replaces VI. Weatherwise 9, 132-135.
24. Schleusener, R.A. and Paul C. Jennings, 1960, An energy method for relative estimate of hail intensity, Bull. Am. Met. Soc., 41, 372-376.
25. Schotland R.M. and Kaplin E.I. 1957, The Collision efficiency of cloud droplets. Peng Press.
26. Strong, Geoffrey S. 1974, The objective measurement of Alberta hailfall, M.S. Thesis, Dept. of Geography, University of Alberta, 182 pp.
27. Towery, Neil G., and Staneley A. Changnon 1974, A review of surface hail sensors, J. Wea. Mod., 6, 304-315.
28. Vento, D., 1972, Rev Ilal Geofis, 21, 73-77.
29. Vittori, O.D. and Suiliano and Caporiccol, 1959. The Density of hailstone – Nibula Anno, Vol. 2 Verona.
30. Vittori, O 1960, On the Effects of pressure waver upon hailstones, Nubila, Anno, and Vol. 3 No. 1.

CHAPTER 4

Convection and Thunderstorm Formation

Ascent of air mass is attributed to the formation of convective clouds. Sufficient literature is available on the formation of these clouds in same air mass and also in frontal processes. In tropical countries one is mainly concerned with air mass process. Due to heating of underlying surface and the unstable stratification of atmosphere, instability is released and cumuliform clouds and often thunderstorms, hailstorms, duststorms, tornadoes etc arise and develop. All these phenomena are connected with powerful vertical air current. Concept of vertical current becomes more conspicuous in the case of hailstorms i.e. thunderstorms which support hail in them. Thus the main foundation of the study of hailstorm lies in the study of the convective motion giving rise to strong vertical currents, powerful enough to support hails. Such a situation, however, has to be supported with the favorable conditions for the formation and growth of ice particles within them. Increasing availability of new observation techniques over the last score has led to a great deal of descriptive material on the structure of hailstorm. In the present chapter we will study the convective process, formation and movement of thunderstorm.

4.1 Convection

The fundament building block of hailstorms, as of all convective motions, is a cell. A cell in this context is the dynamical entity characterized by a compact region of relatively strong vertical current, which can be identified by the radar from its associated volume of relatively intense precipitation. Convective element may takes numerous shapes and forms. Various terminologies are used to describe them. But majority of meteorologists prefer to call it as bubbles. Low (1925) observed that over the heated underlying surface several layers carrying bubbles are noticed. Shipley (1941) had recognized that within cumulonimbus cloud there exists turbulence, vigorous and sustained ascending currents in the central part and the presence of horizontal as well as vertical gusts at the same level surrounding the central core of vertically rising air. Systematic study of vertical motion inside the cloud was first of all attempted by Ludlam (1959) who found that top of the thick cumulus cloud may develop at a rate of 3-5m/sec and sometimes even at the rate of 10m/sec. But values calculated by Shiskin (1960) showed that top of these clouds develop at much slower rate of 1.2-1.5 m/sec. Estimation of the vertical currents with the help of development of the top, however, is likely to be erroneous as vertical current within the cloud may be much stronger than growing top; Zak and Fedora (1958).Over estimation of the rate of vertical development may be also due to the neglect of the effect of horizontal movement of the cloud during radar measurements. Ascending currents of the order of 50 m/sec or more have also been observed inside a matured cumulonimbus cloud; Weisman et al. (1983).

4.1.1 Gutman Model of Cell

Gutman (1963) proposed that moist air enters the cloud through the lower and lateral bases and immediately begins to ascend. The water vapour contained in it condenses and latent heat of condensation is released which heats the central region of the lower part of the cloud. Apparently almost all the water vapour manages to condense in the lower part of the cloud. Hot air, therefore, ascends in the central of the cloud and carries along formed water droplets. In the central part of the cloud air ascent mainly follows dry adiabatic law. As a result temperature difference with the surrounding remains positive but begins to decrease. This difference vanishes at a point on the cloud axis (center), situated approximately two third up the height of the cloud. Any further ascent of air together with the water droplets is by inertia. Temperature difference continues to decrease with increasing height and at this stage it is negative. The rising air, thereafter, gradually slows down vertically and the air moves horizontally. As the particles move farther from the cloud axis, they become warm at slower rate due relatively lesser vertical current as water is condensed and cools faster as the height increases. Hence at a given height the central core of the cloud may be hot while its periphery is cool. In the neighborhood of the cloud top where the vertical velocity vanishes, it will obviously be colder than the surrounding medium and fully saturated with water. As the air in the central part and upper part of the cloud carrying water droplets radially moves away from the cloud axis, it being cooler than surrounding at once begins to descend. This induces evaporation of water droplets which further accentuates the cooling; vis-à-vis intensifies the descending currents. As a result all the liquids moisture evaporates and dry air begins to flow from the middle part of the cloud. This cooling process diminishes rapidly in a horizontal direction as one moves further away from the cloud and then transforms into weak heating.

Gutman's (1963) axisymmetric model had provided the distribution of vertical and horizontal velocity within the cloud at the initial stage of its development. But his concept of dry adiabatic ascent was debatable. His stationary asymmetric model also does not explain the moisture accumulation below the cloud top which leads to the formation of larger droplets.

4.1.2 Lebedev Model of Cell

Lebedev (1964, 1966) gave more realistic explanations of convective cell in which he accounted for the moisture accumulation at the top of the cloud. He further accounted for evaporation at the cloud boundary, which give the convective cells a typical shine in the atmosphere and also included the change in atmospheric stratification in his model. He determined the velocity component u, v, w in the horizontal x, y and vertical z directions respectively. Some of the important theoretical results obtained by him are summarized below:

I. During the whole time before which cloud begins to disintegrate the ascending current velocity rises to a maximum value after which it begins to fall.

II. Ascending current velocity attains maximum value (10.3 m/sec) at a height 3 km a.m.s.l.40 minutes after the beginning of the convection.

III. Only at the moment of intense precipitation fall at the level of maximum ascending current it transforms into descending currents.

IV. Ascending currents are observed only at 4-6 km and do not exceed 1.5 m/sec in magnitude, thereafter.

His results were obtained on the MINSK computer and are subject to observational verification.

4.1.3 Mull and Kulshreshtha Model

To satisfy the slow rate of growth of cumulus top Mull and Kulshreshtha (1969) proposed the model of layered convection. Concept of layer may not be true in cumuliform stage but it provides some directive for the pre-cumuliform stage. Each layer of cloud while transforming from stratiform to cumuliform, could be comprised of pair of vertical vortices such that each pair of these vortices will resolve in similar direction in the region of contact and that even if between each such pair there did exist initially another vortex with opposite sense of rotation, it is probably very short lived and unstable. Thus the rotation and circulation in the case of a cumulus on this consideration may be expected to be shown in Fig. 4.1. This bubbled structure meets the observational facts that a developing cumulus cloud even at its ab-initio stage of formation is not an uniformly ascending mass but a complex system of irregular current. During the distinctly cumulus stage of its growth, however, this model does not explain the observed features of strong ascending currents in the central region of cloud.

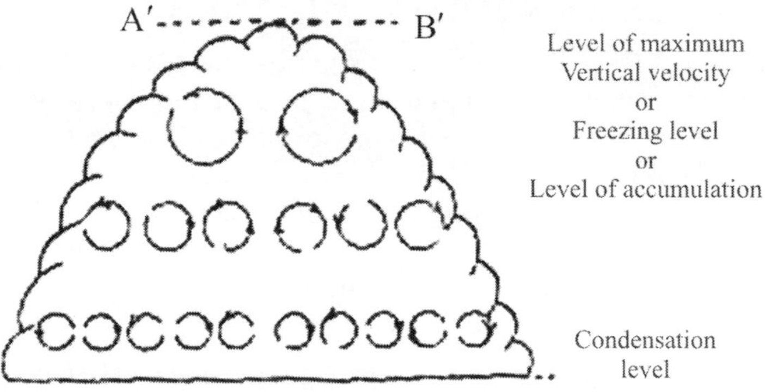

Fig. 4.1 Layered convection by Mull and Kulshreshtha (1969).

4.1.4 Ludlum Model of Cumulus Convection

By early sixties cumulus and cumulonimbus convection were appreciated to be quite different – Ludham and Kamburova (1966). In cumulus convection the primary element is the thermal – a volume of air which expands, thus thoroughly mixing its own warmer air with the ambient colder air. The familiar cauliflower shaped cumulus cloud comprises of many such thermals. Once the cumulus cloud produces shower, usually when cloud top is in mid-troposphere, a marked transformation occurs, particularly in the size and the vigour of the parent cloud. This is due to freezing of droplets in subzero temperature. Mull and Rao (1950) found in one case the rise of 4.5 °C in temperature with (13.5 gm/kg) minimum amount of liquid water at the level of accumulation in the super cooled layer (0-10°C). Within half an hour of precipitation formation, successive towers of cumulus penetrate into the higher troposphere. Prolonged convection then leads to mature Cumulonimbus cloud.

4.2 The Concept of Buoyancy

Vertical momentum equation in its inviscid form is

$$\frac{dw}{dt} = -\frac{1}{\rho}\frac{\partial p}{\partial z} - g \qquad .....(4.1)$$

where w is the vertical velocity component, ρ is the air density, p is the pressure and g is the gravitational acceleration. Multiplying 4.1 by ρ we get

$$\rho\frac{dw}{dt} = -\frac{\partial p}{\partial z} - \rho g \qquad .....(4.2)$$

In hydrostatic balance, $w = 0$. Hence we get

$$0 = \frac{\partial \bar{p}}{\partial z} - \bar{\rho} g \qquad .....(4.3)$$

Subtracting (4.3) from (4.2) we get,

$$\rho\frac{dw}{dt} = -\frac{\partial p'}{\partial z} - \rho' g \qquad .....(4.4)$$

where primed p and ρ are the deviations of pressure and density from the horizontally homogeneous, balanced base state. Rearrangement of the terms in (4.4) gives,

$$\frac{dw}{dt} = -\frac{1}{\rho}\frac{\partial p'}{\partial z} - \frac{\rho'}{\rho} g \qquad .....(4.5)$$

$$= -\frac{1}{\rho}\frac{\partial p'}{\partial z} + B \qquad .....(4.6)$$

where B $= -\frac{\rho'}{\rho}g$ is familiar buoyancy force and $-\frac{1}{\rho}\frac{\partial p'}{\partial z}$ is the vertical perturbation pressure gradient force (VPPGF). The VPPGF arises from the velocity gradient (which imply acceleration of air) and from the presence of density anomalies. VPPG therefore has both nonhydrostatic and hydrostatic components. Generally it tends to offset some of the acceleration induced by buoyancy force. An exception could be in updraft with large vertical wind shear in which VPPGF may act in the same direction as buoyancy force, especially at lower levels, thereby augmenting the vertical acceleration. Hence generally relatively high (low) pressure tends to be located above arising (sinking) warm (cold) bubble and relatively low (high) pressure tends to be located beneath arising (sinking) warm (cold) bubble. Thus an upward directed buoyancy force (associated with warm bubble) tends to be associated with downwards directed perturbation pressure gradient force and downward directed buoyancy force (associated with cold bubble) tends to be associated with upward directed perturbation pressure gradient force. Hence considering the effect of perturbation pressure gradient warm (cold) bubbles tend not to rise (sink) as fast as one would expect based on the consideration of buoyancy force alone. For smaller size bubbles the buoyancy force may be relatively larger than the perturbation pressure gradient force but as the bubble increases in size the vertical perturbation pressure gradient force becomes so large that it completely off sets the buoyancy force and net acceleration is zero. This is hydrostatic limit. In other words the horizontal scale has become very large as compared to the vertical scale.

It is common for the buoyancy force to be written as

$$B = -\frac{\rho'}{\overline{\rho}}g \qquad .....(4.7)$$

where $\rho = \rho(x,y,z,t)$ has been replaced by $\overline{\rho} = \overline{\rho}(z)$ in the denominator (a "base state" density profile). This is the equivalent of making the anelastic approximation. Using the equation of state and neglecting products of perturbations, it can be shown that

$$\frac{\rho'}{\rho} \approx \frac{p'}{\overline{p}} - \frac{T_u'}{\overline{T}_u} \approx -\frac{T_u'}{\overline{T}_u} \qquad \text{.....(4.8)}$$

where T_u is the virtual temperature and we have assumed $|p'/\overline{p}| \ll |T_u'/T_u|$. The buoyancy force therefore can be approximated as

$$B \approx \frac{T_u'}{\overline{T}_u} g \qquad \text{.....(4.9)}$$

Often it is customary to designate the "base state" virtual temperature as that of the ambient environment, and the perturbation virtual temperature as the temperature difference between the environment and an air parcel rising through an updraft, so that

$$B \approx \frac{T_{u_p} - T_{u_{env}}}{T_{u_{env}}} g \qquad \text{.....(4.10)}$$

where T_{u_p} is the virtual temperature of an air parcel rising through the updraft and $T_{u_{env}}$ is the virtual temperature of the ambient environment. When an air parcel is warmer than the environment, a positive buoyancy force exists, resulting in upward acceleration.

4.2.1 Parcel Theory

The simplified approach of predicting vertical velocity of convection is by parcel theory which assumes vertical acceleration to be equal to buoyancy (B) only

$$\frac{dw}{dt} = B \qquad \text{.....(4.11)}$$

If we multiply both sides of (4.11) by $w \equiv \frac{dw}{dt}$ we get

$$w\frac{dw}{dt} = B\frac{dz}{dt} \qquad \text{.....(4.12)}$$

$$\frac{d}{dt}\left(\frac{w^2}{2}\right) = B\frac{dz}{dt} \qquad \text{.....(4.13)}$$

$$dw^2 = 2Bdz \qquad \text{.....(4.14)}$$

Next, we integrate 4.14 from the Level of Free Convection (LFC) to the equilibrium level (EL). Refer Appendix-B for definitions. We will assume that $w = 0$ at the LFC; since the only force considered here is buoyancy force (which does not become positive until above the LFC; by definition). The integration of (4.14) yields,

$$\int_{LFC}^{EL} dw^2 = 2\int_{LFC}^{EL} Bdz \qquad .....(4.15)$$

$$w_{EL}^2 - w_{LFC}^2 = 2\int_{LFC}^{EL} Bdz \qquad .....(4.16)$$

$$w_{max}^2 = 2\int_{LFC}^{EL} Bdz \qquad .....(4.17)$$

$$w_{max} = 2\sqrt{2CAPE} \qquad .....(4.18)$$

where CAPE is the Convective Available Potential Energy; defined by

$$CAPE = \int_{LFC}^{EL} Bdz \qquad .....(4.19)$$

The prediction of w_{max} in a convective updraft typically is too large by parcel theory. This is due to following reasons.

(i) Vertical perturbation pressure gradient force (VPPGF) also sometimes referred as aerodynamic drag are neglected.

(ii) Mixing with ambient environmental air or entrainment was neglected.

(iii) The weight of the condensate or hydrometeor loading with in a parcel of rising moist adiabatically is ignored.

(iv) Compensating subsidence in the surrounding air is ignored.

Therefore w_{max} predicted by (4.14) can be interpreted as the upper limit for the vertical velocity in buoyant convection. w_{max}, therefore, is also known as thermodynamic speed limit.

Entrainment and hydrometeor loading are discussed in 4.2.2 and 4.2.3 respectively. Compensating subsidence is described in section 4.6, separately.

4.2.2 Entrainment

Mixing of environmental air into arising air parcel from its surface boundaries is known as entrainment. By entrainment the parcel is diluted as a result equivalent potential temperature (θ_e) of rising parcel is reduced leading to the realization of lesser CAPE and lesser w_{max}.

Updraft dilution increases with the tilt of an updraft, which increases the surface area of the updraft which is exposed to the ambient environment. Entrainment on the side of the rising parcel also increases with the vertical acceleration, owing to mass continuity. This is referred to dynamic entrainment.

Entrainment may be quantified by the rate of change of temperature excess i.e., how warm a parcel is as compared to environment. It can be reasonably approximated as

$$\frac{dT_u'}{dt} = \left(r - \Gamma_m\right)w - \left[T_u' + \frac{L}{c_p}q'\right]\lambda \qquad .....(4.20)$$

where T_u' and q′ are the virtual temperature and specific humidity excess over the ambient environment, w is the vertical velocity, r is the environmental lapse rate, Γ_m is the moist adiabatic lapse rate, L is the latent heat of vaporization, c_p is the specific heat at constant pressure and λ is the entrainment rate. It is possible to measure the entrainment rate by *in situ* thermodynamic measurements within cloud. In simple updraft model entrainment rate is often parameterized as updraft width.

4.2.3 Hydrometer Loading

As air parcel rises moist adiabatically, condensate is carried along with the parcel. The condensate has mass and weight of the condensate exerts a downward acceleration equal to gq_c where q_c is the mass of the condensate per kg of air. Maximum value of q_c with a strong updraft are typically 8-18 *g/kg*. The effect of condensate or hydrometer loading may be incorporated into buoyancy term B as

$$B = g\left(\frac{T_u'}{\overline{T}_u} - q_c\right) \qquad .....(4.21)$$

Thus a 3 °C virtual temperature excess will be offset entirely by a condensate concentration of 10*g/kg*, assuming $\overline{T}_u \approx 300K$.

4.3 Computation of Ascending Velocity

In all the theoretical and experimental investigation of the cloud formation the knowledge of the ascending current velocities at different levels of atmosphere is of fundamental importance. Hence various methods are presented here for their calculation.

(a) Weickmenn Method: This method is based on parcel theory. Neglecting the effect of entrainment of air into the cloud Weickmenn (1962) calculated velocity of ascending current from the simplified buoyancy equation,

$$\rho\left(\frac{\partial^2 z}{\partial t^2}\right) = g\left(\frac{T_m - T_s}{T_s}\right) \qquad \text{.....(4.22)}$$

where T_m is the temperature of moist adiabatically ascending air and T_s is the temperature of the surrounding environment. For subzero part of the cloud Weickmann applied correction in T_m necessitated due to the increase in temperature resulting from the release of latent heat when super cooled droplets freeze. With this method the value of w at any level in the atmosphere may be computed. Integrating (4.22) from time t_1 to t_2 we get

$$W_{t_1} - W_{t_2} = \left(\frac{g}{\rho}\right) . \left(T_m \, log\left\{\frac{T_{st_1}}{T_{st_2}}\right\} - (t_1 - t_2)\right) \qquad \text{.....(4.23)}$$

where T_{st1} and T_{st2} are the temperature of the surrounding environment on two different times t_1 and t_2. W_{t1} and W_{t2} are the vertical velocities on time $t_1$and t_2 respectively. In his calculation he however neglected the effect of entrainment of air into the cloud of strong convective process. It has been observed that neglecting entrainment, particularly when the average relative humidity in layer 250 – 500 mb is below 60%, it is likely to cause significant error. This is because the drier the surrounding air the lower the temperature of the ascending air in comparison with the one obtained with adiabatic process.

(b) Slice Method: In the parcel theory it is assumed that the displaced air parcel does not disturb the surrounding atmosphere. In reality the vertical motion of the displaced air parcel requires compensatory vertical motion by the surrounding atmosphere. The effect of compensatory motion is treated as the so called 'Slice method' devised by Bjerkness (1938) and Patterson (1939). The ascending current velocity can be measured from the relation,

$$W = \left[2C_p\left\{\left(T_m - T_0\right) - S_{imax}\left(T_m - T_{dry}\right)\eta\right\}\right]^{1/2} \quad \text{..... (4.24)}$$

where $C_p(T_m - T_0) - S_{imax}(T_m - T_{dry}) = \Delta q$ is the thermal energy of the air mass. S_{imax} is the optimum amount of cloud which forms when maximum convective energy is liberated and η is conversion factor of thermal into kinetic energy.

To compute the layer wise velocity Shishkin (1960a) divided the environmental curve (emagram) in to n separate layers at the point of inflexion of the temperature stratification, starting from convective condensation level (CCL). If there is no such stratification and environmental lapse rate is smooth then layers of thickness 100-150 hPa may be taken (see fig 4.2). Dry and moist adiabetes are drawn from the lower boundary of each layer. T_{dry} and T_m are the temperatures at the end of the dry and moist adiabates respectively, where they intersect the nth layer. T_0 is the temperature at the condensation level. $T_{dry\ n}$ is the temperature at the end of the dry adiabate which connects the condensation level to the intersection with the nth layer. T is the temperature of the surrounding air at the lower boundary of the layer. The differences $T_m - T_0$, $T_m - T_{dry}$, $T - T_{dry}$ and $T_{dry\ n} - T_{dry}$ are determined from the aerological diagram(fig. 4.2). Corresponding values are shown in Table 4.1.

Readers are advised to read the temperature values in Emagram and then compare it with the computations in Table 4.1. η and q values are shown in Table 4.1 Col. 7 and Col. 15 respectively.

For the first layer the ascending current velocity is determined from the equation,

$W_1^2 = 2\Delta q_1\eta_1$,[From Table 4.1 $W_1^2 = 2 \times$ Col. 7 $\times$ Col 15 i.e., Col. 17]

And similarly for the ith layer the vertical velocity may be,

$$W_i^2 = 2\Sigma\Delta q_i\eta_i, \quad \text{.....(4.25)}$$

Layer wise computation of vertical velocity is mentioned in the Table 4.1.

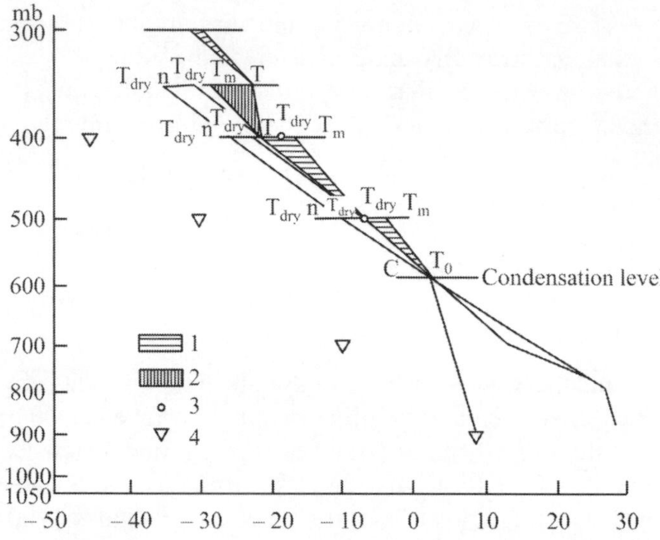

Fig. 4.2 Emagram for calculating the ascending current velocity, taking into account entrainment of surrounding air into the cloud. Instability energy: 1 is positive area; 2 is negative area; 3 is the wet bulb temperature T and 4 is the dew point T_d.

While computing the vertical current it is to be noted that process of mixing environmental air into the ascending cloud mass known as entrainment significantly alters the lapse rate and the liquid water content within the cloud. However, if average relative humidity is above 60% in free atmosphere in layer 850-500 hPa and there exists strong vertical velocity (*e.g.* 20 m/sec or so) the correction for entrainment need not be applied while calculating the vertical velocity. This is because high humidity and strong vertical motion in a developed cloud itself implies large horizontal dimension (10-15 km diameter) of the cloud. And since ascending currents is formed by air mass lifted from the condensation level in the middle part of the cloud where they attain the maximum values, the effect of entrainment may be neglected.

Table 4.1 Example of calculating the ascending current velocity from the Emagram in Fig. 4.2

Layer	T	$T_m - T_0$	$T_m - T_{dry}$	$T - T_{dry}$	$T_{dr} - T_{dry\,n}$	$\frac{T_{dr} - T_{dry\,n}}{T_{dr}}$	$\Sigma(T - T_{dry})$	$\Sigma(T_m - T_{dry})$	$\frac{\Sigma(T - T_{dry})}{\Sigma(T_m - T_{dry})}$
1	**2**	**3**	**4**	**5**	**6**	**7**	**8**	**9**	**10**
n_1	8.0	2.0	3.0	1.0	5.0	17.5	1.0	3.0	0.33
n_2	–9.5	3.5	11.5	8.0	30.0	105.0	9.0	14.5	0.62
n_3	–24	3.0	5.0	2.0	46.0	161.0	11.0	19.5	0.56
n_4	–34	–2.5	2.0	4.5	64.0	224.0	15.5	21.5	0.72

Layer	$\sqrt{\frac{\Sigma(T - T_{dry})}{\Sigma(T_m - T_{dry})}}$	$1 - \sqrt{\frac{\Sigma(T - T_{dry})}{\Sigma(T_m - T_{dry})}}$	4 × 12	3 – 13 = Δq	Σ(14) (ΣΔq)	15 ×7	2×16	w_1 m/sec
1	**11**	**12**	**13**	**14**	**15**	**16**	**17**	**18**
n_1	0.58	0.42	1.2	0.8	0.8	14	28	5.3
n_2	0.79	0.21	2.4	1.1	1.9	199.5	400.0	20.0
n_3	0.75	0.25	1.2	1.8	3.7	595.7	1190	34.6
n_4	0.88	0.15	0.3	–2.8	0.9	201.6	403.2	20.0

(c) Sulakvelidze Method of Computing Maximum Vertical Velocity: When only W_{max} is required and not its distribution with height then the following simplified equation can be used.

$$W_{max} = [2C_p(\Delta T/T_s)\{\log(p_1/p_2)\}(T_s - T_{dry})]^{1/2} \quad \text{..... (4.26)}$$

where

C_p = Specific heat at constant pressure

T_s = Temperature of surrounding

T_m = Lowest temperature attained by the air parcel lifted along moist adiabate from the level of condensation.

$\Delta T = T_m - T_s$ is the maximum temperature difference between the air ascending from the condensation level along a moist adiabate and the surrounding air at the same level.

p_1 & p_2 = atmospheric pressure at the condensation and convection levels respectively.

T_{dry} = Lowest temperature attained by the parcel lifted dry adiabatically from condensation level to the level of convection.

The quantity $\left\{ log\left(p_1/p_2\right)\left(\frac{T_s - T_{dry}}{T_s}\right)\right\}$

is the efficiency of the ascent of warm air mass.

The equation gives good result when $W_{max} \geq 15$m/sec.

(d) Empirical Method: With the help of quick response thermograph velocity of the ascending currents may be related with the temperature difference (ΔT °C) between the ascending and surrounding environment. Ludlam (1952) found that expression for w (the vertical velocity in m/sec) in the cloud may be given as

$$w = 2\,\Delta T \text{ m/sec} \quad \text{.....(4.27)}$$

Vulfson (1954) also derived an expression for the vertical velocity with the thickness of the thermal or ascending jet in the cloud which again depends on ΔT. Thus

$$w = 0.02h^{3/4}\left(1 - \frac{r^2}{R^2}\right)^{1/2} \quad \text{.....(4.28a)}$$

where h is the altitude above the cloud base in meter, R is the radius of the thermal in meter and r the horizontal distance from the center of the thermal in meter. Ascending current increases with height above the cloud base. It reaches maximum beneath the top and reaches maximum value w_{max}, after which it begins to fall rapidly as the cloud top is reached. If origin of coordinate is taken as cloud base and z-axis is taken vertically upward then as a first approximation the magnitude of the ascending current velocity may be represented as follows:

In the lower part of the cloud where $z \leq z_{max}$

$$w(z) = w_0 + (w_{max} - w_0)\, z/z_{max} \qquad .....(4.28b)$$

where w(z) is the ascending current velocity at height z; w_0 is the ascending current velocity at the cloud base for z = 0; w_{max} is the maximum ascending velocity and z_{max} is the level of the maximum velocity w_{max}. In the upper part of the cloud for which $z \geq z_{max}$

$$w(z) = W_{max}\left[1 - \frac{z - z_{max}}{z_a - z_{max}}\right] \qquad .....(4.28c)$$

where z_a is the height at which the ascending current velocity vanishes. It may be noted that cloud top usually lies below the level z_a. Equation 4.28a, b and c describe ascending current velocity fairly well.

(e) Radar Method: The free lift of the Pilot Balloon is defined as the difference between the total lift and the weight of the balloon and its load. The free lift is really the net buoyancy of the balloon. Thus free lift is the force tending to drive the balloon upward, which depends upon the amount of hydrogen gas filled in it. Free lift pilot balloons may be used for observing the vertical velocity in cloud after applying suitable correction for its buoyancy. Three mutually perpendicular planc triangular foils may be attached to the sounding balloon so as to act as reflecting target for radar. The equivalent reflecting surface of such reflection whose planes are triangles of sides a is given by

$$S = \left(\frac{4\pi a^4}{3\lambda^4}\right) \qquad .....(4.29)$$

where λ is the wave length. Edge length by 30-40 cm is taken to ensure reliable reception of the reflected signals from a distance of 15-18 km for λ = 3.2 cm and from 30–40 km for λ = 10.0 cm. The spatial coordinates of the reflector (slant range, elevation and azimuth) are determined for two different times and the vertical velocity of atmosphere is obtained by dividing the vertical displacement of target by the time interval, minus the rate of ascent of balloon due to free lift.

Dispensing aluminum foil pieces in the cloud and tracing them with the help of radar is also helpful to determine the turbulence inside the cloud and their horizontal and vertical dimensions.

4.4 Thunderstorm

A thunderstorm is defined as a storm that contains lightening and thunder which is caused by unstable atmospheric conditions. This can occur singularly, in clusters or in lines. Severe thunderstorms can bring in heavy rains which can cause flash floods, strong winds, lightening, hail and tornadoes. A thunderstorm is called as severe if it contains hail that is ≥ 1.9 cm (≥ 0.75 inch) in diameter or if the wind gusts are reaching to ≥ 93 km/h (≥ 58 mph).

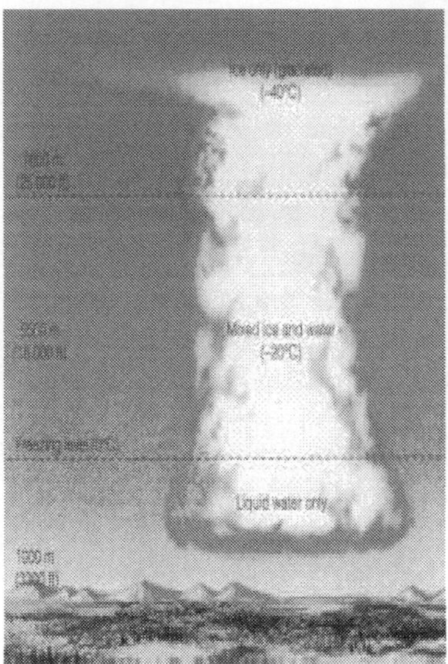

Fig. 4.3 Thunderstorm.

These types of clouds which can span, vertically, even up to the top of the troposphere and can be more than 19 Km (12 miles) high. During extreme storms, the updrafts can reach as high as 16.2 Kmph (100 mph) while the downdrafts can be even higher. Aircrafts totally avoid these types of clouds as the turbulence found inside them are usually on the extreme side. It has been estimated that there are nearly 16 million thunderstorms annually all over the globe. Kampala, Uganda holds the record of about 240 thunderstorm days annually. There are about 1800 thunderstorms occurring at any moment across the world. All thunderstorms produce lightening which often strikes outside of the area where it is raining and is known to fall more than 10 miles away from the rainfall area. A severe thunderstorm can produce winds that can cause as much damage as a weak tornado and these winds can be life threatening.

Lightning in a thunderstorm results from the buildup and discharge of electrical energy between positively and negatively charged areas. The action of rising and descending air within a thunderstorm separates positive and negative charges. Most lightning occurs within the cloud or between cloud and ground. A cloud-to-ground lightning strike begins as an invisible channel of electricity charged air moving from the cloud toward the ground. When one channel nears an object on the ground, a powerful surge of electricity travels from the ground upward to the cloud, producing the visible lightning strike. The average flash of lightning could be a 100-watt light bulb for more than three months. The air near a lightning strike is heated to 50,000° F - hotter than the surface of the sun. The rapid heating and cooling of the air near the lightning channel causes a shock wave those result in thunder. "Heat lightning" actually is lightning from a thunderstorm, too far away from thunder, to be heard.

4.4.1 Formation of Thunderstorm

It is now generally accepted that cumulus cloud build up in the form of individual bubbles or air parcels (thermals). These thermals form due to heat exchange between heated earth's surface and the surface air layer. While moving upward a thermal mixes with the surrounding air and leaves behind it a wake of air, which is colder than the thermal but warmer than the surrounding atmosphere. The penetration level of convection depends on the temperature lapse rate and the time of the day. With considerable atmospheric instability, bubbles reach the condensation level and with the released latent heat fresh surges of bubbles, these air parcels gradually increase in number as the cumulus enlarges into a large cumulus.

Most of the thunderstorms are comprised of ordinary cells which drift with the mean wind speed. A cell, as has been discussed in the beginning of this chapter is a dynamical entity characterized by a compact region of relatively strong vertical air motion which can be identified by its volume of relatively intense precipitation. Ordinary cells are short lived units of convection. They undergo through the following three stages.

(a) Cumulus stage (updraft alone)
(b) Mature stage (updraft and down draft)
(c) Dissipating stage (downdraft alone)

Mature stage of ordinary cell does not last more than 15-30 min. A few ordinary cells under favorable condition set into much larger scale and much intense in airflow go into virtually steady state with updraft and down draft coexisting for a long period of 30 min or more. Super cell lasting for 12 hours have also been reported. Super cells usually drift either to the right or to the left of mean wind, without any new echo forming on the right or left flank. A typical thunderstorm formation is shown in the Fig.4.4.

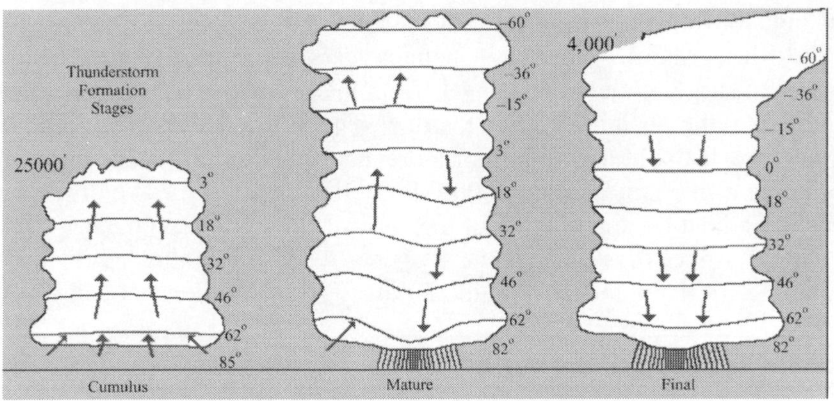

Fig. 4.4 Typical Thunderstorm Formation Stages with atmospheric thermal profile.

1. *Cumulus stage*: Cumulus cloud develops into tall cumulus when rising air currents, reach above 20,000 feet.
2. *Mature stage*: Ascending air reaches height where precipitation occurs. Initially, cloud air is warmer than surrounding air. Falling rain & ice cool air. Downdrafts created.
3. *Final or Dissipating stage*: Entire Cloud is having downdraft. Air is adiabatically warmed. No air is ascending & cooling. Precipitation decreases. Upper winds blow ice crystals into anvil top.

Cumulus Stage: Refer Fig. 4.5. The solar heating of earth's surface during the day warms the air around it. Since warm air is lighter than cool air, it starts to rise. The moisture of the warm air condenses into a cumulus cloud. The cloud will continue to grow as long as warm air below it continues to rise.

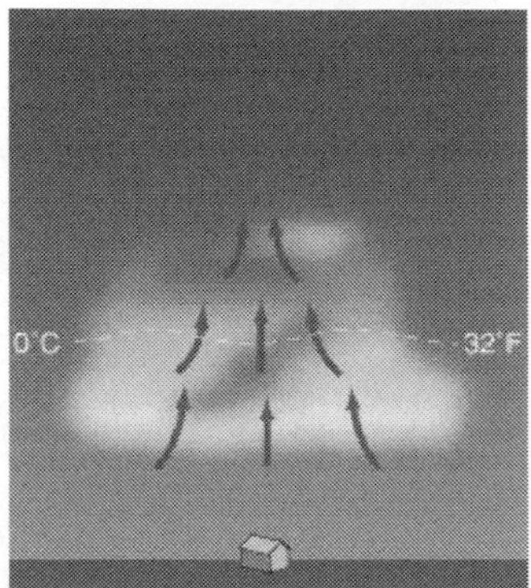

Fig. 4.5 Cumulus stage.

Mature Stage: Refer Fig. 4.6.When the cumulus cloud becomes very large, the water in it becomes large and heavy. Raindrops start to fall through the cloud when the rising air can no longer hold them up. Meanwhile, cool dry air starts to enter the cloud. Because cool air is heavier than warm air, it starts to descend in the cloud. The theory of downdraft is to be explained in detail under in section 4.6.The downdraft further pulls the precipitation downward, making rain, snow or hail. This cloud is known as cumulonimbus cloud (Latin word which translates to 'rain heaps') because it has an updraft, a downdraft and rain. Lightning and thunder also start to occur, as well as heavy precipitation. The cumulonimbus is now a thunderstorm cell.

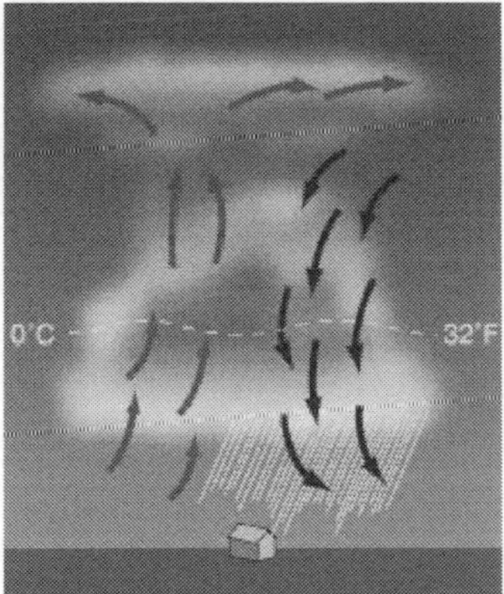

Fig. 4.6 Mature Stage.

Dissipating Stage: Refer Fig.4.7. After about 30 minutes, the thunderstorm begins to dissipate. This occurs when the downdrafts in the cloud begins to dominate over the updraft. Since warm moist air can no longer rise, cloud droplets can no longer form. The storm dies out with light rain as the cloud disappears from bottom to top.

For a single cell the mature and dissipation stages take about one hour in any ordinary thunderstorm.

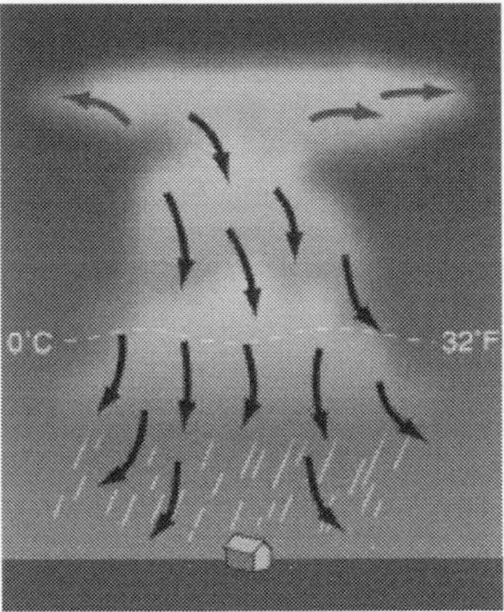

Fig. 4.7 Dissipating Stage.

4.5 Parts of Matured Thunderstorm

(a) Anvil: The Anvil is one of the most impressive features of a severe storm due to its areal coverage and icy texture. Within a severe storm, moisture is transported from the lower troposphere to deep into the upper troposphere. Not all moisture that is ingested into a storm is precipitated out of the storm. Some of the moisture in a strong updraft is lofted so high into the troposphere that it is not able to drop back down immediately. Strong upper level winds move and fan the moisture out over great distances. The temperature of the anvil is frigid cold. The light density of the moisture allows the wind to move easily. A forecaster can note the direction and speed of the upper level winds by noting the anvil's orientation. The moisture within the anvil will be blown downstream.

(b) Overshooting Top: The core of the updraft has the strongest convective upward vertical velocity. This core of rapidly rising air will only slow down and stop when it encounters a very stable layer in the atmosphere. This very stable layer is the tropopause. Air will rise as long as it is less dense and therefore more buoyant than surrounding air. The faster air rises the longer it takes generally to slow down and stop once it encounters a very stable layer. This occurs because a moving object has momentum. That part of the updraft that has the greatest momentum will form the overshooting top on a severe thunderstorm.

(c) **Mammatus:** Mammatus are pouched shaped clouds that protrude downward from the thunderstorm's anvil. They form as negatively buoyant moisture laden air sinks. The cloud remains visible until the air sinks enough that the relative humidity falls below 100%. The portion that has a relative humidity of 100% remains visible. Theories to how they form include:

1. Turbulent eddies mixing down moisture,
2. Evaporative cooling with surrounding air causes pockets of sinking air,
3. Pockets of precipitation falling out of the anvil that produce virga. Mammatus tend to be most prominent in extremely severe storms but can occur when storms are not severe also. Fig. 4.8 shows mammatus clouds.

Fig. 4.8 Mammatus clouds.

(d) **Flanking Line:** The flanking line is produced by convergence along an outflow boundary extending from the storm. This outflow is often air from aloft that is converged into warm and moist air near the surface. It can be seen as a line of developing cumulus clouds extending from the storm. The cumulus closer to the storm tend to be more mature and eventually merge into the parent storm. The flanking line often feeds into the updraft of the storm.

(e) **Rain Core/Hail Core:** The core refers to the heaviest precipitation. The most violent rain and hail in a supercell tend to be on the outer edge of the updraft on the downdraft side of the storm. Extreme turbulence on the edge of the updraft can contribute to significant hail growth. As hail falls into above freezing air it sheds its moisture as rain.

(f) **Wall Cloud:** The wall cloud is located in the updraft region of a supercell. Rising air cools and condenses out moisture once it is saturated. Due to the rapidly rising air and the verticality of the rising air, the cloud base is close to the ground within the wall cloud. The wall cloud will often be witnessed as rotating since directional wind shear acts on the updraft as it rises. Tornadoes can occur under the wall cloud.

(g) **Rain-Free Base:** The updraft region in supercells will often lack precipitation. This is most true for developing supercells and for classic low precipitation supercells. As a supercell matures or has a high moisture content, often precipitation will wrap around the updraft region and eventually fall into the updraft region. The updraft region of a supercell will be tilted with height. This will deposit the precipitation away from the updraft and thus this also results in less precipitation in the updraft region. Being in the rain free base region offers an awe-striking view of the storm.

(h) **Forward Flank/RearFlank Downdraft:** The forward flank downdraft is the outflow from the rain-cooled air of the storm's downdraft. The rear flank downdraft is air from aloft that is transported down to the surface from colliding with the storm. The rear flank downdraft air tends to be dry and warm since the air warms by adiabatic compression as it sinks to the surface. Adiabatically warmed air will also decrease in relative humidity if no precipitation falls into the air. The rear flank downdraft tends to be warmer than the forward flank downdraft since rain the evaporational cooling is not as common in the rear flank. Shear is enhanced along these flanking downdraft boundaries and the shear can be magnified along where the two flanks merge. The right balance of shear and instability release can lead to tornadogenesis.

4.6 Downdraft

4.6.1 Definitions

Downdraft is defined as downward moving air. Fujita and Wakimoto (1983) defined downburst as particularly strong downdraft which induces outburst of damaging wind near ground. They further divided downburst into macroburst and microburst. Macroburst when the horizontal outflow size is greater than four km and the lifetime is of 5 to 20 minutes and microburst when the horizontal outflow size of less than four km and lifetime is of 2 to 5 minutes. Microburst is also of two types; dry and wet. Dry microburst is accompanied by less than 0.01 inches (< 0.25 mm) of rain at surface otherwise it is wet microburst.

Based on the strength of the downward air motion Knupp and Cotton (1985) categorised it as weak (< $5ms^{-1}$), moderate (5-10 ms^{-1}) and strong (> 10 ms^{-1}). Maximum speed of downdraft seems to be limited to 20 ms^{-1}.

Radar Meteorologists define microburst when peak to peak differential Doppler velocity across the divergent centre is greater than 10 ms^{-1} and the distance between these peaks is less than four km (Wilson et al, 1984). Robert and Wilson (1986) characterised the microburst as low reflectivity (< 35dBZ), moderate reflectivity (35-55 dBZ) or high reflectivity (> 55dBZ). Albeit little correlation was found between microburst intensity and peak radar reflectivity.

Joint Airport Weather Studies (JAWS) project (McCarthy et al 1982) and Northern Illinois Meteorological Research on Microburst (NIMROM) project (Fujita 1978, 1985) and FAA/Lincoln Laboratory Operational Weather Studies (FLOWS) project (Wolfson et al 1985) have provided important observational finding regarding microburst. Dry microburst was sometimes found to be as strong as wet microburst. Most JAWS microburst intensified after first reaching the ground and typically obtaining their peak outflow speed 5 minutes after the initial detection of divergence at the surface. The typical JAWS microburst had peak outflow intensity lasting for 2 to 4 min, a peak downdraft speed of approximately 12 ms^{-1}, an outflow depth of less than 1 km and life time of 5 to 10min. They were essentially stationery and most had propagation speed of less than 4 ms^{-1}.

4.6.2 Types of Downdraft

Based on the number of observations and numerical cloud model results Knupp and Cotton (1985), classified the convective cloud downdrafts which are located within or very near cloud edge, as under. Refer Fig. 4.9.

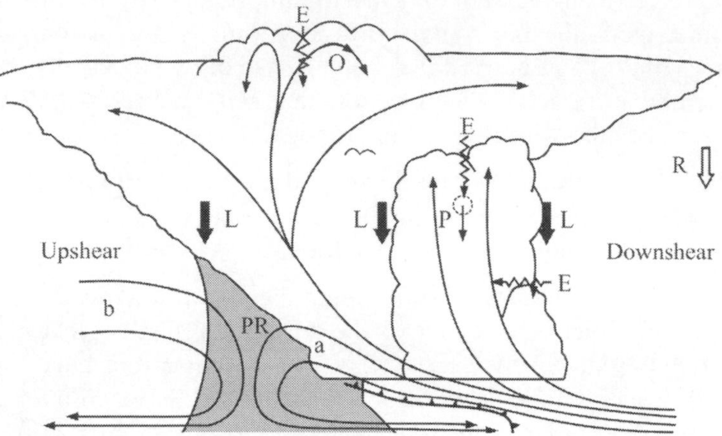

Fig. 4.9 Schematic of updraft, downdraft and entrainment flows within a typical cumulonimbus cloud based on the observational and numerical model studies. E denotes Entrainment, PR denotes Precipitation downdraft, P denotes Penetrative downdraft, L denotes Cloud updraft related downdraft and O denotes Overshooting top downdraft. Large scale downdraft shown as R.

(i) **Precipitation downdraft:** Shown as PR in the figure. This downdraft reaches the ground and causes divergence in the form of gust front.

(ii) **Penetrative downdraft:** This downdraft is generated due to entrainment (E) on the cloud boundaries. Evaporative cooling causes the cool parcel of air to descend. Shown as P in the figure. This does not reach the ground level.

(iii) **Cloud updraft related downdraft:** Shown as L in figure. It is caused due to the return of updraft currents after hitting the top of the cloud, near the cloud internal boundaries. This downdraft also remains elevated and does not reach the ground level.

(iv) **Overshooting top downdraft:** This is a subclass of class (iii) and are the downdrafts caused by the return of prime updraft current inducing overshooting at the cloud top. Shown as O in the figure. This downdraft also remains elevated and does not reach the ground level. Table 4.2 shows the speed, width and depth of these downdrafts.

Table 4.2

Downdraft Types	Typical values			Level*
	Speed ms^{-1}	Width Km	Depth Km	
Precipitation(PR)	1-15	1-10	1-5	l,m
Penetrative(P)	1-15	< 1.0	0.5 – 5.0	m,u
Cloud updraft edge(L)	< 1-5	< 1-5	< 1-5	m,u
Overshooting(O)	1-40	0.5 – 5.0	1-3	u
*Relative cloud levels: l =low, m= medium and u= upper				

Large scale downdraft shown as R in the Fig.4.9, which is far removed from the cloud is induced due to cloud scale circulations. This is having typical speed of < 1 ms^{-1}, width could be 5-25 km and depth of 1-5 km. This downdraft is also limited to middle and upper levels of the cloud. As only precipitation downdraft reaches the ground, only this is discussed henceforth in this chapter.

4.6.3 Downdraft Initiation Hypotheses

Various hypotheses on the initiation of downdraft are described below.

(a) **Entrainment and Precipitation loading:** Byers and Braham (1949) suggested that entrainment within the updraft causes evaporation and cooling. This creates negative buoyancy which is further enhanced by the precipitation drag generated due to precipitation loading. It is however observed that on several occasions reasonably heavy rainfall from convective clouds during Indian southwest monsoon is not accompanied by downdraft. Hence Entrainment and Precipitation loading may not be only initiating factor though they may be contributory factors.

(b) **Drop Breakup:** Drop breakup during rain could be important microphysical mechanism helping evaporative cooling. The breakup produces small drops which can evaporate more rapidly. As the frequency of collisional breakup increases rapidly with the rainfall rate this could be the possible mechanism of the wet downburst that are accompanied with heavy rainfall. But then downburst should follow heavy rainfall-which is not always true.

(c) **Congelation Process:** Mull and Rao (1950) suggested that sudden congelation of super cooled water droplets close to freezing level in sub-zero region causes liberation of some heat. This would then result in a local turbulence which would possibly bring some of the ice crystal in this warmer layer below the freezing level. Melting of the crystals would cool the surrounding air thereby initiating the downdraft. Though the hypothesis locates the region near the freezing level as one from where the downdraft originates but generation of local turbulence against the in situ upward currents is debatable.

(d) Hail as Initiator of Downdraft: When hail fall below the freezing level they melt wholly or partly and cool the immediate environment. The air thus cooled starts descending and downdraft is initiated. Just as the hailstone melt and cool the environment just below the freezing level they also produce warming above the freezing level by virtue of growth into larger sizes. This would add to net negative buoyancy. The travelling downdraft will have to encounter deceleration due to adiabatic warming and viscous interaction with the environment or subsequent rising levels. At the same time its negative buoyancy will get strength with the cooling due to evaporation of water droplets in it as well as cooling due to entrained environmental air from outside the cloud in the lower region of the cloud. Albeit hail and rain both grow by same mechanism but hailstone grows into larger sizes than water drops because of greater forces of cohesion. It is expected that the updraft which could be strong enough to hold up the largest of the water drops would not be able to do so for the hailstone, which would start falling. Hence they are supposed to initiate the downdraft. Once the updraft is weakened at certain places due to above processes rain drops may also begin falling which were held up.

Hail theory of downdraft suffers with one inherent error of assumption that all thunderstorms carry hailstone. Typical finger, scallop or hook types of echoes are not seen in all the thunderclouds. However, it is possible that small size embryonic graupels are essential component of cumulonimbus. Further the graupel particles have larger radius so the transfer of heat from environment to the particle is faster. Also since the melted water can soak into the graupel particle to certain distance hence it presents an ice surface to the environment whose surface temperature is 0^0C. Therefore heat transfer to the graupel particle is faster. For the same reason vapour can condense at a faster rate on the graupel particle, thereby further increasing the rate of melting. Proctor (1989) observed that hail may produce stronger microburst in more stable environment than graupel or snow whereas in the deep unstable environment (lapse rate $\gtrapprox 8^{o}\text{Ckm}^{-1}$) graupel or snow may produce microburst of greater intensity. Thus snow, graupel and hail are all seen to initiate the downdraft which is further accentuated by the liquid raindrops drag and evaporative cooling.

4.6.4 Structure of Precipitation Downdraft

The structure of downdraft is highly dependent on the static stability of the environment wind shear profile, the cloud microphysical processes and precipitation characteristics. Vertical velocity magnitude are typically 5-10 ms^{-1} with in downdraft. But it can reach 20 ms^{-1}. Descend begins close to freezing level and has horizontal scale similar to the low level precipitation region below cumulonimbus cloud. Fig.4.10 shows the radar observed RHI imagery of cross section of a microburst during JAWS (from Fujita and Wakimoto (1983). Note the broad initial shape penetrating as narrower column in the intermediate levels up to its descend till about ≈ 1 km close to surface then diverges as a broader head splash after touching the ground.

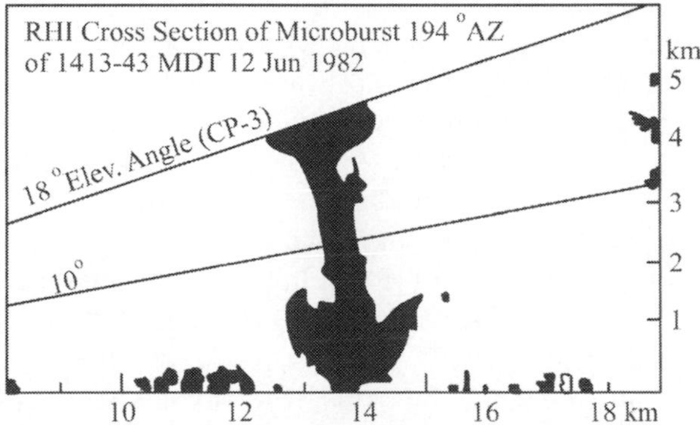

Fig. 4.10 Radar observed RHI cross section of microburst during JAWS (Fujita and Wakimoto, 1983).

It is generally accepted that dry microbursts are associated with unstable lapse rate ≥ 8.0°Ckm^{-1}. In case of evaporative cooling the size of rain droplet is important as smaller droplets evaporate faster. In combination with unstable lapse rate (≥8.0 °Ckm^{-1}) and high rain water mixing ratio (≥2 gKg^{-1}) near the cloud base with minimum downdraft radius of less than 1 km, evaporation generated downdraft could be intense in magnitude. Downdraft tend to be stronger in relatively dry environment because entrained dry air enhances evaporation; however moist environment which has generally more stable temperature profile also induces negative buoyancy due to difference between the virtual temperature of downdraft and its environment. Resulting wet microburst are also as intense as dry.

4.6.5 Numerical Analysis of Downdraft

Eq. 4.30 is the vertical equation of motion(Cotton and Anthese, 1985).

$$\frac{d\overline{w}}{dt} = -\frac{1}{\rho_0}\frac{\overline{\partial p'}}{\partial z} + g\left[\frac{\overline{\theta v'}}{\theta v_0} - \frac{C_v}{C_p}\frac{\overline{p}'}{p_0} - \left(r_c + r_r + r_i\right)\right]$$

term a term b term c term d

$$+ \text{ viscous terms and eddy stress terms} \qquad .....(4.30)$$

term e

Single primed quantities denote departure from the basic state (subscript zero) which varies only in height. The basic state is typically regarded as the undisturbed environment immediately adjacent to the storm. In eq. 4.30 p is pressure; θ_v is virtual potential temperature; and $r_{c,}$ r_r and r_i are the mixing ratios of cloud water, rain water and ice water respectively. Ignoring frictional effects (term e), vertical accelerations are produced by perturbation pressure buoyancy and its vertical gradient (terms c and a), thermal buoyancy (term b) and condensate loading (term d). The pressure term (term a) can be subdivided into buoyant and dynamic component (refer Rotunno and Klemp (1982), Klemp and Rotunno (1983) and Schesinger (1984). In term d we note that condensate loading is detrimental to updraft but helps downdraft. Modelling studies (Klemp and Wilhemson (1978), Schlesinger (1980) have indicated that for scale > 1 km above the subcloud layer positive buoyancy forcing within the updraft is far greater than negative buoyancy forcing within the downdraft. Because of this individual parcel excursion tend to be $\approx < 4$ km in the downdraft and $\approx > 10$ km (or entire tropospheric depth) in the updraft.

Through one dimensional model based on the combined effect of melting and evaporation Srivastava (1987) showed that at any given time concentrated cooling due to melting and evaporation exists just below the level to which the downdraft has penetrated at that time. However as it descends the temperature of the downdraft exceeds that of environment due to compressional warming. In that state though the buoyancy is positive throughout the depth of column but downdraft is driven by the weight of the precipitation which generates small negative buoyancy. Hence albeit downdraft could be initiated by melting and evaporation but latter on sustained by precipitation loading. Comparing the relative roles latent heats of evaporation(≈600 cal/g) and melting (≈80 cal/g) in initiating the downdraft Srivastava, 1987 pointed out that under typical environment conditions millimetric-sized ice particles may melt completely in a fall through a few kilometres above melting temperatures, whereas the raindrops of the same size cannot evaporate completely under similar conditions. Therefore even though the latent heat of melting is much smaller than that evaporation, ice particles are potentially important for generating the downdraft by cooling of atmosphere.

4.7 Types of Thunderstorms

Thunderstorms are classified into the following four categories based on the physical characteristics of the storm:

(a) **Single Cell Thunderstorms:** These storms have a lifespan of about 30 minutes and are not severe. Typically they last 20-30 minutes. Also known as 'Pulse storms' they can produce severe weather elements such as downbursts, hail, some heavy rainfall and occasionally weak tornadoes (Fig. 4.11). The storms of this kind are rare and as these occur randomly, they are very difficult to forecast.

Fig. 4.11 Single cell storm.

(b) **Multi-Cell Cluster Storm:** These are the most common type of thunderstorms and occur in a group. As environmental windshear increases to moderate level multicell storm form (Fig. 4.12). A group of cells moving as a single unit, with each cell in a different stage of the thunderstorm life cycle. Multicell storms can produce moderate size hail, flash floods and weak tornadoes. Storms of this kind can persist for a few hours.

Fig. 4.12 Multicell Cluster Storms.

(c) Multi-Cell Line Storm: It consists of a long line of thunderstorms that can produce hail the size of golf balls and weak tornadoes. Multicell line storms consist of a line of storms with a continuous, well developed gust front at the leading edge of the line (Fig. 4.13). These storms can produce small to moderate size hail, occasional flash floods and weak tornadoes. Storms of these kinds can easily be predicted with the help of radar. Multicell line storms are better known as squall lines.

Fig.4.13 Multicell line storms.

(d) Super-Cell Storms: These are although rare but highly organized thunderstorms (Fig. 4.14). Storms of this kind pose a high threat to life and property. Also known as thunderstorm with deep rotating updraft, these storms can produce strong downbursts, large hail, occasional flash floods and weak to violent tornadoes.

Fig. 4.14 Super cell storm.

The flanking line of the supercell behaves differently than that of the multicell cluster storm, in that updraft elements usually merge into the main rotating updraft and then explode vertically, rather than develop into separate and competing thunderstorm cells (Fig. 4.15). In effect, the flanking updrafts "feed" the supercell updraft, rather than compete with it.

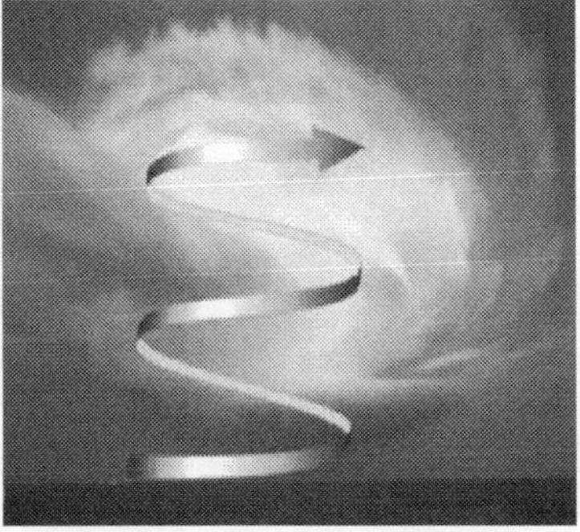

Fig. 4.15 Rotating updraft.

Based on precipitation and wind characteristics a severe thunderstorm is defined as one having large hail, at least 1.9 cm (0.75 inches) in diameter, and/or damaging winds, at least 93 kmh^{-1} (26 ms^{-1}) or 50 knots.

4.8 Movement of Thunderstorm – Stochastic Approach

Radar can be applied in predicting the direction and speed of movement of cumulonimbus by extrapolating the sequence of radar pictures by the method of curve fitting. For developing precise forecasting tool for the motion of storm without any radar aid detailed theoretical studies are lacking. Marwitz, 1972 studied the motion of the severe storm. Forecast of direction and movement of thunderstorm is challenge to any meteorologist. Operational forecasters may use stochastic approach of vectorial wind addition method for a quick computation of direction of movement to a reasonably good degree of accuracy.

(a) **Direction of Movement:** In this method all the level winds beginning from the cloud base up to the top of the cloud is vectorially added, as shown in the Fig.4.16. Reasonably good estimate of the level of top of the cloud is important to achieve the accuracy. This could be achieved either by Radar observation or by analyzing temporally and spatially the nearest T-Φ gram plot. In Fig 4.16 the base of cloud is at 910 hPa and the top at 200 hPa. Available winds at various heights are indicated on the left hand side of the figure. Right hand side of the figure shows the vectorial addition of wind. Resultant addition vector HO provides the direction of movement of the thunderstorm. Accuracy of prediction is in the range of ± 5^0. This method also explains why some supercells move leftward or rightward to the direction of mean wind. Vertical extents of supercells reach great heights. Hence the contribution to the resultant vector of higher level winds predominates, causing the steering of the cell biased with the largest wind vectors. The method is based on the hypothesis that isolated cumulus cloud mass is a floating balloon in atmosphere and it is steered in the direction of resultant wind.

Base of cloud at 910 hPa		
Wind Direction (Degree)	**Wind Speed Kt**	**Height (m) a.m.s.l**
340	03	717
320	08	1442
030	08	3070
070	10	5780
030	10	7500
060	10	9560
040	20	10820
020	20	12270
Top of the cloud at 200 hPa		
Winds at different levels		

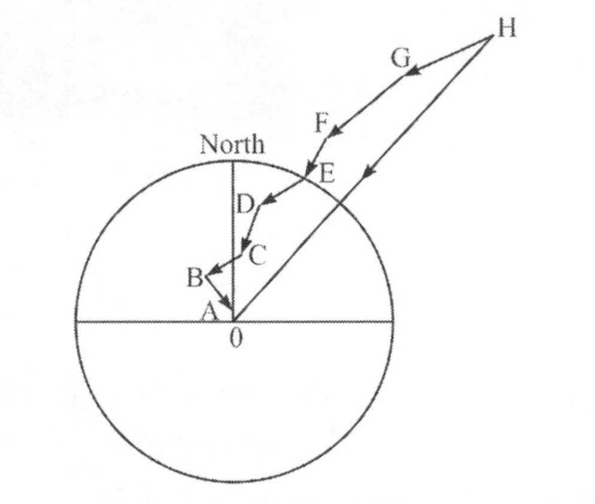

Vectorial addition

Fig. 4.16

(b) Speed of Movement: Out of sample of 60 thunderstorms in India the Speed of cumulonimbus has been observed to be ranging from 5 m/s to 22.2 m/s. Mean speed of sample was 11.7 m/s with $\sigma = \pm 3.9$ m/s (Kumar and Pati, 2016).

Higher the top of the cloud more would be its volume. Higher volume cloud would be offered higher resistance by the atmosphere, slowing the speed of movement. Speed of movement of cumulonimbus, therefore, could be empirically anticipated based on the Table 4.3.

Table 4.3

Top of Cumulonimbus	Speed indicated by the Percentage of the Resultant Addition Vector
9 Km	50%
10Km	45%
11Km	40%
12Km	35%
13Km	30%
14Km	25%

Eq. 4.31can be used for the velocity of single cell thunderstorm.

$$V = h\,[P_L\,(1 - T_A + T_L) + P_U\,(T_A - T_L)]/100 \qquad \text{.....(4.31)}$$

where

V = The predicted speed of the movement of Cumulonimbus in kt.

h = Modulus of HO in kt. (Refer Fig. 4.16)

P_L = Lower limit of % in table 4.3(Col.2)

P_U = Upper limit of % in table 4.3(Col. 2)

T_A = Actual height of the top of the cumulonimbus in kt.

T_L = Lower limit in the bracket of actual height of the top as per col.1 of Table 4.3.

The accuracy of the prediction speed is in the range of ± 20% of the computed speed. In Fig. 4.16 the cloud height reached to 12270 m. Hence T_A = 12.27 and T_L= 12. P_L is 35% and P_U is 30%. Resultant vector's modulus (h) is 74 kt. It indicated movement with the speed of 25 kt (20% of 25 kt is ± 5 kt).

$$74[35(1 - 12.7 + 12) + 30(12.27 - 12)]/100 \approx 24.9 \approx 25 \text{ kt}$$

The method described above is appropriate for isolated cumulonimbus cells with average 'Total Precipitable Water (TPW)' at 500 hPa level of the order of 5.5cm. For multicell storms the hydrodynamics of air motion around the storms and in between adjacent cells becomes highly complex and hence it may not be applicable. Vertical tilt, vertical shear, water loading or Total Precipitable Water content and vertical currents in the cloud and formation of new cells ahead would also influence its speed of movement. Hence detailed theoretical studies and numerical simulations are need to develop a model for the movement of severe storms.

References

1. Beers N.R. 1946, Bull Am Met Soc., 27, 2, p-25.
2. Byers H.R. and Braham jr R. R.(1949) The thunderstorms, U.S. printing Office, Washington DC, pp287.
3. Bjerkness J (1938): Saturated ascent of air through a dry adiabatic descending environment, QJRMS, 65.
4. Cotton, W. R. and Anthes (1985) The dynamics of clouds and mesoscale weather systems, Part I. The Dynamics of Clouds. Academic, Orlando, Fla.
5. Fujita T. T.(1978) Mannual of downburst identification for project NIMROD. SMRP Res. Pap. 156, University of Chicago, Dept. Geophysical Sciences, Satellite and Mesometeorology Research Project, pp 104.
6. Fujita T. T and Wakimoto R.M. (1983), JAWS microburst revealed by triple Doppler radar, aircraft and PAM data. Preprint, 13th Conf. on Severe Local Storms, Tulsa, Amer. Met. Soc., pp 97-100.
7. Fujita T. T. and Wakimoto R., (1983), Microburst in JAWS depicted by Doppler radars, PAM and aerial photographs, preprints 21st Conf. Rad. Met., Edmonton, AMS, 19-23.
8. Fujita T. T.(1985) The downburst, Microburst and Macroburst. Satellite and Mesometeorology res. Project, Dept. Geophysical Sciences,, University of Chicago, pp122.
9. Gutman L.N. 1963, Stationary axial – symmetric model of cumulus clouds DANSSR, Vol. 150 No. 1.
10. Harrison L.P. 1940, Report on Max Possible Liquid Water content per unit volume of air in heavy rain and cloud, U.S. weather Bureau.
11. Klemp, J.B and Wilhemson R.B. (1978,a) The simulation of three dimensional convective storm dynamics, J. Atm. Sc., 35, pp 1070-1096.

12. Klemp, J.B and Wilhemson R.B. (1978,b) The simulation of right and left moving storms produced through storm splitting, 35, 1097-1110.
13. Klemp, J.B. and Rotunno, R.(1983) A study of tornadic region within a supercell thunderstorm, J. Atm. Sc., 40, pp 359-377.
14. Knupp R. Kelvins and Cotton R. William (1985), Convective Cloud Downdraft structure: An interpretive survey, Reviews of Geophysics, Vol. 23, No. 2, May, pp 183-215.
15. Kumar P and Pati Debaprasad (2016) Pre Hail Detection Algorithm for Reaction Time in Hail Control, Communicated to Meteorology and Atmospheric Physics (MAAP).
16. Lebedev S.L. 1964, Nonstationary one Dinensional model of convective cloud DANSSR, Vol 156 No. 4
17. Lebedev S.L. 1966, Formation and Evaluation of convective clouds, TundyVGI No. 3(5).
18. Low A.R. 1925, Nature, 115, P 299.
19. Ludlam F.H. 1952, The structure of cumulus cloud, proc OSTIV, Madrid.
20. Ludlam F.H. & Scorer R.S. 1953, Bubble theory of penetrative convention, QJRMS, Vol. 79.
21. Ludlam F.H. 1959. Thunderstorm studies, Weather, Volume 14, Issue 5,pages 177–180, May 1959.
22. Lublam F.H. and Kamburova Petia L (1966) Rainfall evaporation in thunderstorm downdraughts, QJRMS Volume 92, Issue 394, October, Pages 510–518.
23. Marwitz, J.D., 1972a, The structure and motion of severe hailstorms. Part I: Supercell storms. J. Appl. Meteor., 11, 166-179.
24. Marwitz, J.D.,, 1972b, The structure and motion of severe hailstorms. Part II: Multicell storms. J. Appl. Meteor., 11, 180-188.
25. Mull, S., and Rao Y.P., (1950). Origin of downdrafts in a thunder storm and the circulation in full fledged thunder cell, IJMG, Vol.1. No.4, pp 291-297.
26. Mull, S and Kulshrestha S.M. 1969, on the formation of hailstone vol. 13, splNo. Mar 62, IJMG, pp 95-103.
27. Petterson, S., (1939): Contributions to the theory of convection. Geofys. Publ, 12, 5-23.
28. Proctor H. Fred (1989), Numerical simulation of an isolated microburst. Part II: Senstivity experiments, J. of Atm, Sc., Vol. 46, No. 14, pp 2143-2165.

29. Robert R.D. and Wilson J.W., Nowcasting microburst events using single microburst data, (1986). preprint, 23rd Radar Met. Conf., Snowmass, CO, AMS, 45, pp3137-3160.

30. Rotunno, R. and Klemp, J.B. (1982), The influence of shear induced pressure gradients on thunderstorm motion, Mon. Weather rev.,110, pp136-151.

31. Schlisinger, R.E. (1980). A three dimensional numerical model of an isolated thunderstorm, II, Dynamics of updraft splitting and mesovortexcoplet evolution, 37, pp 395-420.

32. Schlisinger, R.E. (1984). Mature thunderstorm cloud top and dynamics: Three dimension numerical simulation study, J. Atm. Sc., 41, pp 1551-1570.

33. Shipley J.F.1941, QJRMS, 67,292, p 340.

34. Shiskin N.S. 1960(a), Investigation of Disintegration of convective clouds with unstable atmospheric stratification, Turdy GGO No. 102.

35. Shiskin N.S. 1960(b) Calculation of vertical development rate of convective clouds, Turdy GGO No. 104.

36. Srivastava R.C. 1962, Eddy structure of updraft in cumulus clouds and observed depth of first radar echo, Vol. 13, Spl No. March, pp 109-116.

37. Srivastava R.C. (1987), A model of intense downdrafts driven by the melting and evaporation of precipitation, J. Atm. Sc., 44, No. 13, pp 1752-1773.

38. Sulakveledze G.K. Rainstorm and hail, IPST Press, Jerusalam.

39. Weisman, M.L., Klemp, J.B. and Miller L.J., Modelling and Doppler analysis of the CCOPE August 2 Supercell storm, in Preprint, 13th conf. on Severe storm, AMS, Boston,, pp 223-226.

40. Wilson J. W. Roberts R.D., Kessinger C. and McCarthy, (1984), Microburst wind structure and evaluation of Doppler radar for airport wind shear detection. J. Climate Appl. Met., 23, pp 898-915.

41. Wolfson M.M., DiSttefano J.T. and Fujita, (1985), Low-altitude in the Memphis, TN Area based on mesonet and LLWAS data. Preprints, 14th Conf. on Severe Local Storms, Indiapolis, Amer. Met. Soc., pp 322-327.

CHAPTER 5

Hail Growth and Hailstorm

The convective clouds are composed of liquid water droplets. If the temperature is below 0 °C liquid water droplets may be in the thermodynamically unstable supercooled state or as ice crystals. Top of cumulonimbus cloud's altitude reaches even upto tropopause and since it has strong updrafts, cold air regions and sufficient ice nuclei and supercooled water, squall line thunderstorms and supercell thunderstorms form the most frequent hail producers. Supercells have the greatest potential for damaging hail since they live and physically taller than most thunderstorms. Hailstones generally begin forming on seeds of small frozen raindrops or soft ice particles known as graupel which are hardened conglomerates of snow flakes. For the frozen raindrops or graupel to grow into true hailstones, they must accumulate additional ice by spending time in cloud regions, rich in supercooled water, where temperatures are below the 0°C (32 °F) level. For the hail embryos to grow, they must remain in a layer of supercooled water for sometime —

the longer the residency, the larger the hailstone's potential size. For the smallest hailstone to form, an updraft of at least 36 km/h (24 mph) is required. Larger stones those termed golf-ball size (1- 3/4 inch diameter) require updrafts of around 88 km/h (55 mph) to form. Softball-size hail involves updrafts of over 160 km/h (100 mph).

5.1 Precipitation Process Chain

Growth of hailstones in atmosphere may be considered as a sequence of events, not necessarily independent; which begin with the synoptic environment and undergo a "precipitation process chain" as shown is Fig. 5.1. The chain extends from macro-scale (synoptic) processes to cloud micro-physical processes, ending with precipitation at the ground. The links in the chain represent basic scales of meteorological phenomena that differ primarily in terms of the relative importance of various dynamic, thermodynamic, turbulent and micro-physical processes on different temporal and spatial scales. For example, the coriolis effect is of paramount importance on the synoptic scale but usually considered unimportant at the cumulus and smaller scale. Therefore, forecasting methods of hail have to relate processes at the extreme ends of the chain. Hence it depends on the success in forecasting mesoscale characteristic from synoptic characteristics, storm environment from mesoscale, cloud characteristics from the storm environment and hail growth from the cloud characteristics.

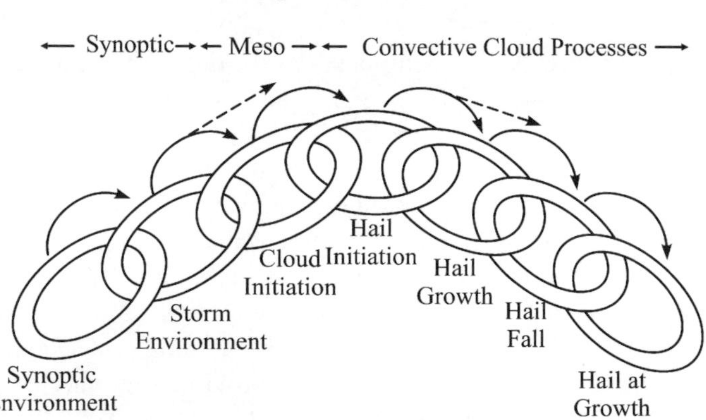

Fig. 5.1 "Precipitation Process Chain" illustrating the sequence of events, not necessarily independent, which lead to precipitation at the ground. Processes within each 'link' or event often occur on different spatial or temporal scales.

It has been observed that on synoptic scale presence of low pressure area or upper air cyclonic circulation in the lower levels in association with following situation normally give rise to hailstorm.

(a) Higher value of low level moisture.

(b) Middle layer (700 hp or above) humidity decrease

(c) Approach of trough in the upper troposphere over the region.

(d) Presence of westerly jet maxima to the southwest or northeast or the station in northern hemisphere.

(e) Presence of vertical wind shear of more than 30 kt between 1.5 to 9.0 km amsl in general. In exceptional cases in easterly wind regime in south peninsular India wind shear of less than 20 km have also been seen during hailstorm.

Detailed regionwise favourable conditions for the formation of hailstorms over India will be presented in the chapters VI to IX. Present chapter, therefore, deals with the basic physics of the hail growth.

5.2 Mechanism of Hail Growth

Under the backdrop of precipitation process chain a discussion of the microphysical processes in the hail growth is imperative. Hail growth process takes place in two stages.

Stage I: Growth of hail embryos.

Stage II: Growth of hail embryos into hailstones.

Stage I: Size of embryos may vary from a few millimeter to a centimeter. The growth takes place in relatively weak updraft region of the cloud. Studies regarding hail embryo sources and growth trajectories have yielded the following information on their formation (Knight and Knight, 1999):

- Time-developing updraft: daughter cloud*.
- Flanking line of cumulus congestus: feeder cells*.
- Core of a long-lived updraft: broad droplet spectra producing frozen drops.
- Merging mature cell.
- Mature cell positioned upwind: transfer by mid-level winds.

- Edges of main updraft: re-entry into main updraft via cyclonic flow.
- Melting and shedding of water drops from growing hailstones.
- Entrainment of stratiform cloud: debris, anvils.
- Embedded small-scale updrafts and downdrafts.
- Recirculation/multiple-passes through updraft core.

(* defined in section 5.4)

A large body of the evidence at hand suggest that in an open system, where embryo growth takes place in a distinctly different location with the embryos subsequently transported into the large, main updraft, most likely under the influence of the mid-level winds, the final growth to hail size takes place in a single pass through the sloping updraft. Doppler radar studies and deuterium to hydrogen ratios indicate that most of the growth occurs between -10°C and -25°C. It is conjectured that this mode is dominant for most hailstorms on the plains. It must also be mentioned that a single, specific embryo source is still not specified, and that in any storm, concepts may overlap or change with time. Federer and Waldvogel (1978), on the basis of hailstone sample from five storms in Switzerland, had concluded that these embryos develop through following processes.

(i) ***Ice Process*****:** In storms with relatively warm cloud bases, a big drop zone exists which contains large droplets that freeze and become efficient hail embryos.

(ii) ***Coalescence Process*****:** Embryos also grow with the riming and freezing of unstable super cooled droplets.

Stage II: Hail embryos are transported into main cloud to be carried by the strong updraft to grow into large hail. Though theoretically it is simple to foresee two different stages of growth of hail, however, it may be noted that such a clear distinction may not be always available in all the hailstorms. A conceptual idealized model of hail growth presented by Krauss (1982) is presented in Fig.5.2.

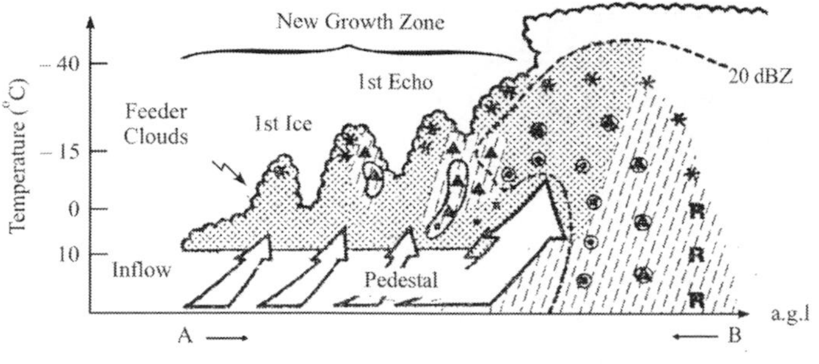

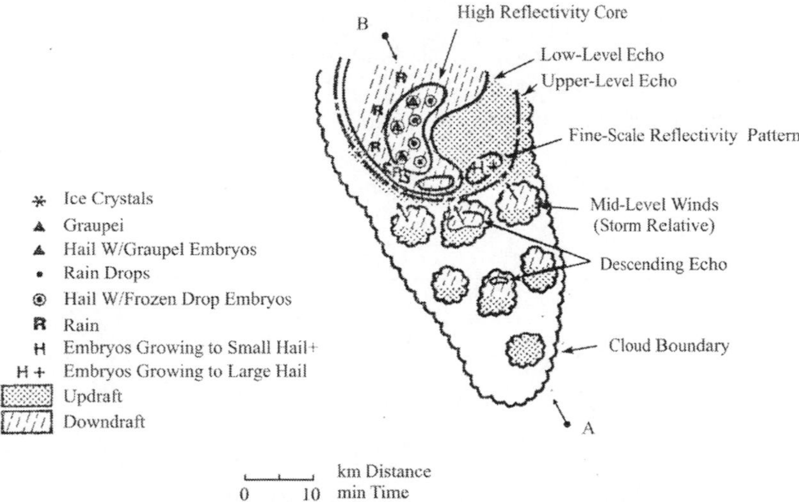

Fig. 5.2 Vertical and plan views of a conceptual model of the precipitation processes leading to the formation of hail in Alberta hailstorms (from Krauss, 1982).

5.3 Vertical Wind Shear and Tilt

It has been recognized that sustainability of the thunderstorm convective motion significantly depends on the vertical wind shear in the environment. But vertical wind shear of the environmental wind must not be so strong as to shear off the growing cumulus top and inhibit its vertical growth. But some amount of vertical wind shear is essential for the development of severe convective storm like hailstorm. The tilt provided by the vertical shear is helpful in following aspects:

(a) Preventing the high concentration of precipitation that might overload the updraft and cause it to decay by limiting the time of residence in the region of growth.

(b) The updraft tilt in relation to environmental wind may determine whether it is possible for downdraft to be maintained and coexist with updraft, without interference.

(c) It may also help ascertain whether precipitation particle grown during the first ascent will be subsequently able to re-enter it during the second ascent for their further growth.

It has been noted that the higher the top of a cloud the larger will be tilting angle with respect to cloud. For cloud top reaching 200 hPa the tilting angle needed to sustain the updraft buoyancy against the loading effect of the rain water is of the order of 25^0 (Cheng- 1989). Browning and Ludlam (1960) developed the concept of a combination of wind shear and tilted updraft enabling the updraft and downdraft to be maintained continuously without much interference, thereby enhancing the overall energy of storm. Fig.5.3.

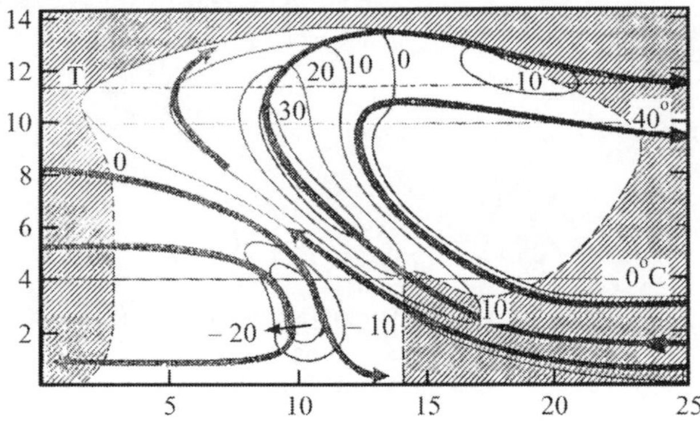

Fig. 5.3 Tentatively inferred streamlines relative to storm and isopleths of vertical velocity (ms^{-1}) within a vertical section along the direction of movement of the Wokingham supercell storm. Storm motion is from left to right. The unshaded area corresponds to radar reflectivity in excess of 30 dbz. (after Browning and Ludlum, 1960).

Yet other consequences of the vertical wind shear are:

(a) Taking away the detrained air from the top of the cloud which in turn helps in removing the heat of fusion caused due to the hail formation. This, prevent rise of temperature within the cloud inhibiting melting of hail.

(b) Horizontal velocity associated with tilted updraft also has the tendency of size sorting of precipitation with the largest particles falling on the edge of the precipitation-free portion of strong updraft.

Although strong vertical wind shear is a favourable feature for the formation of hail storm but it is not necessary condition. In South Indian region weak vertical wind shear has also produced hailstorm. Refer sec. 9.2.

5.4 Commonly used Nomenclatures Related to Hailstorms

(a) **Feeder clouds:** These are the weak updraft cumulus cloud associated with the flanking line of a large isolated cumulus cloud (or super cell), which some time merge with them or seems to intensify them. These cloud feed hail embryos in the main cloud. Refer Fig. 5.2(Top).

(b) **Daughter cloud:** Multicell storms are characterized by successive formation of discrete new cells usually in the right flank sometimes arranged in lines, at a distance up to 30 km from the hailstorm core. These cloud do not feed the mature cloud instead they grow and become the mature cloud themselves. Therefore they are known as daughter cloud.

(c) **Shelf cloud:** Shallow cumuliform cloud usually located on the right flank of the main hail cloud. Refer Fig. 5.10(b)

(d) **Pedestal cloud:** These clouds are situated just below the most intense updraft part of the mature cloud and are very low. Its base is lower than the shelf cloud. Refer Fig. 5.10(b).

5.5 Hail Size on Surface

Factors associated with surface hail size are the initial hail size in storm, humidity, thermal profile and vertical wind shear. Large stones have better chance to reach ground than small stones which melt faster on their descent. Melting is always faster in wet environment than in relatively drier, hence environmental relative humidity (RH) has strong effect on the size of the stones reaching the ground. This is the reason that hails falling within rain melt at a much greater rate than hails falling separate from rain (especially for small hail). Greatest rate of melting occurs for small stones in wet environments (e.g. high RH and within rain). Drier air in ~800-500 hPa layer is very important to produce evaporative cooling and limit hail melting during descent. Giant stones (≥ golf ball) in dry environments outside of heavy rain shaft, even in warm environment with high freezing level will have least rate of melting. Vertical wind shear also plays critical role in surface hail size by separating the updraft and downdraft and limiting the duration of hail falls in heavy rain. The lesser the duration the lesser is the melting. Hence vertical wind shear is as much important for hail size on the surface as are the thermal and moisture profiles.

Equal radar reflectivity in two storms does not imply equal hail size due to these multiple reasons. Equal values of reflectivity can be generated from a fewer of large stones with low density/size distribution, and ensemble of small stones with high density/size distribution.

5.5.1 Fawbush and Miller Technique of Possible Hail Size

The method is illustrated in tephigram shown in Figure 5.4. Two temperature differences *c*–*B* and *b*–*B* are obtained by starting from the environment temperature *C* at the convective condensation level (CCL). From this point ascend up along saturated adiabatic to reach the point *b* which is at the pressure level of the −5 °C environment temperature *B*. From the point *B* descend dry adiabatically to reach *c* at the CCL. The temperature difference *b*–*B* is an indirect measure of the amount of CAPE gained, and hence vertical velocity acquired, by a parcel ascending in the lower part of the cloud below the −5 °C level.

The −5 °C level is discussed by Fawbush and Miller (1953) (FM) and will often be well within a region of cloud with strong hail production. The greater the vertical velocity then the larger the size of the hail stone that can be developed by accretion processes. The difference *c*–*B* is an indirect measure of the depth of cloud below the −5 °C level. Provided the cloud top is sufficiently cold for ice to form (<−20 °C) then a relatively large cloud layer below the −5 °C level will provide ideal conditions for hail growth. However, the relationship between having sufficient upward motion to support large hail and favourable conditions for hail growth is such that both are needed for really large stones to develop.

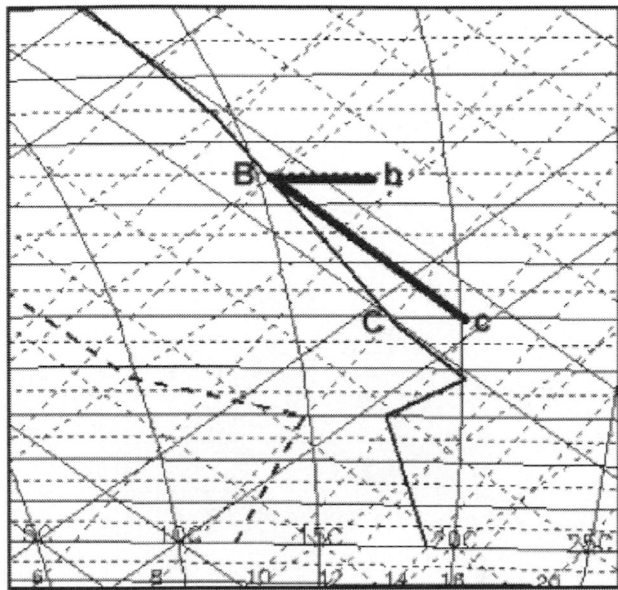

Fig. 5.4 T- φ gram constructions required to determine possible hail size.

This can readily be seen in Table 5.1 which is then used for the next step in the process. In the original diagrams provided by Fawbush and Miller, Table 5.1 was in the form of a nomogram thus providing hail diameter as a continuous function of temperature differences *b–B* and *c–B*. To help automate the process, the nomogram has been discretized in Table with hail sizes converted to millimetres. The temperature difference *b–B* has been specified to the nearest half degree (although only whole degree values are shown in the table for clarity). The temperature difference *c–B* is calculated to the nearest 5 °C. Thus, a look-up table provides the hail size once *b–B* and *c–B* are known.

If the wet-bulb freezing level is higher than 3350 m then stones are allowed to melt by an amount according to thc height of the wet-bulb freezing level and diagnosed hail size, see Table 5.2 derived from Fawbush and Miller (1953). All stones of any size are deemed to melt completely before reaching the ground if the wet-bulb freezing level exceeds 4400 m. As a last check, if the final hail size is greater than or equal to 10 mm and CAPE is less than 400 J Kg^{-1} then the hail size is set to 5 mm since peak vertical velocities implied from the CAPE would not be large enough to support large stones (also from Fawbush and Miller).

Table 5.1 Table showing hail size (mm) as a function of temperature differences b–B (along bottom with half values not labelled) and c–B (along left-hand side) as indicated in Figure 5.4

c–B	50	5	10	15	20	20	25	30	35	40	45	50	50	55	60	65	75	80	85	90	100	105	110	115	120	120	120
	45	5	10	15	20	20	25	30	35	40	45	45	50	55	60	60	65	70	80	85	90	95	100	105	110	110	110
	40	5	5	10	15	20	20	25	30	35	40	40	45	50	55	60	60	65	70	75	80	80	85	90	100	100	100
	35	5	5	10	15	20	20	25	30	35	40	40	45	45	50	55	55	60	60	65	70	70	75	75	80	80	80
	30	5	5	5	10	15	20	20	25	30	35	40	40	40	45	45	50	50	55	55	60	60	60	60	65	65	65
	25	2	5	5	5	10	15	20	20	25	30	30	35	40	40	40	40	45	45	45	50	50	50	50	50	55	55
	20	2	2	2	2	2	5	10	15	20	20	20	25	25	30	30	35	35	40	40	40	40	45	45	45	45	50
	15	0	0	2	2	2	2	5	10	10	15	15	20	20	20	25	25	25	25	30	30	30	35	35	35	35	35
	10	0	0	0	0	0	2	2	2	5	5	5	10	10	15	15	15	15	15	15	20	20	20	20	25	25	25
	5	0	0	0	0	0	0	0	2	2	2	2	2	5	5	5	5	10	10	10	10	10	10	10	15	15	15
		0		1		2		3		4		5		6		7		8		9		10		11		12	
	b–B																										

Table 5.2 Corrected hail size (mm) after melting as a function of raw diagnosed hail size from Table 5.1 (along bottom) and height of wet-bulb freezing level in metres (along left-hand side)

4400	0	0	0	0	0	0	0	0
4150	0	0	0	0	0	0	5	5
3950	0	0	0	0	5	10	10	10
3750	0	0	0	5	10	15	15	15
3550	0	0	5	10	20	20	25	30
3350	0	5	10	15	25	50	65	75
	< 5	< 10	< 20	< 25	< 50	< 75	< 100	< 125
	Hail diameter (mm)							

5.6 Numerical Simulations Hailstorm

Farley, (1987) attempted the numerical modelling of hailstorms and hailstone growth and also presented the role of low-density riminggrowth in hailproduction. Farley et al., (1989) subsequently provided numerical simulation of graupel/hail initiation via the rimimgof snowin bulk water microphysical cloud model. Orville (1996) attempted one dimensional numerical simulation of hail in weather modification. With this one could learn some thing, especially relating to the detailed microphysical growth of the particles. The interaction between the microphysical effects and dynamical responses could be sorted out only by using three-dimensional cloud-field models Barth et.al. (2007). Thus more detail three-dimensional, non-hydrostatic, time dependant, compressible system was developed, which was based on the Klemp and Wilhelmson (1978) dynamics, Lin *et. al.* (1983) microphysics, and Orville and Kopp (1977) thermodynamics.

Global and most Regional NWP models have an average grid spacing of order tens of kilometres and are unable to resolve explicitly the cumulus-scale convective systems that provide the conditions for hail production. Instead, the average effects of these and other sub-grid-scale phenomena are represented at the grid scale by parameterization schemes. These schemes are primarily designed to account for the transfer of heat, moisture and momentum, not for specifically forecasting convection.

5.6.1 Technique of Hailstorm, Prediction by Convective Diagnostic Procedure (CDP)

Convective Diagnosis Procedure (CDP) (Hand and Cappellutib, 2011), through NWP was initially aimed to provide additional diagnoses from NWP outputs of the probability of convection occurring in the UK. The outputs also provided an indication of the expected intensity of precipitation, accumulation and lightning activity. The system was operational in the Met Office from 1999 to 2005 when it was replaced by similar outputs from very high resolution models. However, a version of the CDP has been retained in order to provide a post-processing tool for site-specific diagnoses of convective hazards from gridded model data where surface conditions at a site may be different from those at a particular representative model grid point. Hail potential based on Fawbush and Miller (1953) was included in the updated version.

The convection diagnosis process begins by calculating the mean potential temperature (T) and humidity mixing ratio (T_d) in a surface layer. In this application global model grid point orography was used to define the ground surface height with temperatures and humidities from screen height and model levels 1 and 2. Screen height in the model is 1.5 m above ground, level 1 is 20 m and level 2 is 80 m. Using these mean values, (labelled T, T_d in Figure 5.5), a parcel is lifted adiabatically to find the convective condensation level (CCL) (U in Figure 5.5). At the CCL the parcel is then lifted up a saturated adiabat to the next model level. If it is stable at this level, (parcel temperature less than that of the environment), calculations cease and the parcel's convectively available potential energy (CAPE) is set to zero. If the parcel is unstable then the convective cloud base height (CCB) is set to the CCL plus 50 m and saturated ascent continues. (The addition of 50 m to the CCL for cloud base is consistent with standard forecasting practice in the United Kingdom to allow for upward transport and entrainment in convective up-draughts). Provided the parcel temperature exceeds the environment temperature, saturated ascent continues level by level until the parcel temperature lapse rate equals the environment lapse rate. The ascending air parcel is diluted by entraining environmental air from the current grid point at a fixed rate of 0.5% at each model level. This will produce convective cloud tops at the 'slice' level (C in Figure 5.5). However, ascent will continue to the 'parcel' level (W in Figure 5.5) for more vigorous convection if V–C in Figure 5.5 is greater than 1 °C (a value which makes additional computation worthwhile). The convective cloud-top height (CCT) is set to the height of the last model level where ascent was taking place. The CAPE is then calculated by integrating all the positive contributions during saturated ascent from CCL to CCT. Convective cloud-top temperature (CTT) is set to the model environment temperature at CCT.

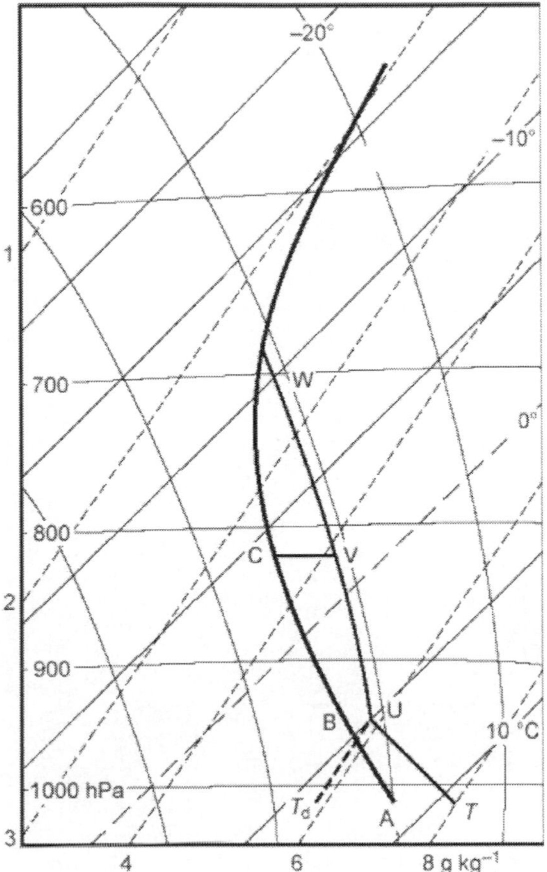

Fig. 5.5 Tephigram illustrating the parcel construction TUVW to determine convective cloud bases and tops. ABC is an environment temperature curve.

T is a layer mean potential temperature and T_d a layer mean humidity mixing ratio. The environment dew point curve has been left out for clarity.

If the parcel CAPE is greater than 50 J Kg^{-1} (equivalent to a peak vertical velocity of 10 m s^{-1} at cloud top) and CTT is less than −20 °C with a cloud depth greater than 3500 m and a base below 5000 m above ground level, then the potential for hail is assessed. These conditions will provide an increased likelihood of cumulonimbus clouds with strong updrafts, liquid water, and with a good portion of the cloud layer below freezing (0 °C). Possible hail size is determined using the technique of Fawbush and Miller (1953) (FM) which has provided useful guidance to Met Office forecasters in recent years.

5.7 Cave Channel Hypothesis

Observed size sorting during the hailfall led Fengqin Kang et al (2007) to propose the concept of Cave Channel hypothesis of hailstone formation with in a thunderstorm. Based on their 3-D computer simulation of a typical hailstorm on Qinghai Tibetan plateau they proposed formation of high humidity cave channel regions in hail clouds to explain the varied sizes of hailstones i.e. large, medium and small, by the same hailstorm. The information of the large-scale circulations for the cloud model was provided by the MM5V3 model. The results showed that the water content of each hailstone bin was significantly large in the "cave channels" (CC). Refer Fig. 5.6 which points the location of cave channel in the core of the high vertical velocity and the region of zero horizontal velocity.

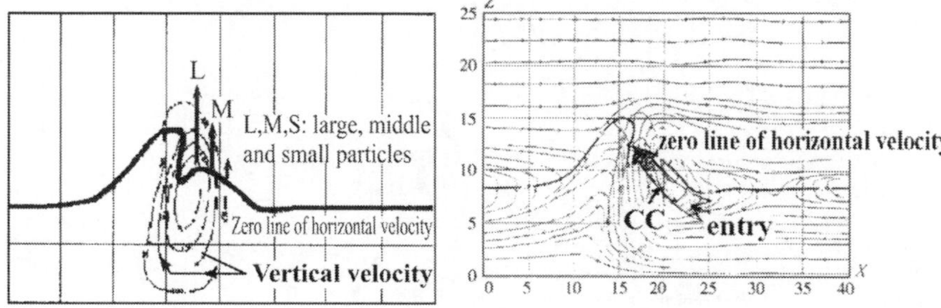

Fig. 5.6 Location of cave-channel (CC) indicating maximum vertical velocity region and zero line of horizontal velocity. Formation zones large, medium and small hailstorms are indicated.

At the initial stage of hail formation, there was also high water content region consisting of small ice particles (d < 1 mm), graupel and hail embryos (1 mm < d < 5 mm), as well as small hailstones (5 mm < d < 10 mm), around the altitude of –30 to −50°c, above the high water content center associated with the "cave channels". Between them there was a gap of lower water content, which indicated that the main mechanisms of hail formation were different in those two regions. As the hail and rain fell, the maximum center at higher level dropped until it merged with a lower equivalent. The larger the hail particles, the earlier the maximum centers merge with each other. During the hailstorm dissipation period the downdraft occurred in the region of "cave channels" and the "cave channels" gradually fade, thereafter. However, it remained still the center of high hail water content, even though all updraft airflow turned to downdraft airflow. "Cave channels" are not the only regions of hailstones formation, but are nonetheless effective in the size sorted growth of hailstones. It should be, therefore, also the main region of suppressing hail growth from small to large.

5.8 Echoes Associated with Hailstorm on Radar

Radar is extremely useful tool for the detection of any hydrometeor in atmosphere. Meteorological unit of measurement of reflectivity is dBZ. It stands for decibels of Z. It is a meteorological measure of equivalent reflectivity (Z) of a radar signal reflected off a remote object. The reference level for Z is 1 mm^6 m^{-3}, which is equal to 1 μm^3. It is related to the number of drops per unit volume and the sixth power of drop diameter. Reflectivity of any cloud is dependent on the number and type of hydrometeors, which includes rain, snow and hail and the hydrometeors' size. A large number of small hydrometeors will reflect the same as one large hydrometeor. The signal returned to the radar will be equivalent in both situations, so a group of small hydrometeors is virtually indistinguishable from one large hydrometeor on the resulting radar image. A meteorologist can determine the difference between one large hydrometeor and a group of small hydrometeors as well as the type of hydrometeor through knowledge of local weather condition contexts. On dBZ-scale of rain could be categorized as, heavy, moderate or light as per the following criteria:

40 dBZ as heavy

24-39 dBZ as moderate

8-23 dBZ as light

Conventionally with increasing intensity of precipitation different colour schemes are used e.g. light (green), moderate (yellow), and heavy (red). The latest radars use five colors, including two shades of red and magenta to indicate the heaviest returns. To assess three dimentional extention of any cloud system radar images are obtained in two forms. If antenna rotates making small constant angle with horizontal then the picture is termed as Plan Position Indicator (PPI). If the antenna scans in a fixed vertical plane then the image produced is termed as Range Height Indicator (RHI). PPI and RHI images are shown below:

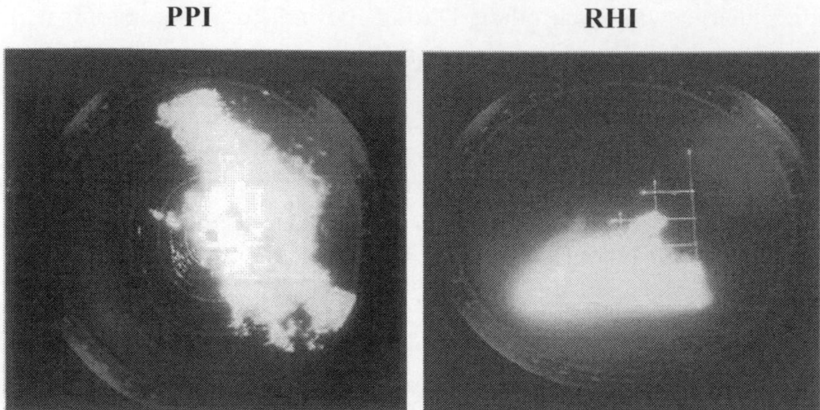

Fig. 5.7 PPI and RHI cloud photographs from the storm detecting radar at Mumbai (India) on 26 July 2005. (Left) PPI at 1130 h IST having radar range at 100 km with each range ring of 20 km scale. (Right) RHI at 1430 h IST having radar range at 50 km with scale height as 5 km each in vertical, while horizontal scale is 10 km each. (Colour Plates Pg. No. 354)

Hailstorm Echo

Radar echoes, associated with hailstorms, are normally associated with reflectivity (Z) of the order of 45dBZ or more. When radar reflectivity of 45 dBZ or higher are observed above the freezing level where large amounts of supercooled or solid water is present at high altitudes, the presence of hail is highly probable. With 45dBZ or more of radar reflectivity, if the images indicated on the screen are typically those as are explained below, then the prospects of hails are almost certain.

(a) ***Fingers:*** Protrusion like fingers, of the order by 1.5 to 8 km, in length.

(b) ***Hook echo:*** A wall is seen on RHI during the most intense phase of storm. When such an echo is viewed on PPI by horizontally scanning radar with slightly narrow beam, the wall surrounding the vault (refer subsection (g) below) is generally seen as hook echo. Refer Fig. 5.8. Fig 5.9 shows the hook echo in; forming stage in the lowest echo picture; Formed stage in the middle echo picture and dissipating stage on the top echo picture.

Fig. 5.8 Typical Hook echo and Bounded Weak Eecho Region (BWER). (Colour Plates Pg. No. 354)

Hooks or finger shapes are usually appended to core portions of a storm cell. They may show steep gradients and heavy returns within their narrow bands. Many times, hooks prove to be tornadoes, and fingers may represent hail shafts.

(c) ***Scallops:*** Blunt protuberances 1.5 to 5 km in length from the edge of the thunderstorm. These are other signs of actively growing aggressive storms. Scallops can turn into hooks.

(d) ***Forward Overhang:*** It is extending outward, downwind from the main echo mass of the storm. Its length may be 100 to 300 km. It is comprised of thick cirrostratus / altostratus cloud indicating outflow near the top of the strong updraft. Close to the area of rain this is thick and relatively lower in base.

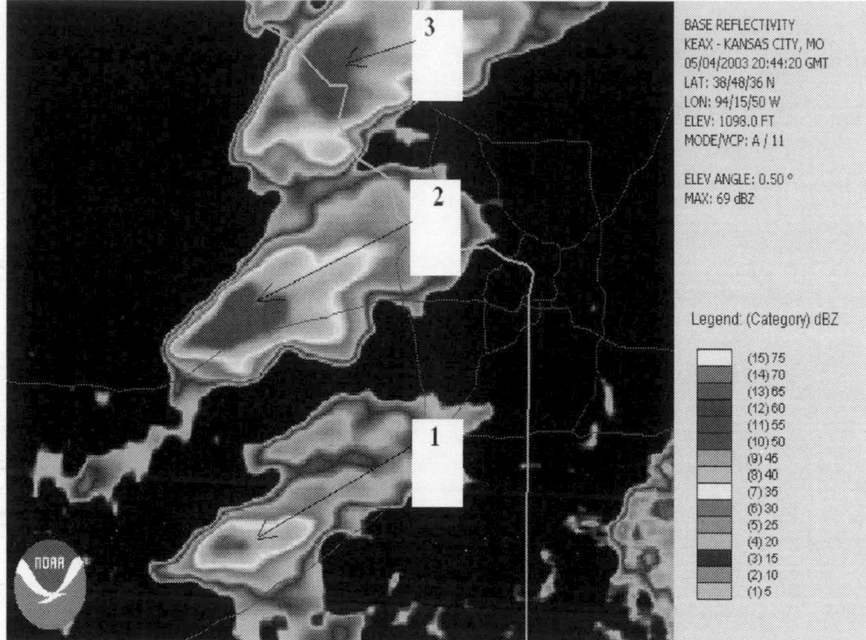

Fig. 5.9 Forming (1), formed (2) and dissipating (3) hook echoes from bottom to top. (Colour Plates Pg. No. 355)

(e) ***Weak Echo Region* (*WER*):** An early step in a storm organizing into a tornado producer is the formation of a Weak Echo Region (WER). This is an area within the thunderstorm where precipitation should be occurring but is "pulled" aloft by a very strong updraft. WER is the most intense part of the updraft region of the mature cloud showing lowest reflectivity. Although it is filled with cloud particles but updraft is so strong that it does not permit them to grow to large enough size to be clearly detected by radar. The weak echo region is characterized by weak reflectivity with a sharp gradient to strong reflectivity above it and partially surrounding the sides. The region of the precipitation lofted above the WER is the echo overhang consisting of precipitation particles diverging from the storm's summit that descend as they are carried downwind. Wall Boundary of Weak Echo Region (WER) as seen on RHI is also known as Wall.

(f) ***Bounded Weak Echo Region* (*BWER*):** Weak echo region in severe storm surrounded by precipitation are termed as Bounded

Weak Echo Region (BWER). Refer Fig. 5.8 which shows BWER based on the colour scheme.

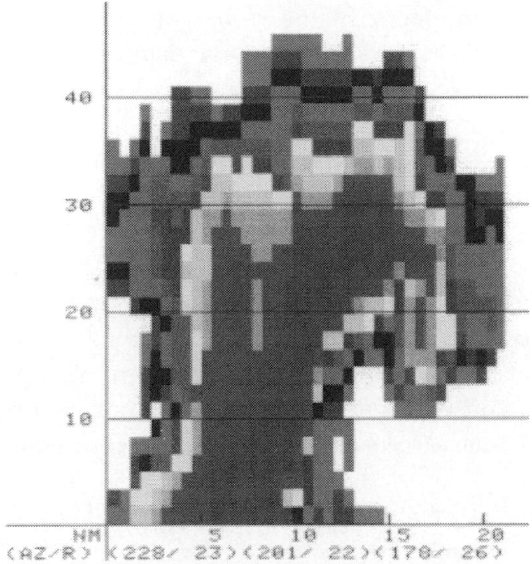

Fig. 5.10 Bounded weak echo region (BWER). (Colour Plates Pg. No. 355)

(g) ***Vault*:** Those weak echo regions in severe storms which are wholly or partially bounded by precipitation are termed as vaults. These vaults may contain cloud particles or even graupel with diameters less than 2 mm resulting in reflectivity values between – 30 and –50 dBZ which are too weak to be detected by most of the weather radars.

(h) ***Notch*:** Echo free vault in RHI when seen in PPI is known as notch. Notch may also be the result of dissipation of cloud with the release of hails (Slot and Hiser – 1955).

(i) ***Hole*:** On a PPI when the vault is seen for a storm which is located at a little distance from the radar station, it looks as a darker hole embedded in a brighter echo surrounding. It is also termed in some of the books as 'Dry-Hole'. Name is apparently to identify the region which is comprised of only small cloud droplets in it. Large precipitation particles do not develop owing to the high velocity with in the hole rendering insufficient time period for any droplet to grow to radar detectable size.

5.9 Satellite observation of Hailstorm

Rosenfeld,et al(2008) had presented an indirect conceptual model that facilitates the inference of the vigour of severe convective storms and hence those of tornadoes and large hail, by using satellite-retrieved vertical profiles of cloud top temperature (T) and particle effective radius (r_e). The driving force of these severe weather phenomena was the assumption that the high updraft speed can sustain the growth of large hailstones and provide the upward motion that is necessary to evacuate the violently converging air of a tornado. Stronger updrafts could be revealed by the delayed growth of r_e to greater heights and lower T, because there is less time for the cloud and raindrops to grow by coalescence. Fig. 5.11 shows satellite observed hailswath on the ground. More direct exclusive method of hailstorm was presented by Cecil, (2009). His concept of satellite observation of hailstorm is based on the assumption that large ice hydrometeors scatter upwelling microwave radiation away from a satellite's field of view, causing brightness temperature (TB) depressions (refer Appendix-B for definition) far below the thermodynamic temperature in the atmosphere. The comparison of surface hail observation with TB were made in U.S.A. and it was noted that the probability of large hail increased with decreasing brightness temperature. The largest hail size categories were associated with progressively lower brightness temperatures.

Albeit the assumption was a significant step towards monitoring of hailstorm by satellite but there are several questions which makes it not unambiguous e.g. ensemble of smaller size hails in resonance with the upwelling radiation instead of large hail may give the same brightness temperature depression. The upwelling brightness temperature represents column-integrated effects, more responsive to middle and upper layers of the atmosphere in these storms (not directly seeing hail near the surface). A given brightness temperature does not uniquely map to a particular vertical profile of hydrometeors having particular distributions of particle type, size, and density. There are several combinations that can yield the same brightness temperature. Hence local conditions of melting during fall and local climatology has to be kept in mind before interpreting hailstorm.

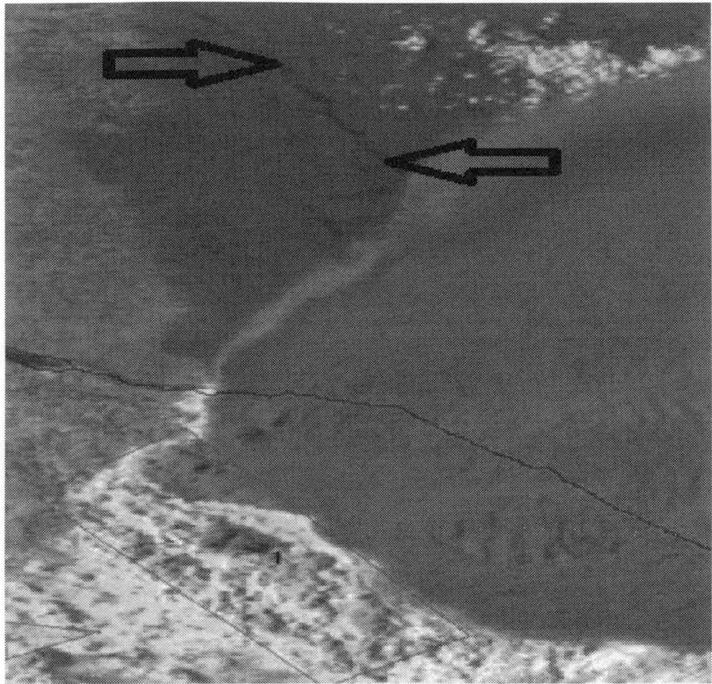

Fig. 5.11 A tornadic storm with 4.5 inch hail. The image is based on the NOAA-AVHRR overpass on 29 June 2000, 2221 UTC, over a domain of 282 264 AVHRR 1-km pixels. The cloud occurred in southwestern Nebraska. A hail swath (marked with arrow) on the ground can be seen as the dark purple line emerging off the north flank of the storm, oriented nw-se. Two hail gushes are evident on the swath near the edge of the storm. The precipitation swath appears as darker blue because of the cooler wet ground; Rosenfeld, et al (2008). (Colour Plates Pg. No. 356)

The 37-GHz channel seemed most useful for identifying severe hailstorms. The 85-GHz channel sometimes had extremely low brightness temperatures without reports of large hail, presumably because a deep column of large graupel or small hail is sufficient to scatter the shorter wavelength radiation. The precipitation fraction of hail as observed by Tropical Rainfall Measuring Mission (TRMM) Microwave Imager (TMI) in 37 and 85 GHz channels is shown in Fig 5.12. It may be noted that minimum brightness temperature (TB) between 50 to100 indicated 100 % hail fraction of precipitation in 37 GHz channel.

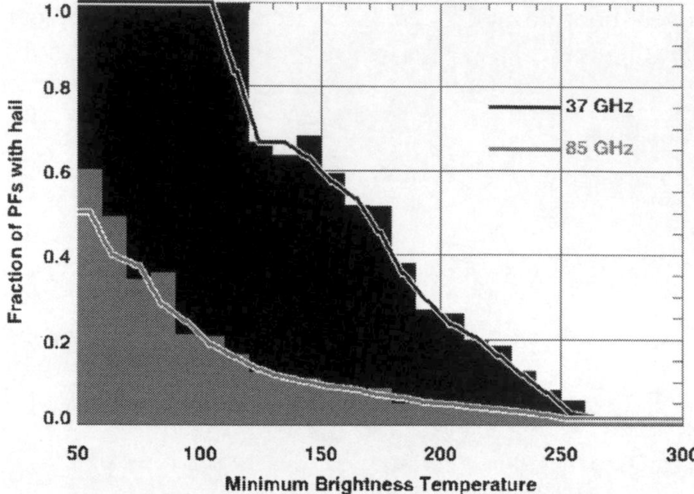

Fig. 5.12 Fraction of TRMM precipitation features (PF) in south-central and southeast United States with *Storm Data* reports of large (1 in. diameter/2.5 cm, or greater) hail, as a function of precipitation feature's (PF) minimum 85-GHz PCT (gray) or 37-GHz PCT (black). (Polarization-corrected brightness temperature = PCT). Methodology as in Cecil (2009).

It is important to emphasize that satellite observation technique do not unambiguously identify the presence of large hail, but instead identify radiometric signatures consistent with observed hailstorms.

5.10 Structure of Hailstorm

Most simplified structure of hailstorm is shown in Fig 5.13. Main cloud could be tornado with protruding funnel or just a cumulonimbus cloud.

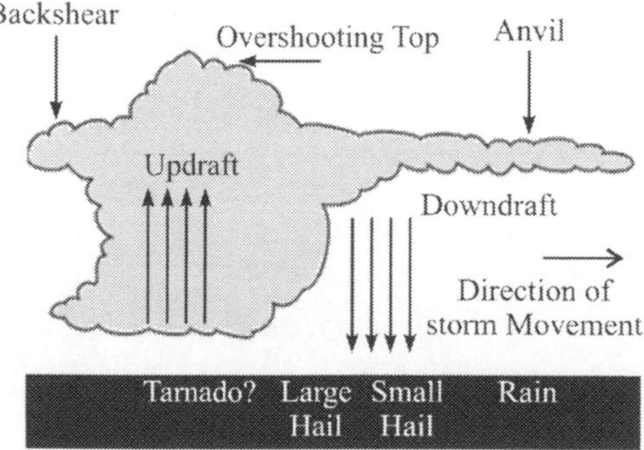

Fig. 5.13

For over-simplification downdraft is shown separately, out of the cloud. But they co-exist within the mature cumulonimbus cloud as has been explained in 4.6. Hailstorm structure is described in the following sections.

(a) **Super Cell Storm Structure:** The model of super cell presented by Browing (1977) for middle latitudes may be helpful in developing a model for tropics, too, with one remark. While the anvil of the middle latitude severe thunderstorm generally extends ahead of the thunderstorm in the direction of its movement, in tropics they sometimes extend to the rear also of the direction of movement (Asnani-1992 b). In tropics, sometimes, ahead of the squall-line the airflow is directed towards the squall-line at all levels. The boundary layer air enters at the front of the squall-line, rises in the updraft and leaves the cloud in the rear at the upper level. As a result anvil trails behind the storm. In contrast anvil in mid latitude is always ahead of the storm. Salient features of Browing model are presented in Figs. 5.14(a) and (b) which represents the schematic plan view and vertical cross section of the storm respectively. Various details of the super cell are as under:

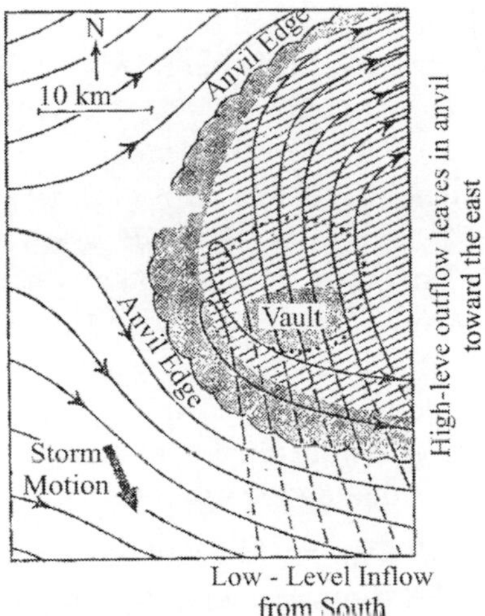

Fig. 5.14(a) Schematic model of the Fleming supercell hailstorm, taken from Browning and Foote (1976).

Plan view in Fig. 5.14(a) is showing the principal features of the airflow within and around the Fleming storm. Regions of radar echo are shown hatched. Areas of cloud devoid of detectable echoes are stippled.

The dotted circle represents the extent of intense updraft in the middle troposphere. The thin lines are streamlines of airflow relative to storm. Some of the streamlines represent the strong westerly environmental flow at the middle-level being diverted around the main updraft. Others represent the low level southerly inflow towards the updraft (dashed lines) and also part of the high level outflow.

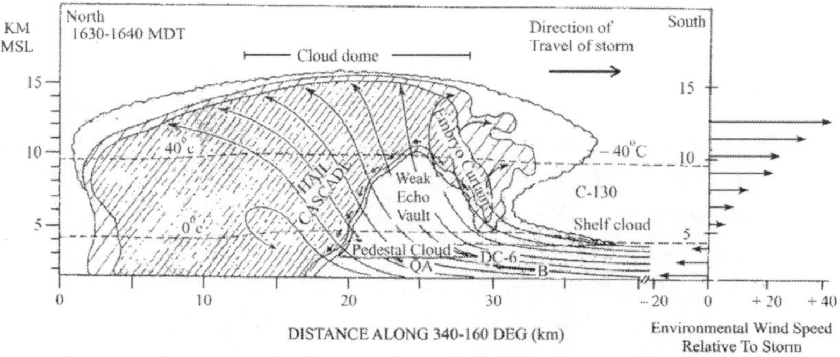

Fig. 5.14(b) Vertical section showing features of the visual cloud boundaries of the Fleming storm superimposed on the radar echo pattern.

The vertical section in Fig. 5.14(b) is oriented along the direction of travel of the supercell storm, through the centre of the main updraft. Two levels of radar reflectivity are represented by different densities of hatched shading. Areas of cloud devoid of detectable echo are shown stippled. The location of four instrumented aircraft are indicated, viz C-130, QA (Queen Air), DC-6 and B (buffalo). Bold arrows denote wind vectors in the plane of diagram as measured by two of the aircraft (scale is only the half of that of the wind plotted on the right side of the diagram). Short thin arrows skirting the boundary of the vault represent a hailstone trajectory. The thin lines are streamlines of airflow relative to the storm drawn to be consistent with other observations. To the right of the diagram is a profile of wind component along the storm's direction of motion, derived from a sounding, 50 km south of storm.

(i) ***Dimension and Shape*****:** A super cell storm is approximately 20-30 km in horizontal extent. Its horizontal cross section is approximately circular or elliptical. Vertically it is about 12 – 15 tall. Its anvil may have horizontal length of 100 – 300km. The main storm top generally looks as smooth or rounded dome.

(ii) ***Anvil Overhang*****:** It is cirrostratus or higher altostratus cloud indicating the outflow near the top of strong updraft. It extends

100-300 km downwind in the direction of upper level flow. Precipitation particles fall from the anvil overhang. These precipitation particles cool by conduction and evaporation. They generate weak downdraft. This downdraft air gets entrained into the outer edges of the cloud updraft. And thus it forms nearly a closed circulation.

(iii) ***Embryo Curtain*:** It is a part thick anvil overhang near the area of rain and hail. The embryo curtain consist of graupel particles several millimeter in diameter having terminal fall speed ($\approx 10 ms^{-1}$) comparable to the vertical velocity in their vicinity. The lower tip of the embryos curtain is not far above the cloud base. Embryo curtain is a favoured region of up and down cycle for hail embryos which grow into hailstorm of substantial size. Refer Fig. 5.14(b).

(iv) ***Base of the Cloud*:** The super cell has two bases. The principal base is known as pedestal cloud base and the secondary base known as shelf cloud base.

(aa) ***Pedestal Cloud Base*:** Strongest updraft enters the storm through this cloud base. It is the condensation level for the ascending air. Level of free convection is situated 1- 2 km above this base. Hence the pedestal cloud are formed by lifting due to outer forcing and not by buoyancy of the parcel itself. This base is well defined and smooth.

(bb) ***Shelf Cloud Base*:** It is at a higher level on the forward side of the thunderstorm cloud than pedestal cloud. It has a diffused appearance. It is the condensation level for the air originating at slightly higher level than that which is feeding the pedestal cloud. Updraft below this cloud is much weaker than that below the pedestal cloud.

(v) ***Weak Echo Region (WER)*:** This is situated just above the pedestal cloud. It is the region of maximum updraft speed of cloud ($\approx$30-40 ms^{-1}). It has horizontal dimension of 5-10 km. Vertically, it extends to half or two third the height of the storm.

(vi) ***Hail Cascade*:** Due to strong updraft in the WER the trajectory of the larger hailstones are roughly concentric with the edge of the weak echo region of the vault. Larger hails fall close to WER on the surface.

(b) Multi Cell Storm Structure: A typical storm of Raymer (Colorado – USA) is shown in Fig. 5.15. As already stated these storms are comprised of ordinary cells each undergoing the life cycle of developing, maturing and dissipating stages. They are arranged in sequence. The daughter cloud arranged in a line, up to a distance of even 30 km subsequently merge with the hailstorm and mature. Young daughter clouds are the one in which hail embryos are believed to form in ordinary multicell storm.

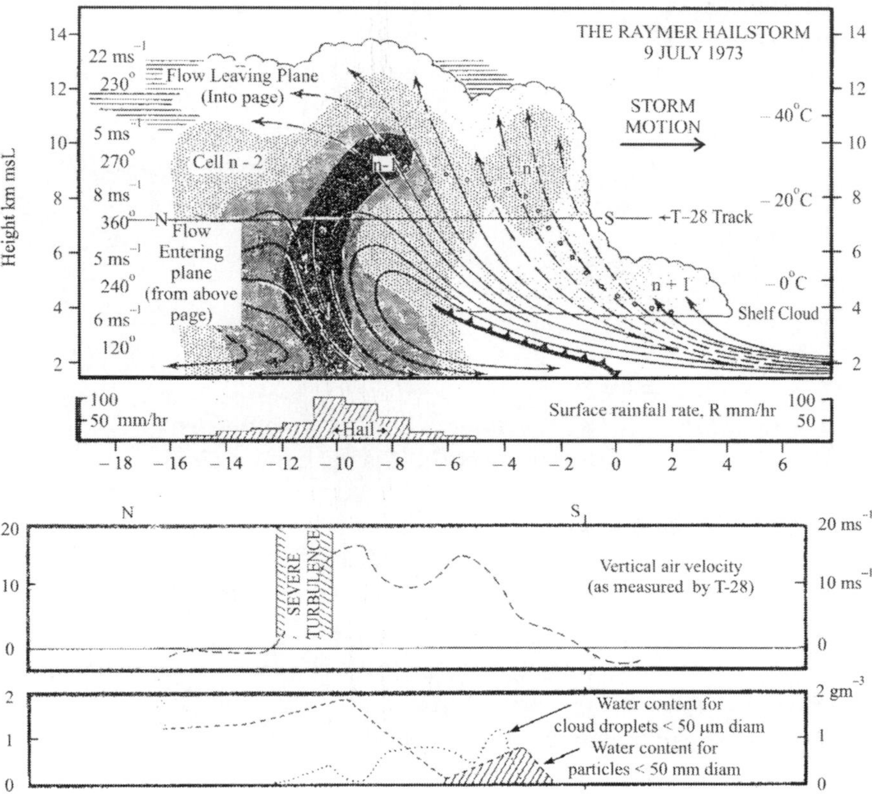

Fig. 5.15 Schematic model of an ordinary multicell hailstorm near Raymer in northeast Colorado, taken from Browning et al (1976)

Fig. 5.15 shows a vertical section along the multicell storm's direction of motion, through a series of evolving cells. The solid lines are the streamlines of flow relative to the moving system; they are shown broken on the left side of the figure to represent flow into and out of the plane and on the right side of the figure to represent flow remaining within a plane, a few kilometers closure to the reader. The chain of open circles represents the trajectory of a hailstone during its growth from a small particle at cloud base. Actually airflow in each cell has been drawn relative to the individual cell and since the developing cells n+1 and n traveled more slowly (5 ms^{-1}) than either the mature cells (7ms^{-1}) or the storm as a whole (10 ms^{-1}), the streamline in the young cells would have had a stronger component from the south relative to the storm as a whole. This explains why in the model the trajectory of the growing hailstone crosses over the streamlines during its early growth as shown in the figure. Lightly stippled shading represents the extent of cloud and the three darker grades of stippled shading represent radar reflectivities of 35, 45 and 50 DbZ. The temperature scale on the right side represents the temperature of the parcel lifted from the surface. Winds (ms^{-1}, deg) on the left side are the environmental winds relative to storm based on soundings behind the storm. Surface rainfall rate averaged over two minutes interval during the passage are plotted below the section. The horizontal lines N-S through the section at about 7 km shows the track of T-28 penetration aircraft. Smoothed data of which are plotted at the foot of the figure. Although the T-28 data was not quite synchronous with the data in the vertical section a comparison of the T-28 updraft velocity measurement with the flow shows the agreement is reasonably good; however the aircraft narrowly missed the core of the updraft, which was as strong as 15ms^{-1} at places. Salient feature of multicell raymer storm are mentioned below:-

(i) ***Developing Stage:*** The cells marked n + 1, n, n-1 and n-2 are a different stages of evolution but are so aligned that they give appearance of one single storm not very different from supercell storm. n-1 cell is seen in its mature stage with vigorous updraft in front and vigorous downdraft in rear. The first echo is seen only at cell n. n + 1 is the daughter cloud which has merged with the main cumulonimbus and does not give echo. They only small precipitation droplets and even hail embryos but not enough to give rain. Time interval in the development of successive cells is about 15 minutes i.e. it takes 15 minutes for a cell to evolve to the stage of n –1 and similarly for n-1 to go to the stage of n-2. In more severe storm the time interval has been observed to be even less (≈5 min.).

(ii) ***Updraft:*** Updraft enters the storm in front and leaves it in the rear. Updraft is drawn from a layer close to ground 500 m deep and 20 km upwind of the storm. It enters through the shelf cloud tilting backward with height. Average updraft velocity at the cloud base is 4 ms^{-1}. Maximum of 8 ms^{-1} is also seen at 7 km level, just above the level of maximum parcel buoyancy, velocity of 20 ms^{-1} is noticed.

(iii) ***Downdraft:*** In Raymer storm the downdraft originated partly in the mid tropospheric (~6 km) level. Part of the vigorous updraft of n-1 cell, turns into vigorous downdraft. It originated close to the level of lowest equivalent potential temperature and descended almost unmixed to surface. Maximum measured downdraft velocity was 15 ms^{-1} close to cloud base. Downdraft velocity greater than 10 ms^{-1} occurred in a region 2 km wide and between levels 2 and 6 km amsl. The depth of the surface outflow was more than 1 km ahead of the storm whereas it was less than 0.5 km in the rear. Gustfront entered about 5 km ahead of the leading edge of the surface precipitation. Maximum horizontal velocity convergence at the interface of inflow and outflow was 1 to 2x $10^{-3}s^{-1}$. Maximum surface horizontal velocity divergence below the strongest downdraft was 4 x $10^{-3}s^{-1}$.

(iv) ***Speed and Movement:*** The Raymer storm travelled roughly with the wind in mid troposphere. Newton and Katz (1958) have found that at the average large convective rainstorms moved about 25^0 to the right and 7 kt slower than mean wind between 850 – 700 hpa layer. Hook echoes (Fujita – 1958) moved 25^0 to the right of the direction of movement of echoes located in the vicinity. Neumann (1965) observed a deviation of 10^0 toward right of the path of hailstorm when it got intensified. Fujita (1965) noted that whenever echo becomes rotational its direction changes abruptly and thus intensifies the storm producing heavy hail and rain. The path of non-rotational echo which maintains the normal course diverges when lying to the left of the path of rotational echoes and converges when lying to the right.

(v) ***Precipitation:*** Rate of rainfall was 100 mm hr^{-1}. Maximum diameter of the hailstones was 15 mm. 5% of the ground precipitation was in the form of hail. 25% of the hailstone embryos were frozen droplets and 75% were graupel. Growth time of hailstone was of the order of 13 min for 5 mm graupel to grow into 15 mm hailstone. Growth time for embryos to grow in to 5 mm graupel was of the order of 10-15 min. Level of hail growth was 8-10 km amsl between the temperature range of -20^0 to -30^0 inside cloud. With the severity of the storm the size of the hailstone increases and minimum time required for growth of embryos to hailstone will also decrease. Also depth of the height of the first echo would also increase.

5.11 Stability Indices

The basic procedure to study any convective cloud is embodied in various stability indices. They are presented below. For details refer Appendix B.

Galway Index or Lifted Index (LI): Raise the parcel moist adiabatically without entrainment from mixing condensation level (MCL) (or lifting condensation level (LCL)) then continue further up to 500 hPa. Subtract the parcel temperature from that of the environment at 500 hPa.

Mixing Condensation Level (MCL) is the lowest height at which saturation may occur if the near surface layer is mixed completely by wind action. Mixing occurs in a layer when vertical windshear or vertical direction wind shear occurs. Shear causes turbulence due to updrafts and downdrafts. These updrafts and downdrafts produce the mixing action. The moisture present in a layer tends to become evenly distributed through the mixed layer.

Showelter index (SI): This is similar to LI, except that the parcel being raised to 500 hpa is from 850 hPa and not from MCL

George Index Or K-Index (KI): It is defined as

$$KI = T_{850} - T_{500} + Td_{850} - dd_{700}$$

where dd_{700} is the dew point depression at 700 hpa, Td_{850} is the dew point at 850 hpa, T_{850} and T_{500} are the environmental temperature at levels indicated by suffix.

Total Total Index (TTI): It is defined as,

$$TTI = T_{850} + Td_{850} - 2\,T_{500}$$

Modified JeffersonIindex (I_{mj}): It is defined as,

$$I_{mj} = 1.68\,W_{850} - 0.5dd_{700} - T_{500}$$

where w_{850} is wet bulb potential temperature at 850 hPa. dd_{700} is dew point depression at 700 hpa. T_{500} is the environmental temperature at 500 hPa.

Potential Instability Index (PI) : It is defined as,

$$PI = T_{w800} - T_p$$

Where T_{w800} is the wet bulb temperature at 800 hPa and T_p is the temperature obtained by lifting parcel from surface wet bulb temperature along saturated adiabate up to 800 hPa.

Rackliff Index (RI): It is defined as,

$$RI = W_{850} - T_{500}$$

where W_{850} is the dry bulb potential temperature at 850 hPa. T_{500} is the dry bulb temperature at 500 hPa.

5.12 Pressure and Humidity Dip

(a) Pressure Dip: This phenomenon is noticed in the barograph trace Fig. 5.16. Dip is of the order of 2 hPa or so. It is sometimes observed that during the first few minutes of large hailstone reaching the ground there is drop in pressure. This is because terminal velocity of the large falling hydrometeor is large and there is quite less time of contact between hail and the surrounding air. Thus the fall of large hail would be accompanied by little exchanges of heat between hydrometeor and the surrounding air. Hence little change is caused in temperature in and outside the cloud. Thus quasistatic perturbations pressure may be controlled transiently only by hydrometeor load shedding factor and drop in pressure is recorded. Theoretically (Asnani-1992 (b)) it has been worked out that for every 1 gm of hailstone per sq cm the barograph would register a fall of 1.8 hpa. But soon after the release of large hailstones from air column smaller hailstones and rain would cool the air. This is because small hailstone would have lesser terminal velocity and hence more resident time in the air. This would cause them to melt partially or wholly by absorbing heat from the ambient atmosphere. Similarly the rain or water on the small hail would evaporate by absorbing heat from the atmosphere. The resulting cooling of atmosphere induces rise in pressure and resume positive perturbation in pressure leveling the dip.

(b) Hygrograph Dip: Often humidity dip is noticed in the hygrograph trace (Fig. 5.16). Thunderstorm project (USA) reported that in the region of divergence in the surface wind i.e. in the out flowing downdraft during heavy rain, the humidity dip is observed in the hygrograph trace. When a downdraft has been initiated and the parcel has started its downward journey many intensely cold drops of rain produced from fully or partially melting hailstone themselves may pass through it. If the descending parcel is saturated and comparatively warmer than these drops, which is normally expected – some water vapour may condense on the cold drops of melting hailstones. These could unsaturate the parcel and cause humidity dip.

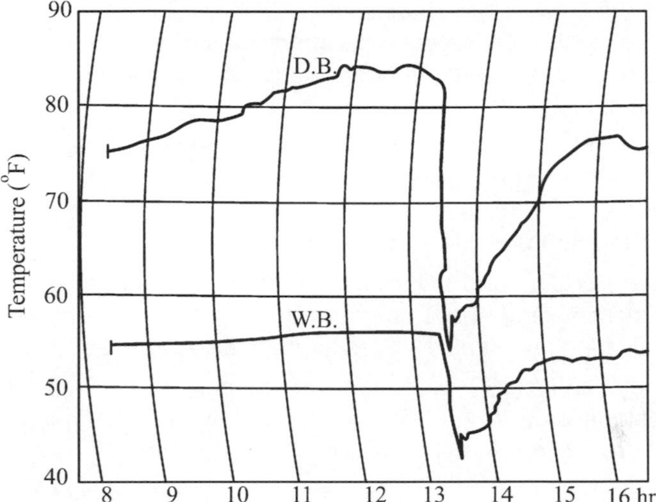

Fig. 5.16 Sharp drop in wet bulb temperature (W.B.) may be noted at the time (13.40-13.50 IST) of hail fall (11 March 1957, Begumpet, India).

References

1. Asnani (1992), tropical meteorology, Indian institute of tropical meteorology Pune. P 153.
2. Asnani (1992 b), 'Tropical Meteorology', Indian Institute of Tropical Meteorology Pune. Pp 10.37.
3. Barth, M.C., S.-W. Kim, C. Wang, K. E. Pickering, L. E. Ott, G. Stenchikov, M. Leriche, S. Cautenet, J.-P. Pinty, Ch. Barthe, C. Mari, J. H. Helsdon, R. D. Farley, A. M. Fridlind, A. S. Ackerman, V. Spiridonov and B. Telenta, 2007: Cloud-scale model intercomparison of chemical constituent transport in deep convection Atmos. Chem. Phys., 7, 4709–4731,
4. Browning k.a. & Ludlam, F.H. (1960), 'Radar analysis of hailstorm', Tech. Note 5, Dept. of met, Imperial College, London pp-106.
5. Browning K.A and Foote G.B. (1976), QJRMS, 102, 4, 499-533.
6. Browning K.A.(1977), The structure and mechanism of hailstorm and hail : A review of hail science and hail suppression, met monograph, American Met Society, Vol. 16, 38. pp 1-43.
7. Byers H.R. & Braham R.R.(1948),'Thunderstorm structure and circulation, j met, 5,3, pp 71.

8. Cecil, D. J., 2009: Passive microwave brightness temperatures as proxies for hailstorms. J. Appl.Meteor. Climatol., 48, 1281–1286.
9. Cecil, Daniel J.And Blankenship Clay B., 2012, Toward a Global Climatology of Severe Hailstorms as Estimated by Satellite Passive Microwave Imagers, Journal of Climate Volume 25 15 January 2012, 687-703.
10. Cheng M.D. (1989), 'Effect of cloud downdraft and mesoscale convective organization on the heat and moisture budgets of tropical cloud cluster. Part i : a diagnostic cumulus ensemble model j. Atmos Sci. 46 1517 –1538.
11. Farley, R.D. and Orville, H.(1986). Numerical modeling of hailstorms and hailstone growth. Part I: Preliminary model verification and sensitivity test. J.Appl. Meteorol., 25, 2014-2035.
12. Farley, R.D..(1987). Numerical modeling of hailstorms and hailstone growth. Part II: The role of low density riming growth in hail production. J. Appl. Meteorol., 26, 234-254.
13. Farley, R.D, Price, P. E., Orville, H. and Hiresh, J.H.(1986). On the numerical simulation of graupel/hail initiation via the riming of snow in bulk water microphysical cloud models J. Appl. Meteorol., 28, 1128-1131.
14. Fawbush EJ, Miller RC. 1953. A method for forecasting hailstone size at the earth's surface. Bulletin of the American Meteorological Society 34: 235–244.
15. Federer, B & Walduegel, A (1978), 'Time- resolved hailstorm analysis and radar structures of Swis storm', QJRMS, 104, pp 69-90.
16. Fengqin Kang, Qiang Zhang and Shihua Lu, (2007), Validation and development of a new hailstone formation theory: numerical simulations of a strong hailstorm occurring over the Qinghai-Tibetan plateau, journal of geophysical research, vol. 112. Also in book "Nucleation and Atmospheric Aerosols", 172–176, © Springer 2007.
17. Fujita T (1958), J.Met, 15 Pp 288 – 296.
18. Fujita T (1958), Monthly Weather Review, 93.2.67-78.
19. Hand WH. 2002. The met office convection diagnosis scheme. Meteorological Applications 9: 69–83.
20. Hand, W.H. and Cappellutib, G. (2011), A global hail climatology using the UK Met Office convection diagnosis procedure (CDP) and model analyses, Meteorol. Appl. 18: 446–458.

21. Rosenfeld, D, William L. Woodley, Amit Lerner, Guy Kelman, and Daniel T. Lindsey, (2008) Satellite detection of severe convective storms by their retrieved vertical profiles of cloud particle effective radius and thermodynamic phase, J. Geophys. Res., Vol. 113.
22. Klemp, J. B. and Wilhelmson, R. B.: The simulation of three-dimensional convective storm dynamics. J.Atmos.Sci., 35, 1070-1096, 1978.
23. Krauss, T.W. (1982), Precipitation process in the new growth zone of Alberta hailstorm, (unpublished).
24. Knight, C. A., and N. C. Knight 1999: Hailstorms. Met. Monographs, Amer. Meteor. Soc., Severe Storms Monograph.
25. Lin, Y. L. Farley, R. D. and Orville, H. D.: Bulk water parameterization in a cloud model. J.Climate Appl. Meteor., 22, 1065-1092, 1983.
26. Mukherjee A.K. Mukopadhyay B.(1983). 'Hail as an initiator for downdraft', Vayumandal – vol. 3, no. 1 & 2.
27. Orville, H. D. and Kopp, F. J, 1977.: Numerical simulation of the history of a hailstorm. J.Atmos. Sci., 34, 1596-1618.
28. Orville (1996) Areview of cloud modeling in Weather Modification, Bull. Am. Met. Soc., 77, 7, 1535-1555.
29. Shishkin NS. 1961. Forecasting thunderstorms and showers by the slice method. Tellus 13: 417–424.

CHAPTER 6

Hailstorm over North-East India

Assam, Arunachal Pradesh, Meghalaya, Mizoram, Nagaland, Tripura, Sub Himalayan west Bengal and Gangetic west Bengal have been categorized under NE India. Northeast India is among the major tea and rice producing areas of the country. In substantially large acreage, the Aman (winter) and Boro (summer) rice crops are raised from December to June and this is the period of maximum hailstorm activity. Hence the loss due to hail to these agricultural produce is quite substantial in NE India.

Ramdas at al (1938) presented the statistics of hail report from Assam and Bengal as shown in Table 6.1.

Although hail fall have been reported in this region mainly from late December to May but the maximum occurrence is experienced in the month of March and April. This may be reason that all the documented case studies of hailstorm for NE India exist only for the months of March and April. Distribution of hailstorms and diurnal variations during 1971-78 over NE India are shown in Fig.s 6.1(a), (b), (c) & (d) (Chowdhury and Banerjee-1983).

Table 6.1 Mean Frequency of days with hail-storm over various areas in North-East India
(The Figures represent number of occasions in 100 years)

Name of the Districts	No. of Stations	JAN	FEB	MAR	APR	MAY	JUN	JUL	AUG	SEP	OCT	NOV	DEC	YEAR
Dihrugarh	34-35	9	9	27	28	-	6	-	-	-	3	-	6	88
Sibsagar	38	8	21	39	47	3	-	-	-	-	-	-	-	118
Tezpur	34-35	8	3	12	27	12	-	-	-	-	-	-	-	62
Gauhathi	83-83	-	18	42	70	27	-	3	-	-	-	-	-	160
Silchhar	38	-	5	21	42	8	3	-	-	-	-	3	-	82
Cox's Bazar	27-28	-	-	4	4	-	-	-	-	-	-	-	-	8
Chittagong	38	3	11	8	24	5	-	-	3	-	-	-	-	54
Narayangunj	38	-	5	39	63	11	-	-	-	-	-	-	-	118
Barisal	38	-	-	18	24	5	-	-	-	-	-	-	-	47
Jeasore	38	8	8	24	53	18	3	-	-	-	-	-	-	114
Calcutta	38	-	5	18	37	3	-	-	-	-	-	-	-	63
Saugor Island	38	3	5	18	58	26	-	-	3	-	-	-	-	113
Burdwan	38	3	11	18	58	26	-	-	3	-	-	-	-	119
Berhampore	38	-	16	18	39	29	11	-	-	-	-	3	-	116
Mymensignh	38	8	-	11	21	13	8	-	-	3	-	-	-	64
Bogra	38	-	-	16	29	16	-	-	-	-	3	-	-	64
Dinajpur	38	-	3	8	39	20	-	-	-	-	3	3	-	76
Jalpaiguri	38	3	5	3	37	24	-	-	-	-	-	-	-	72
Shillong	33-34	-	9	36	55	24	-	-	-	3	-	-	-	127
Cherrapunji	33-34	-	18	64	127	76	6	3	-	-	3	3	-	300
Darjeeling	38	3	26	96	105	89	5	-	-	-	3	3	5	335

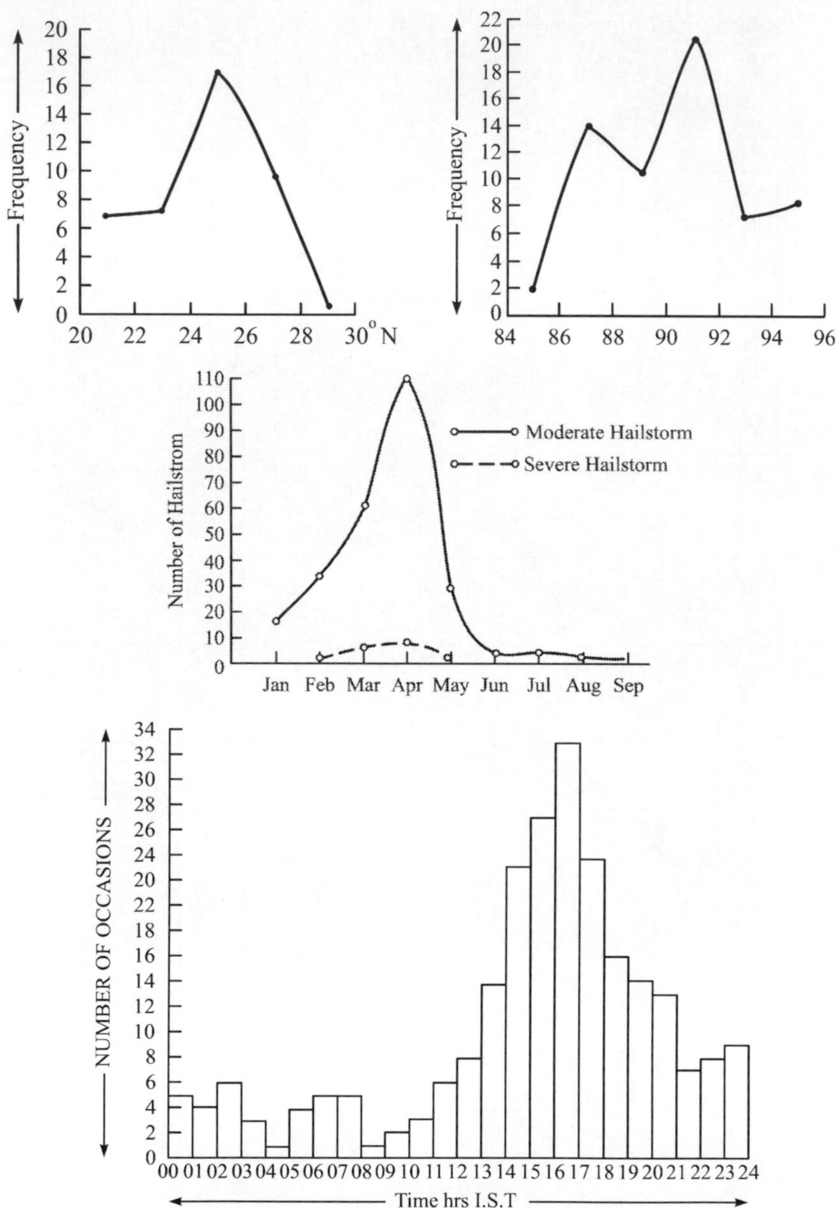

Fig. 6.1 Latitudinal and longitudinal distribution of hailstorm with their monthly and diurnal variations over NE India is shown in (a) Latitudinal Distribution of Hailstorms (top left), (b) Longitudinal Distribution of Hailstorms (top right), (c) Monthly Distribution of Hail Incidences (middle) and (d) Diurnal Variation of Hailstorms (bottom).

(a) Latitudinal Variation

Number of hailstorms which occurred in each of the two-degree latitudinal strip of North-East India between 20^0-30^0 N latitude is shown in Fig. 6.1(a). As portions of some of the latitudinal strips lie in Bangladesh from where no hailstorm reports were available, due weightage was given to the loss of area of these latitudinal strips while computing the number of hailstorms. It may be noted that between 24^0 & 26^0N maximum number of hailstorms occur which is nearly 33 percent of the total number occurring in the region.

(b) Longitudinal Variation

Number of hailstorms which occurred in each of the two-degree longitudinal strip of Northeast India commencing from 80^0E longitude was worked out in the same manner as done for latitudinal strips mentioned above. The longitudinal distribution of hailstorm is depicted in Fig. 6.1(b). Two maxima occur in the distribution curve, the principal one between 90^0 & 92^0E and the secondary between 86^0 & 88^0E.

Considering both the latitudinal and longitudinal distribution of hailstorm, the area between 24^0 & 26^0 N and 90^0 & 92^0E in Northeast India experiences the maximum number of hailstorms. This area roughly covers the mountainous districts of Garo hills and united Khasi Jaintia hills and adjoining areas where about one third of the total hailstorms in the region occur.

6.1 Favourable Conditions for Hailstorm Occurrence

Various synoptic situations on different occasions in general commonly indicated incursion of moisture in lower levels, middle level dryness (with a few exceptions) and pronounced wind shear in vertical. Eastward or northeastward moving upper air cyclonic circulation in the lower levels, over the plains or foothills were also reported in general causing hailstorm – Mukherjee at al (1962), Rakshit and Barman (1963), Sharma (1965), Ray (1971) and Arora (1988). Two types of hailstorms are experienced in NE India, Inter-airmass hailstorms and Intra-airmass hailstorms.

Inter-Airmass Hailstorm

Interacting air masses of contrasting thermal and humid characteristics have been described as the potential region of cyclogenesis, where marked instability frequently develops. Active thunderstorms often develop in this frontal zone of the two air masses – Roy (1939), Chowdhary and Banerjee (1983). Fig. 6.2 shows one such situation in which moist airmass from Bay of Bengal is inducted east of the trough line at 0.9 km. level to meet drier air mass from northwest. Hailstorm was reported over the region on 24th April 1975. It has been obsereved (Chowdhary and Banerjee - 1983) that the boundary zone of two air masses are well marked at 0.9 km and 1.5 km and some times utmost up to 2.1 km. But above that they are ill defined. Convection develops when warm moist air mass glides over the cold and denser current from northwest. The buoyancy is inhanced by the effect of insolation.

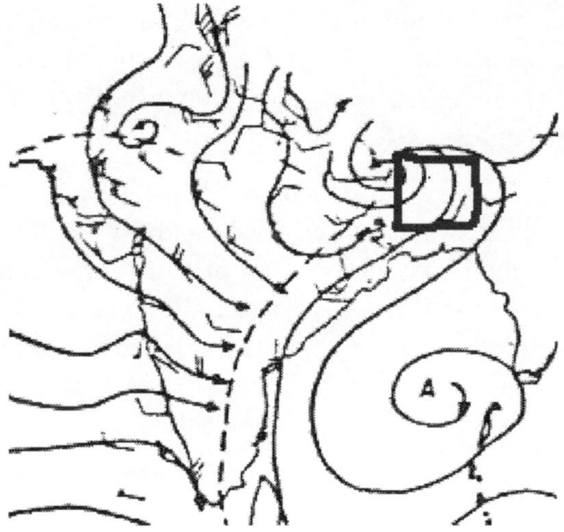

Fig. 6.2 Flow pattern on 24 April 1975 (00 UTC) 0.9 km amsl areas of hailstorm occurrences reported in rectangle

Intra-Airmass Hailstorm

Intra-Airmass or airmass type hailstorms have also been reported by Chowdhary and Banerjee (1983) over northeast India and also by Arora (1988) in his case study on Hasimara. Favourable situations are being mentioned hereunder.

Synoptic Situation

Typical regional conditions which are favourable for the hailstorm formations over NE India are described below:

(a) **Presence of upper air cyclonic circulation upto 850 hPa** Fig. 6.3-6.6 refer case study of hailstorm over Hasimara (Long 89.22^0 E / lat 26.43^0 N) on 8 and 9 March 1987. On 7 March 1987 on 03 UTC a feeble low pressure area with odd isobar could be marked over Punjab, Haryana, Himachal Predesh and adjoining west Uttar Pradesh (Fig. 6.3(a)). Another feeble low pressure was located over north MP and adjoining UP region. At 850 hPa a northsouth trough was also running over Bihar and adjoining Gangetic West Bengal. Anticyclone at 850 hPa was located over Orissa.

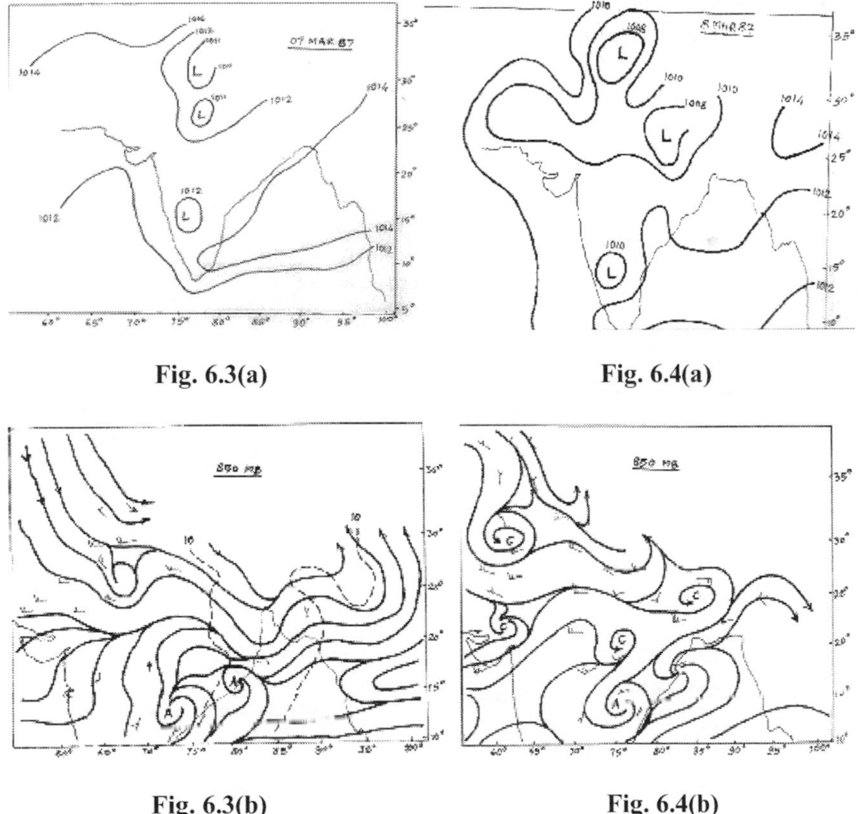

Fig. 6.3(a)

Fig. 6.4(a)

Fig. 6.3(b)

Fig. 6.4(b)

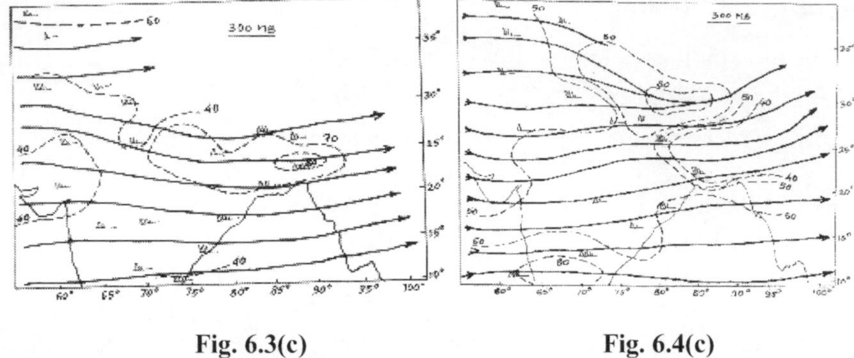

Fig. 6.3(c)

Fig. 6.4(c)

Fig. 6.3 (a) 7 March 87, Surface, 03UTC, **Fig. 6.3 (b)** 7 March 87, 850hPa, 00UTC, **Fig. 6.3 (c)** 7 March 87, 300hPa, 00UTC,

Fig. 6.4(a) 8 March 87 Surface, 03UTC, **Fig. 6.4 (b)** 8 March 87, 850hPa, 00UTC, **Fig. 6.4 (c)** 8 March 87, 300hPa, 00UTC

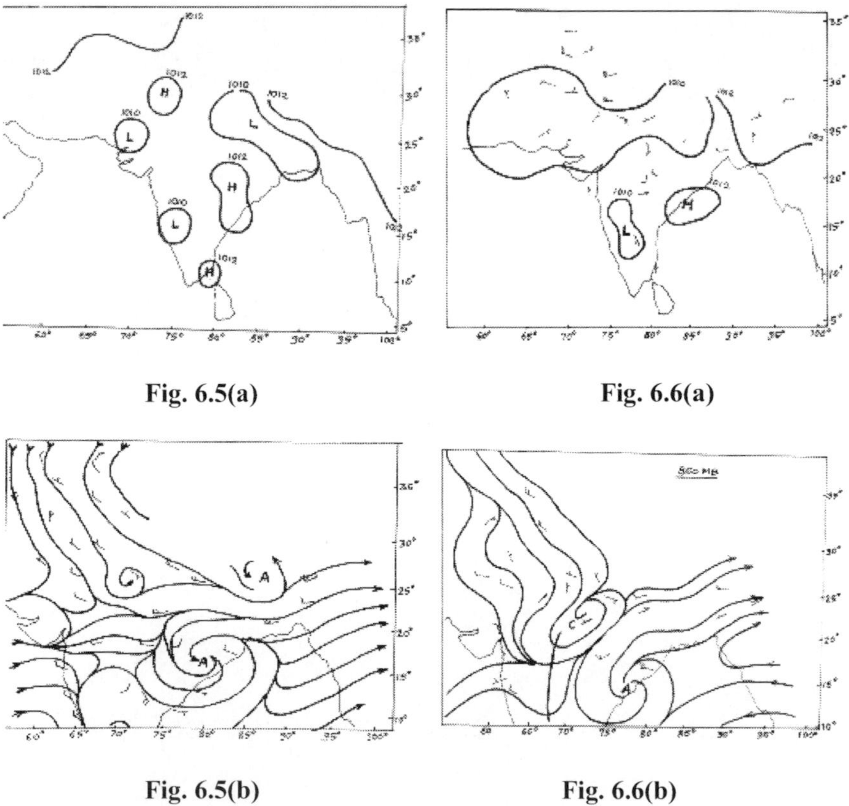

Fig. 6.5(a)

Fig. 6.6(a)

Fig. 6.5(b)

Fig. 6.6(b)

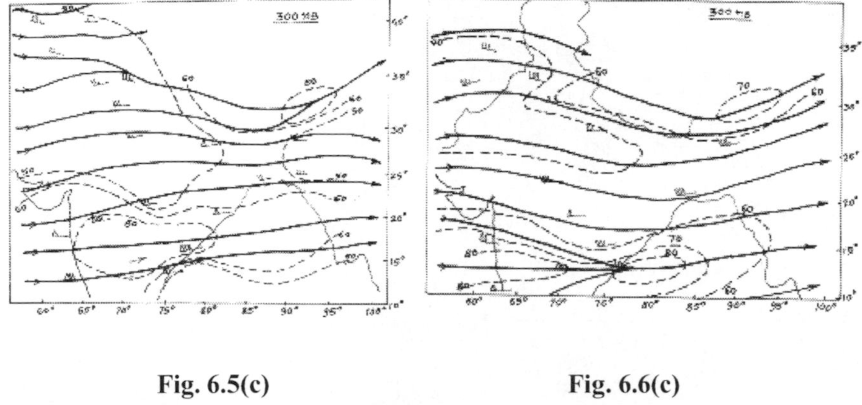

Fig. 6.5(c) Fig. 6.6(c)

Fig. 6.5(a) 9 March 87, Surface, 03UTC, **Fig. 6.5(b)** 9 March 87, 850hPa, 00UTC, **Fig. 6.5(c)** 9 March 87, 300hPa, 00UTC

Fig. 6.6(a) 10 March 87, Surface, 03UTC, **Fig. 6.6(b)** 10 March 87, 850hPa, 00UTC, **Fig. 6.6(c)** 10 March 87, 300hPa, 00UTC

Location of anticyclone is important determinant to know the type of air mass at the place of cyclogenesis. At 300 hPa a westerly jet maximum of 80 Kt was located east of Hasimara over Meghalaya and neighborhood. The low pressure area on surface had shifted eastward only up to east UP and adjoining Bihar region on 8th March 1987 (Fig. 6.4(a)). It persisted only to the west of 87^{0}E on this day, whereas longitude of Hasimara is 89.2^{0}E. There was an upper air cyclonic circulation at 850 hPa on 8 March 1987 over Sub Himalayan west Bengal and adjoining Bihar which persisted on 9 March also (Fig. 6.5). At 300hPa jet maxima over Meghalaya had shifted westward and persisted over Nepal, Sikkim and neighbourhood. It shifted slightly eastward on 9 March providing upper air divergence on both the days. Fig. 6.6 shows the pattern on ceasation of hailstorm activity on 10 mar 87.

Vergence: Vergence patterns for 7, 8, 9, and 10 March 1987 are shown in Fig 6.7 (a), (b), (c), (d) and 6.8 (a), (b), (c),(d). Convergence of -2×10^{-5} Sec^{-1} at 850hPa on 7 March 87 was not supported by the

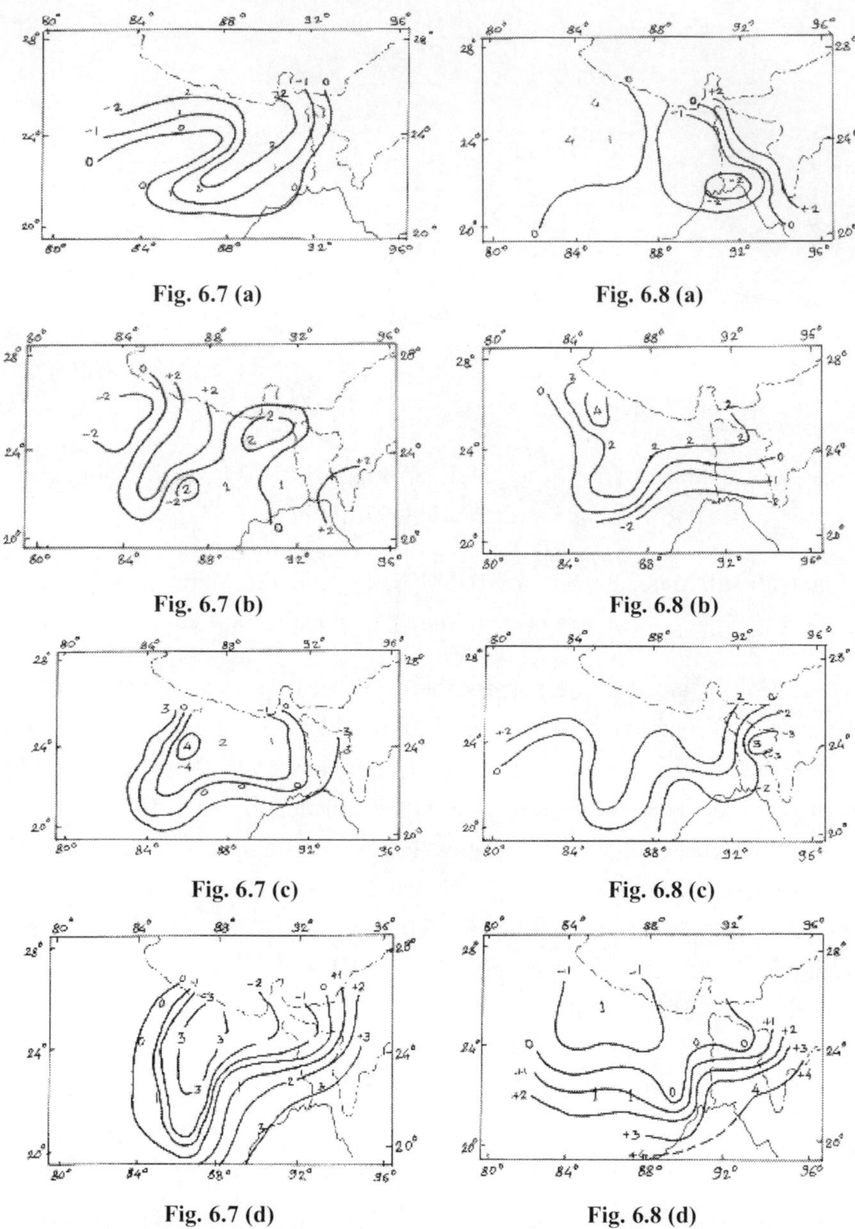

Fig. 6.7 (a) Fig. 6.8 (a)

Fig. 6.7 (b) Fig. 6.8 (b)

Fig. 6.7 (c) Fig. 6.8 (c)

Fig. 6.7 (d) Fig. 6.8 (d)

Fig 6.7(a)($\times10^{-5}$Sec^{-1})850hPa, 7 March 87, **Fig 6.7 (b)**($\times10^{-5}$Sec^{-1})850hPa, 8 March 87, **Fig 6.7 (c)**($\times10^{-5}$Sec^{-1})850hPa, 9 March 87, **Fig 6.7 (d)**($\times10^{-5}$Sec^{-1})850hPa, 10 March 87, **Fig 6.8 (a)**($\times10^{-5}$Sec^{-1})300hPa, 7 March 87, **Fig 6.8 (b)**($\times10^{-5}$Sec^{-1})300hPa, 8 March 87, **Fig 6.8 (c)** ($\times10^{-5}$Sec^{-1})300hPa, 9 March 87, **Fig 6.8 (d)**($\times10^{-5}$Sec^{-1})300hPa, 10 March 87 divergence aloft at 300 hPa. But the wind maxima at the confluence zone indicated in Fig. 6.4(c) provided the suitable divergence aloft of the order of $2x10^{-5}Sec^{-1}$, favouring hailstorm formation. Cloud formation ceased on 10 March 87 with the weakening of the wind maxima (Fig. 6.6(c)) and disappearance of divergence aloft – refer fig. 6.8 (d). In other four case studies by Arora (1988) for the month of April when hailstorm occurred over Hasimara (17 April 84, 26 April 85, 29 April 86 and 10 April 87), on three occasions presences of low level cyclonic circulation up to 850 hPa over sub Himalayan west Bengal and adjoining Bihar region was commonly noticed on the days of occurrence of hailstorms.

(b) Presence of feeble trough at 850 hPa: Rakshit and Barma (1963) reported the passage of feeble upper air trough at 850 hPa across Gauhati on 13 April 62. Hailstorm occurred at 1600 IST. Vertical time section chart (Fig. 6.9) also reported a trough at 300 hPa to the west of Gauhati on the same day. This provided the necessary upper air divergences giving rise to hailstorm. In this case also only a feeble trough at 850 hPa was observed close to station along 87^0E the upper level divergences was provided at 300 hPa by jet maxim with speed of 120 kt, which was situated over Sikkim, sub Himalayan West Bengal and neighborhood.

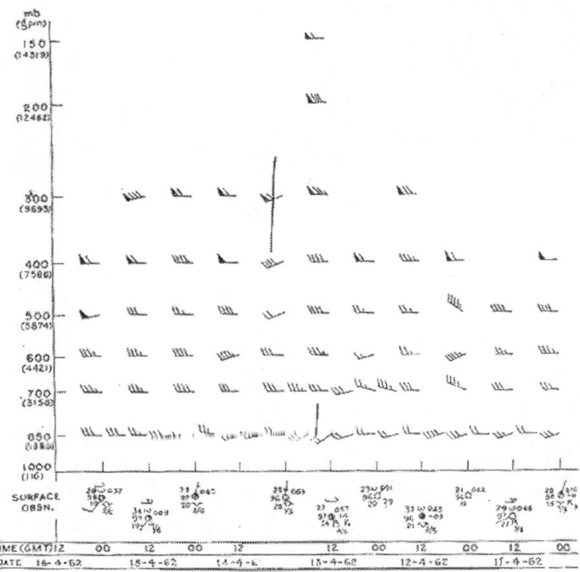

Fig. 6.9 Vertical Time Section Chart

(c) Rapidly eastward moving trough at 0.6 km over taking the upper tropospheric trough: Hailstorm occurred over Gauhati on the evening of 11 Mar 1970; Ray (1971). Refer Fig. 6.10. Interacting air mass types cyclonic circulation was present over Assam at 0.6 km level at 00 UTC on 11 Mar 70. Another quasi stationary upper air trough persisted at 87^0 E at 300 hPa.

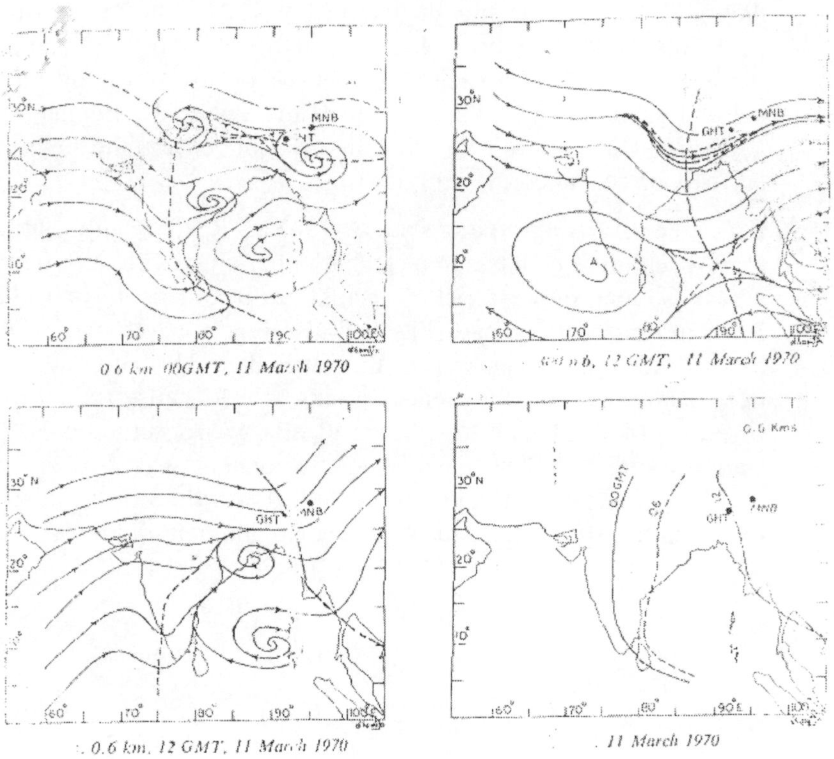

Fig. 6.10

The trough at 0.6 km rapidly moved eastward by more than 5^0 longitude within a period of 6 hours (00 UTC to 06 UTC), overtaking the upper tropospheric trough at 300 hPa. Hailstorm started at 1528 UTC. In 1200 UTC chart there was no cyclonic circulation at 0.6 km. It was only a trough in the westerlies.

(d) Wind discontinuity induced due to the passage of surfaces low pressure area: On 17 April 62 at 1520 IST a hailstorm occurred over Gauhati airport. The surface weather chart of 03 UTC on 17 April showed a low pressure area over Nepal and north

Bengal. It moved by 12 UTC over north Bengal and Northwest Assam (much west of Gauhati).

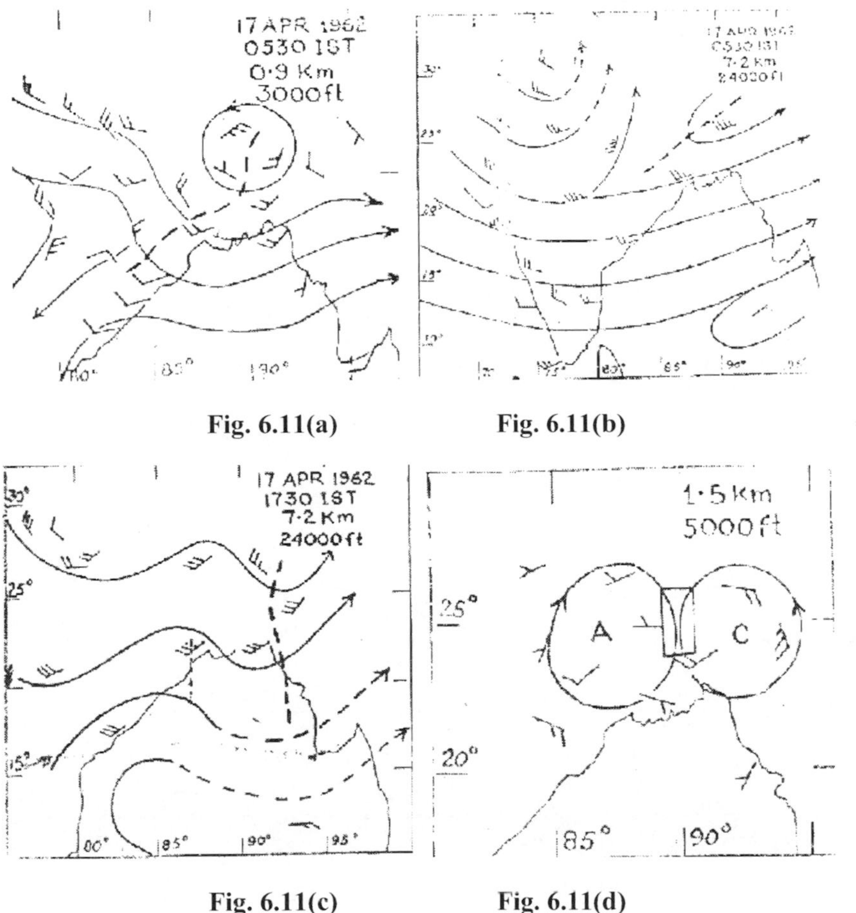

Fig. 6.11(a) Fig. 6.11(b)

Fig. 6.11(c) Fig. 6.11(d)

However, 00 UTC upper air chart showed a wind discontinuity at 0.9 km passing along 90^0 E of long. (Fig. 6.11 (a)). Vertical time section of the winds at Gauhati showed the presences of strong wind field above 3.0 km at 00 UTC, increasing in speed with height, Fig. (6.12). Upper air divergence was provided by the upper air trough at 7.2 km in 12 UTC west of Gauhati (Fig. 6.11(c)) which originated from the NE-SW orientation in 00UTC chart at 7.2 Km (Fig. 6.11(b)). Wind tendency (Fig. 6.11(d)) at 1.5 km indicated that zone of hailstorm was located in between cyclonic and anticyclonic tendencies.

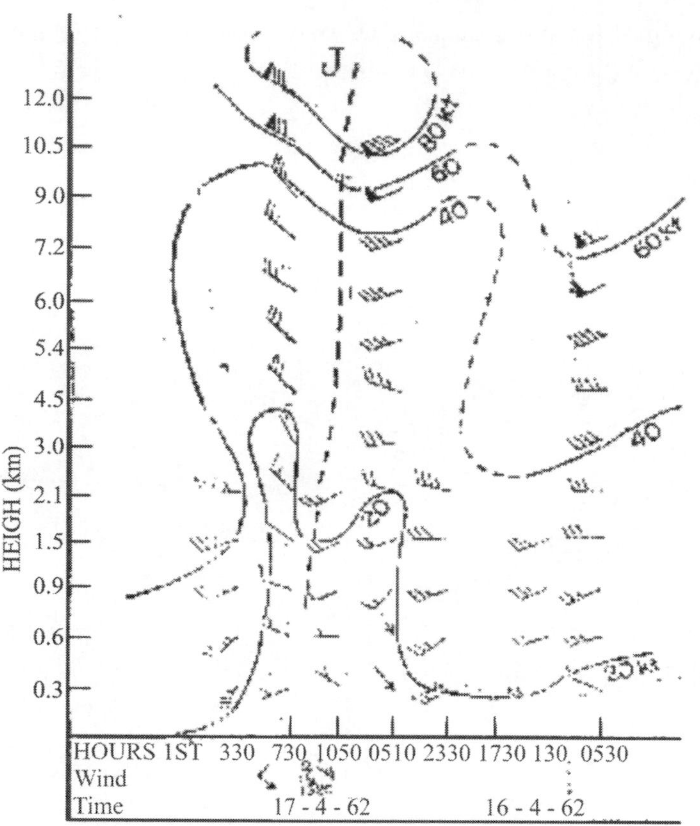

Fig. 6.12 Vertical time Section of Guahati.

6.2 Vertical Wind Shear

The wind shear in all the five cases considered by Arora (1988) were more than 40 kt between 850hPa and 300hPa and more than 50 kt between 850hPa and 200hPa. Table 6.2 gives the wind shear values obtained by Arora (1988), Ray (1971) and Chowdhury and Banerjee (1983) in their studies.

Table 6.2

Vertical Shear Between Levels	Pre-Monsoon				Winter	
	Normal	A	R	CB	Normal	CB
850 – 300 hPa	10 kt	43 kt (Max 90)	45 kt	42 kt	24 kt	69 kt
850 – 200 hPa	33 kt	50 kt (Max 100)	70 kt	61 kt	31 kt	82 kt

A = Arora (1988); R = Ray (1971); CB = Choudhury and Banerji (1983)

High value of vertical wind shear (≥ 40 kt) between 850-300hPa and ≥ 50 kt between 850-200hPa are commonly noted during the premonsoon season hailstorm days. Similarly during winter months it is very high(≥ 69 kt) for the corresponding atmospheric levels.

6.3 Humidity Parameter

(a) Mean mixing ratio (Surface – 850 hPa)

A marked increase in low level moisture 24 to 36 hours prior to the occurrence was observed. During March it became 11.5 gm/kg and during April it rose to 14 gm/ kg on 00 UTC on the days of occurrences. Another study by Ray (1971) of hailstorm over Gauhati on 11 Mar 70 indicated rather much drier lower level.

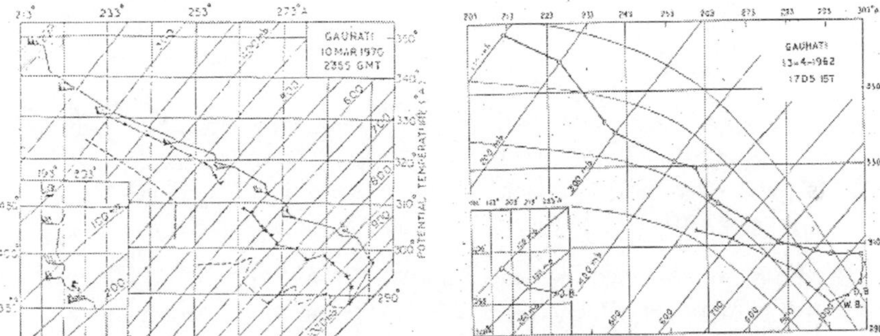

Fig. 6.13 T-Φ gram of Guahati, 10 Mar 1970, 2355 UTC

Fig. 6.14 T-Φ gram of Guahati, 13 April 1962, 1205 UTC

But comparatively more drier than lower level atmosphere prevailed in his study between 850 and 700 hPa; Fig. 6.13. An unusual case of rapid depletion of moisture with height right from the surface layer itself was reported by Rakshit and Barma (1963) on the day of brisk hailshower on 13 April 1962 over Gauhati. In their case the layer between surface to 850 hPa was much more drier than that between 850 to 700 hPa. Refer Tephigram of Gauhati on 13 April 62 (Fig. 6.14). Liquid water content showed value of 3.0 gm kg^{-1} between 2.1 and 3.0 km and 1.0gm kg^{-1} between surface and 2.1 km.

(b) Mean mixing ratio between 850 and 700 hPa

Mean T-Φ grams for the days of occurrence of hailstorm for the month of March and April of Bagdogra are shown in Fig. 6.15 and 6.16 (Arora 1988) respectively. It was found to be 7 gm/ kg during March and less than or equal to 9.0 gm/kg during the month of April. Mean mixing ratio computed only in one case on 26 April 85 was found out to be 2.7 gm/kg. Note the depletion of moisture in the middle level.

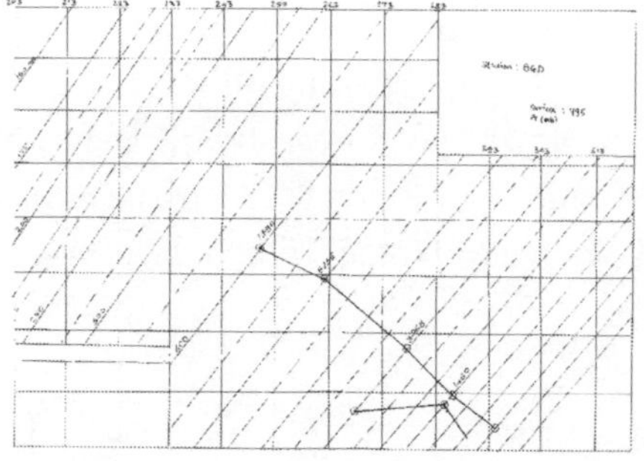

Fig. 6.15 Mean T-Φ gram (00UTC) of Hailstorm days in March over Bagdogra, Mean Surface Pressure 995hPa.

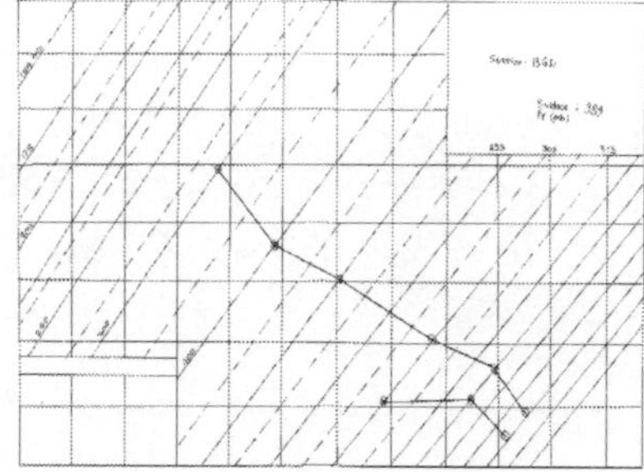

Fig. 6.16 Mean T-Φ gram (00UTC) of Hailstorm days in April over Bagdogra, Mean Surface Pressure 989hPa.

(c) ***Value of indices*:** Table 6.3 shows computation of indices on the day of occurrence (D) and a day prior to the occurrence (D-1) over Hasimara(89.22^0 E Lon./26.43^0 N Lat.). It may be noted that Racliff Index and modified Jefferson Index in all the cases on the day of occurrences at 00 UTC were lying between 29 to 35 and 28 to 39 respectively.

Table 6.3 Value of Indices on the day of occurrence (D) and a day prior to the occurrence (D-1)

Days/ Indices	7 Mar 87	8 Mar 87	9 Mar 87	17 April 84		26 April 85		29 April 86		10 April 87	
	D-1	D	D	D-1	D	D-1	D	D-1	D	D-1	D
Rackliff Index	23	30	31	-	35	28	33	38	31	28	29
Modified Jefferson Index	25	28	34	-	39	-	34	36	36	31	31

6.4 Radar Study of Hailstorm over Gauhati

Two hailstorm observations over Gauhati are described below:

(a) 10 April 1960

Two hailstorm of moderate intensity were experienced at Gauhati air port on the evening between 1815-1819h and between 2022-2025h (Rakshit and Barman-1963). Radar picture of the first system are shown in Fig. 6.17-6.18. Echo of isolated cumulonimbus was observed at 1600 IST with top more than 9.0 km just south of base. At 1741 the cell was found overhead with line type build ups from WNW to ESE direction as shown in Fig. 6.17. At 1800 IST the center portion of radar scope was found covered with a bright – well- edged patch, more towards north east than towards south in Fig. 6.18. The system incessantly moved in northeasterly direction and dissipated.

Hailstorm at Gauhati Airport (NORTH)

Fig. 6.17 - 1741 IST

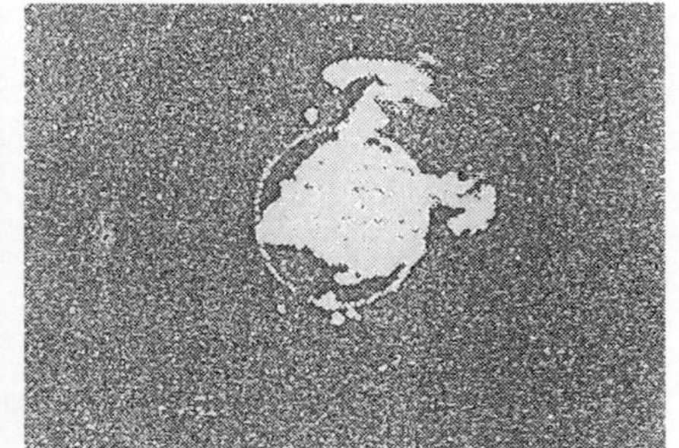

Fig. 6.18 - 1810 IST

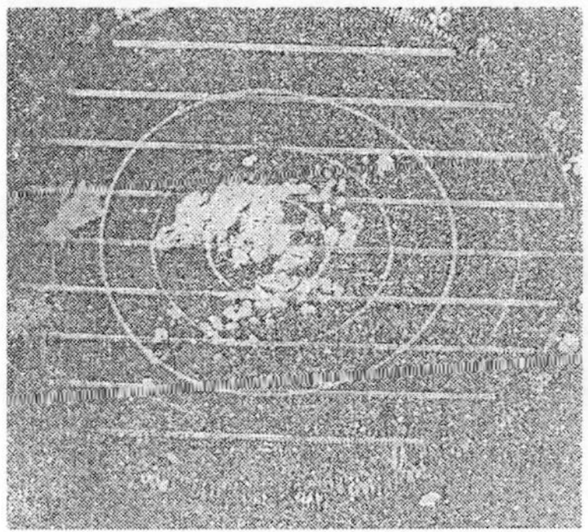

Fig. 6.19 - 1446 IST

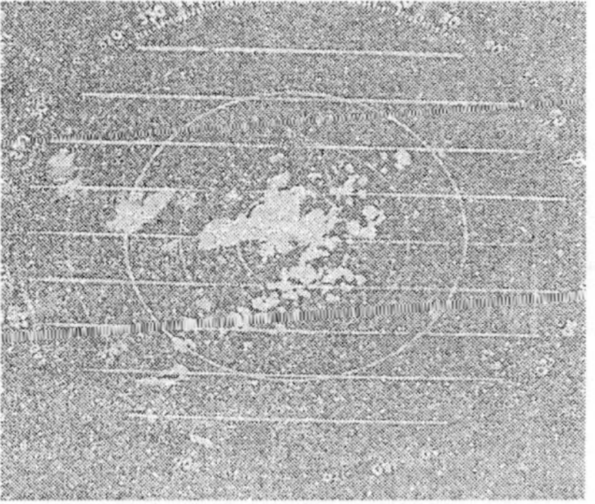

Fig. 6.20 - 1456 IST

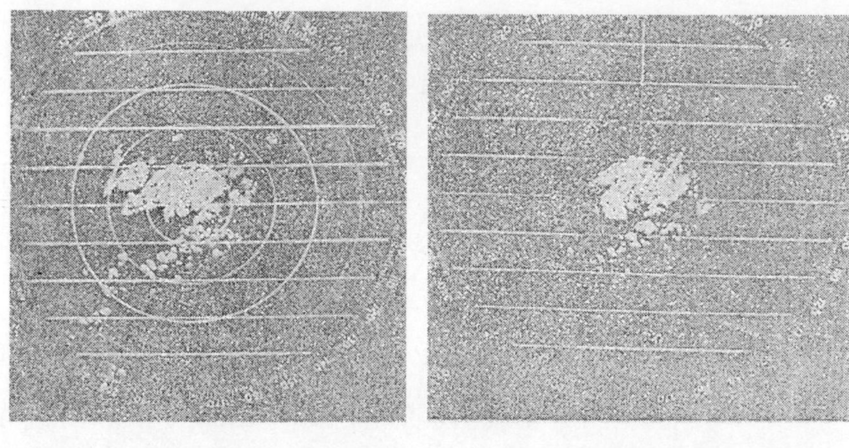

Fig. 6.21 -1507 IST

Fig. 6.22 -1517 IST

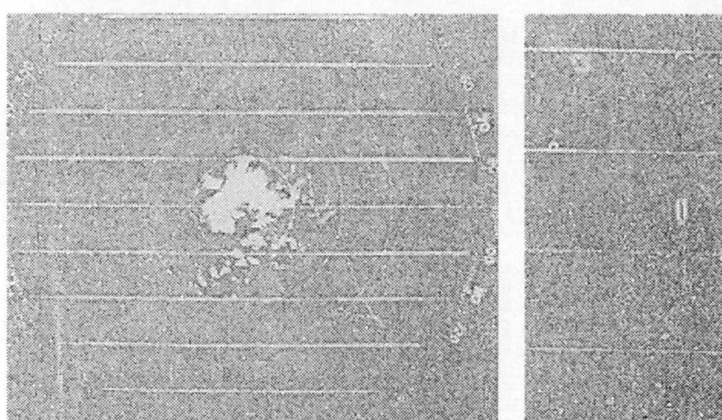

Fig. 6.23 -1518 IST

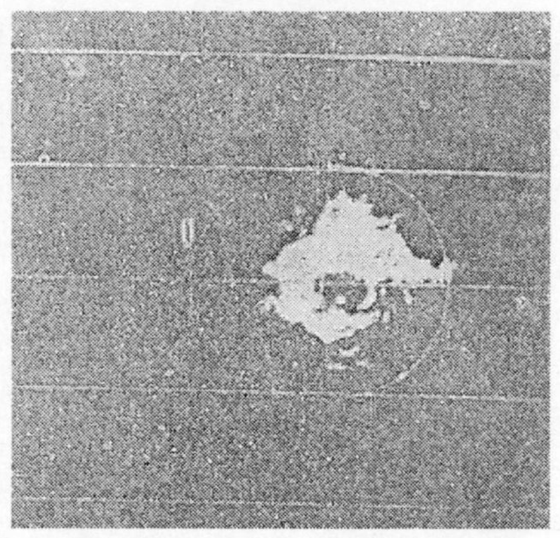

Fig. 6.24 -1521 IST

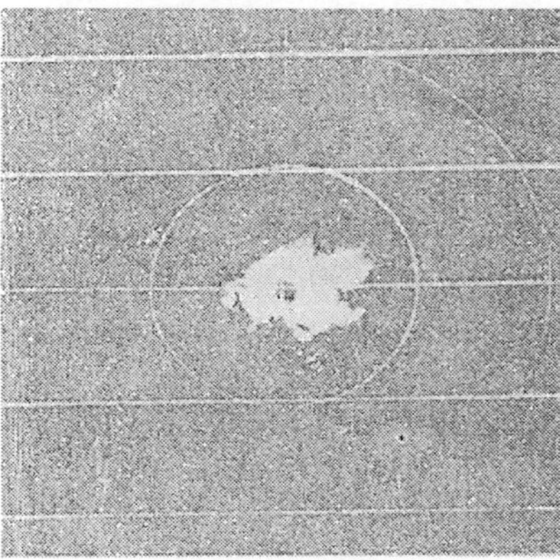

Fig. 6.25 -1525 IST

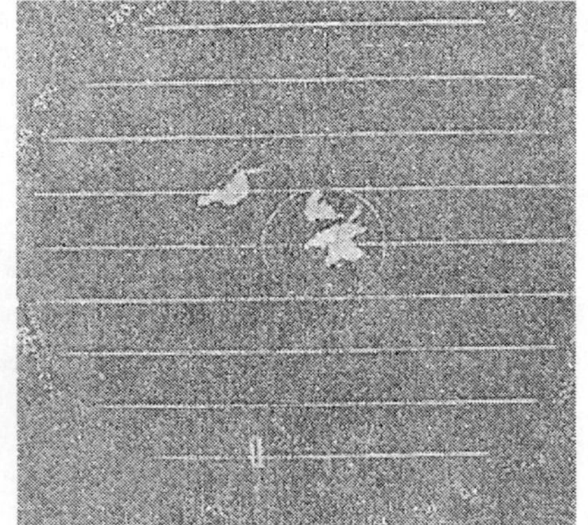

Fig. 6.26 -1529 IST

(b) 17 April 1962

Hailstorm was experienced over Gauhati at 1520 IST on 17 April 1962. The picture periodically recorded at 1446h IST onward till 1529 are shown in Fig.s 6.19-6.26. Fig.s 6.19-6.22 show the daughter cells following the main cell with higher speeds and merging with the main cell. Sharma (1965) noted the varying speed of the daughter cells depended upon their distance from the main storm.Farthest one moved fastest. Speed retarded with the growth in size of the daughter cells. At 1442 IST two very strong patches closely located to each other were seen near 10-15, 15-20 N miles towards west. Another comparatively less intense was located 15-20 miles behind these two towards west and yet other i.e. fourth patch could be seen 20-25nmiles towards further west in 285^0 azimuth. The first one, nearest to the station, was found to have a speed of about 12 n. miles per hour (Fig. 6.19 to 6.21) the second one of about 20 n. miles per hour (Fig. 6.19 to 6.20), the third one of about 35 n. miles per hours (Fig. 6.19 and 6.20), and the fourth the farthest one from the station, of approximately 45nmiles per hours (Fig. 6.20 & 6.21). Thus the speed of these echoes decreased with distances from the station. These four echoes (Fig. 6.20) seemed to merge into one another near about 1507 IST (Fig. 6.21 & 6.22) about 4 n.miles to northwest of the station and after a few minutes transformed into a clear-cut round edged hook at 250^0 and 2 to 4 n.miles from the station (Fig. 6.22) This hook later on, moved with an approximate speed of 35 nmiles per hours. At 1518 IST, the leading edger of the hook was approximately 2 n. miles from the station (Fig. 6.23). As soon as it touched the station 2 minute later, a hailstorm commenced with a northwesterly surfaces squall of 45 kt. It was followed by heavy rain and thunder. It may also be pointed out that rain at 1500 hrs had preceded the hailstorm and it can be seen that the farthest echo (Fig. 6.21) became very weak, due to attenuation. At 1521 IST, at higher elevation of the antenna and low gain, these echoes, which had merged into one another near about 1512 hrs, gave the appearance of a prominent V-shaped hook at its near edge with a clear notch or hole i.e. an open area towards the radar just after one minute of the commencement of the hailstorm (see Fig. 6.24). At 1525 h the hook or the notch began to fill up (see Fig. 6.25). The hailstorm lasted for 8 minutes. After echo a line type character was seen at 10 n. miles in the northwest which had a tendency to move towards northeast (Fig. 6.26).

From the above observations it is seen that the echo associated with the hailstorm, had clear-cut well-defined sharp edges of strong intensity. They are visible even at reduced gain with little change in brightness and did not disappear at 12^0 tilt of the radar antenna (12^0 is the limit for the tilt of this radar) suggesting that the echo was of considerable vertical extent.

References

1. Arora, A.S. (sqn ldr), (1988), on some ynoptc and thermodynamic aspects of hailstorm occurrence over Hasimara during pre-monsoon, SFC thesis, Air forces Admin College, Coimbatore.
2. Chowdhary A. Banerjee A.K., (1983), Hailstorm in NE India Vayamandal.
3. Mukherjee A.K. Gosh, Sk Arunachalam G., (1962) Hailstorm at Gauhati on 18th March 1961, IJMG Vol. 13, No. 2.
4. Rakshit D.K. Barman C.M., (1963), Hailstorm over Guahati air port on 13th April 1962, IJMG, Vol 14. No. 2.
5. Ramdas L.A. Satakopan and Gopal Roa, S., (1938), Frequency of days of hailstorm in India Vol. VIII, Part VI.
6. Ray T.K. (1971), Occurrences of an unusual hailstorm over Brahmaputra Valley, Vayumandal, Oct-Dec.
7. Roy A.K., (1939) Proc. Nat. Inst. Sc, India.
8. Sharma B. L., (1965), Radar Study of hailstorm over Gauhati on. 17th April 1962, Ind. J. Met. Geo. 16, 3 pp 459-466.

CHAPTER 7

Hailstorm over Northwest India

Jammu and Kashmir, Punjab, Haryana, Delhi, west Uttar Pradesh and Rajasthan are categorized under northwest India. Major wheat producing belt of the country falls within NW India region. There is colossal loss of crop every year due to the lashing hailstorm over this region. Most of the hail shower takes place during winter and premonsoon months. A few of them are also experienced during post monsoon period i.e. October and November over Jammu & Kashmir. No hailstorm is normally experienced during monsoon months. Mean frequency of days of hailstorms over various parts of NW India are presented in Table 7.1.

Table 7.2 by Kulshrestha and Jain (1967), shows number of hail reports (within 320 km) of Delhi, observed during December 1957 to June 1960.

Table 7.1

Mean Frequency of day with hail-storm over various areas in northwest India

(The figures represent number of occasions in 100 years)

Name of the Dist	No. of Stations	JAN	FEB	MAR	APR	MAY	JUN	JUL	AUG	SEP	OCT	NOV	DEC	YEAR
Jhansi	38	-	13	8	5	3	3	3	-	-	-	3	5	43
Agra	38	-	16	11	3	5	3	-	-	-	3	-	5	46
Srinagar	38	-	3	13	29	26	13	-	5	5	-	-	-	94
Bikaner	38	3	3	-	8	13	5	-	-	-	-	-	-	32
Jodhpur	38	-	3	-	8	13	-	-	-	-	-	3	-	27
Jaipur	38	5	13	13	13	11	13	-	-	-	-	-	8	76
Ajmer	38	3	5	13	3	8	3	-	-	-	-	-	3	38
Kotah	38	3	3	5	16	8	5	-	-	-	-	5	5	50
Mukteswar	29	48	117	224	217	245	65	3	-	17	62	24	31	1053
Simla	38	68	108	229	211	208	63	3	5	18	58	45	58	1074
Mount Abu	38	8	11	3	8	5	5	3	-	5	3	-	3	54

Table 7.2

Total Number of Occurrences of severe weather over Delhi and neighborhood during the period December 1957 to June 1960

Period of year	Maximum heights attained by echo tops in the area (thousands of ft)*	Number of Days	No. of severe weather reports within ranges of 200 miles around Delhi.					Total No. of reports
			Lightning	Dust storm	Thunder Storm	Squall	Hail	
December to April	30 - 34	3	2	3	18	1	1	25
	35 – 39	16	17	14	99	3	10	143
	40 – 44	24	38	20	179	11	9	257
	45 – 50	3	5	13	33	2	2	55
May To November	30 – 34	10	5	6	24	1	-	36
	35 – 39	36	30	48	137	11	-	226
	40 – 44	46	58	30	206	12	1	307
	45 - 50	117	245	62	629	27	2	965
	Above	34	81	17	204	6	3	311

* During the period December to April echo tops are confined to heights below 50,000 ft.

Monthwise hail reports over Delhi region based on six years data (1958-63) is shown in Fig. 7.1.

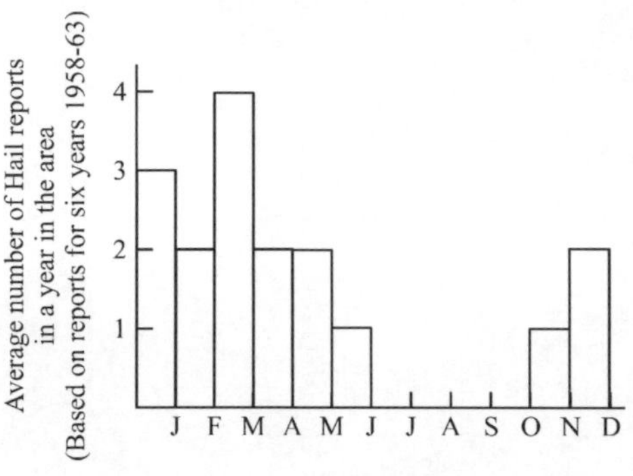

Fig. 7.1 Number of hail reports in the area during the various months.

Fig. 7.2 Preferred time of occurrence of hail.

Most of these hailstorms have been observed towards afternoon, evening or early night hours as shown in (Fig.7.2). Tabib (1977) had produced the mean Tephigram (Fig.7.3) of the days of hailstorm occurrence over Delhi and adjoining regions. He took Tepigram of Delhi and Lucknow as the representative of the atmospheric condition causing hailstorms. Mean profiles of all the hailstorm days during 1966 to 1969 in the region were prepared and compared with mean Tephigram of Delhi for the month.

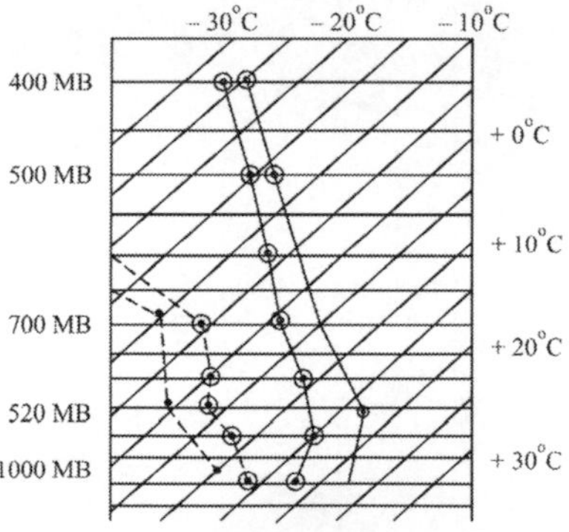

Fig. 7.3 (a) Mean T-Φ gram for the months of February and March (00UTC). Lines with circles correspond to mean T-Φ gram for hailstorm days

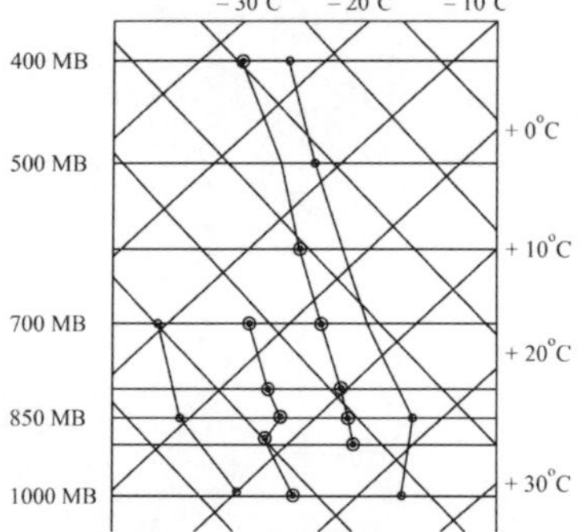

Fig. 7.3 (b) Mean T-Φ gram for the months of February and March (12UTC). Lines with circles correspond to mean T-Φ gram for hailstorm days

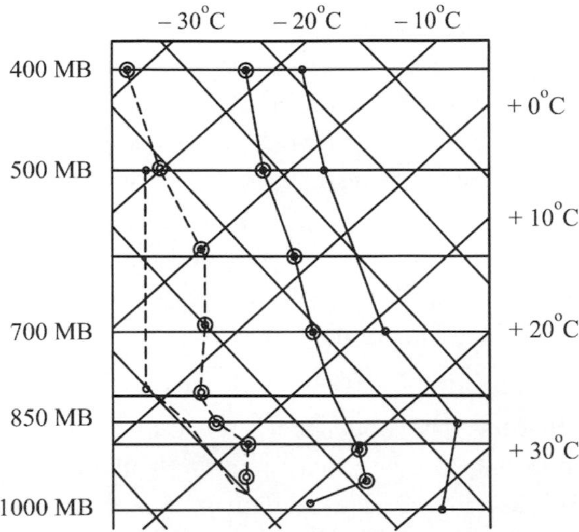

Fig. 7.4 (a) Mean T-Φ gram for the months of April and May (00UTC). Lines with circles correspond to mean T-Φ gram for hailstorm days

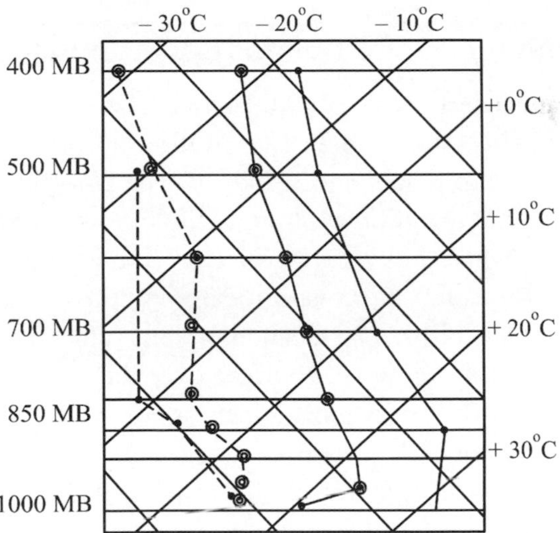

Fig. 7.4 (b) Mean T-Φ gram for the months of April and May (12UTC). Lines with circles correspond to mean T-Φ gram for hailstorm days

Fig. 7.3(a) and (b) show the mean for the month of February a March and Fig. 7.4 (a) and (b) show mean for the month of April and May. It is noted that on hailstorm days freezing level of dry bulb and wet bulb temperature are in general 40 – 90 hPa lower than the normal. However, lowering of freezing level cannot become a predictor for hail storm since lowering or rising by 40 – 90 hPa may be insignificant in the overall genesis of hail formation. On several occasions freezing level has also been observed to rise on the day of hailstorm than the previous day (Ravikiran and Kumar – 1994). Important difference to be noted is comparatively much humid lowest layer and pronounced inversion in the lower layers during April and May for hailstorm days in 00 UTC and 12 UTC mean Tephigrams of these months.

Fawbush and Miller (1953) had also found that almost every hailstorm situation was associated with a low level inversion in the morning radiosonde ascent. Inversion layer may be presumed to be helping in the horizontal accumulation of moisture in the lowest layer without any vertical mixing. Highly moist lowest layer might then further help the buoyancy in the convective process after the inversion is broken due to insolation or any other triggering mechanisms.

7.1 Favourable Condition for Hailstorm Occurrence

Various case studies [(Rao & Mukherjee (1958), Raju (1982), Joshi (1987), Sathe (1991) and Singhal (1992)] have commonly revealed the presence of western disturbances and induced low pressure area as the cause of hailstorm over northwest India. Cut of low has also been referred (Tyagi-1983) to cause severe convective activity over north India in general though none were specifically studied with reference to occurrence of hailstorms. The location of anticyclone at 850 hPa causing inflow of moist Bay of Bengal current over the region and its meeting with the comparatively less moist air mass form west induces hailstorm actively even during early or mid-winters. Singhal (1992) discussed such an inter-airmass hailstorm over Delhi on Dec 90. Inter air mass types hailstorm are more common in northwest India than northeast India [Rao and Mukherjee (1958), Mishra and Prased (1980)].

(a) Western Disturbance as well marked surface low pressure: During the mid winters all the hailstorm cases showed a well marked low pressure area over northwest India displacing anticyclone at 850 hPa over coastal Orissa and adjoining west Central Bay. Hailstorm occurred on 29 Dec. 90 and 06 Jan 89 over Delhi. See fig. (7.5 (a) (i) – (vi)) and (7.5 (b) (i)–(iv)). Note in Figure 7.5 (a) (iii) the moist air mass from Arabian Sea at 850 hPa is intermixing with the dry northwesterly airmass over the region, whereas in figure 7.5 (b) (i) it is the moist air from Bay of Bengal which is intermixing. Upper level divergence was provided by the Sub Tropical Jet ($\approx 2 \times 10^{-5}$ Sec^{-1}) in both the cases. On 6 Jan 89 trough in the westerlies at 300hPa (fig. 7.5(b)(iv)) to the west of Delhi provided further strength to the divergence field ($\approx 5 \times 10^{-5}$ Sec^{-1}).

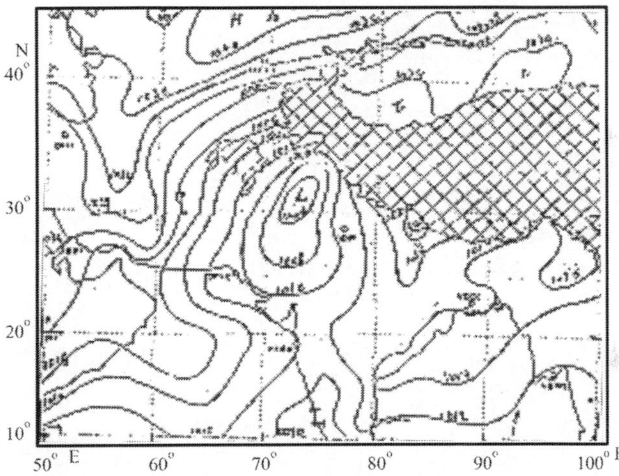

Fig. 7.5 (a) (i) Surface pressure pattern, 1200 UTC of 28 Dec 90

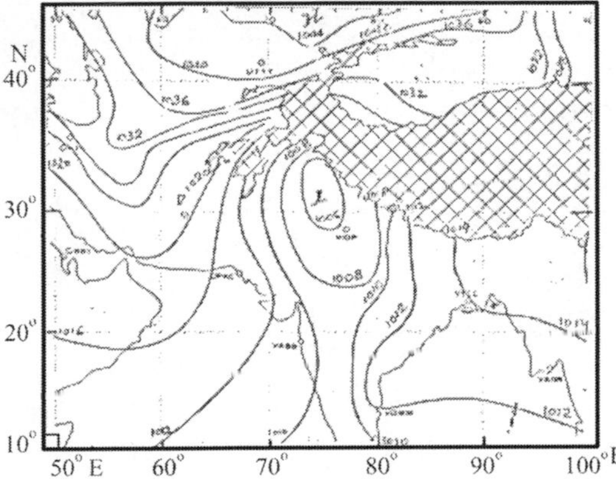

Fig. 7.5 (a) (ii) Surface pressure pattern, 00 UTC of 29 Dec 90

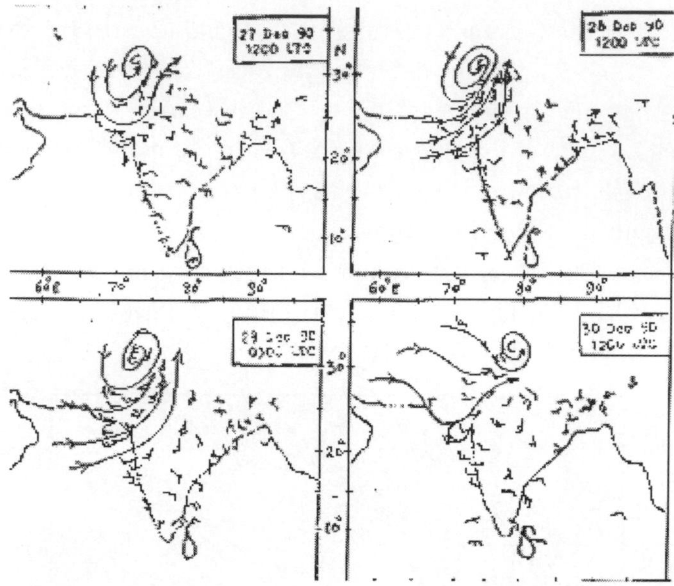

Fig. 7.5 (a) (iii) 850hPa winds from 27 Dec 90/1200UTC (Top left), 28 Dec 90/1200UTC (Top right), 29 Dec 90/0300UTC (Bottom left) to 30 Dec 90/1200UTC (Bottom right).

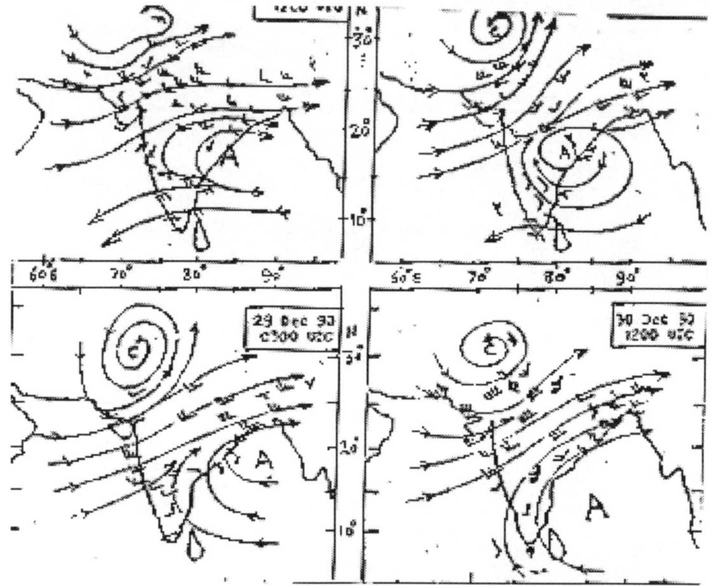

Fig. 7.5 (a) (iv) 700hPa winds from 27 Dec 90/1200UTC (Top left), 28 Dec 90/1200UTC (Top right), 29 Dec 90/0300UTC (Bottom left) to 30 Dec 90/1200UTC (Bottom right).

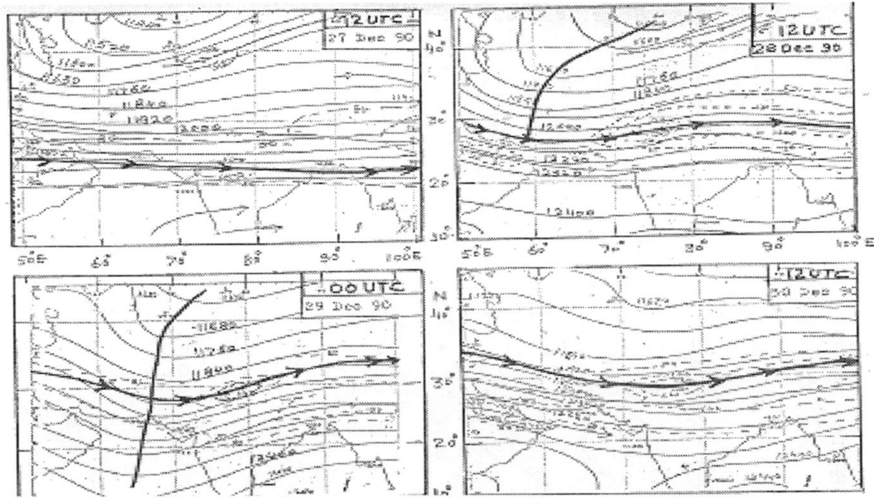

Fig. 7.5 (a) (v) 200hPa winds from 27 to 30 December 1990.

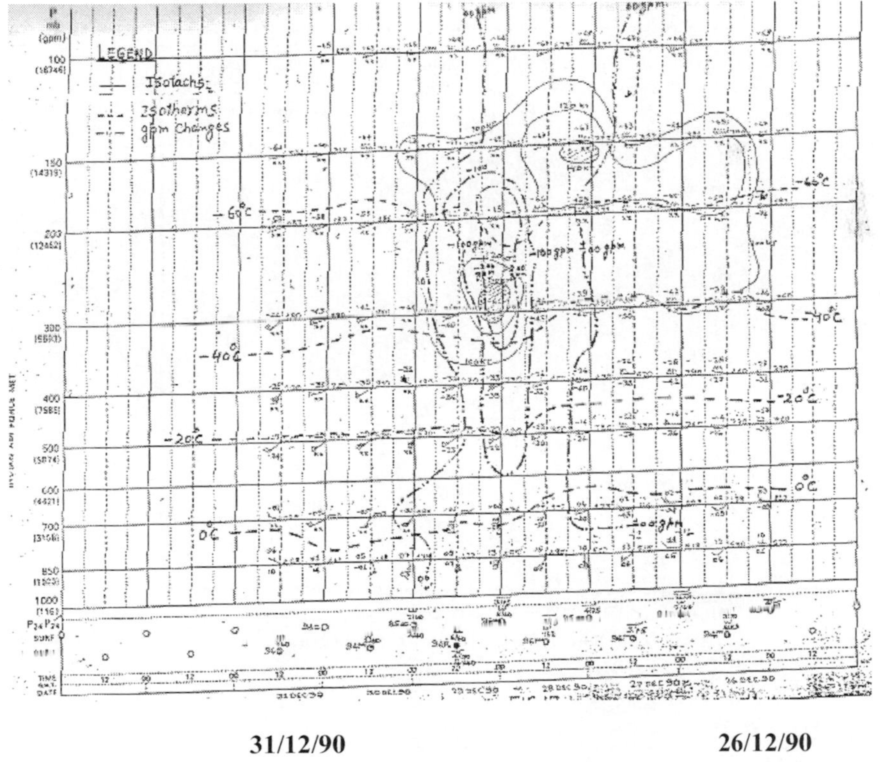

31/12/90 26/12/90

Fig. 7.5 (a) (vi) Vertical Time Section 26 to 31 December 1990.

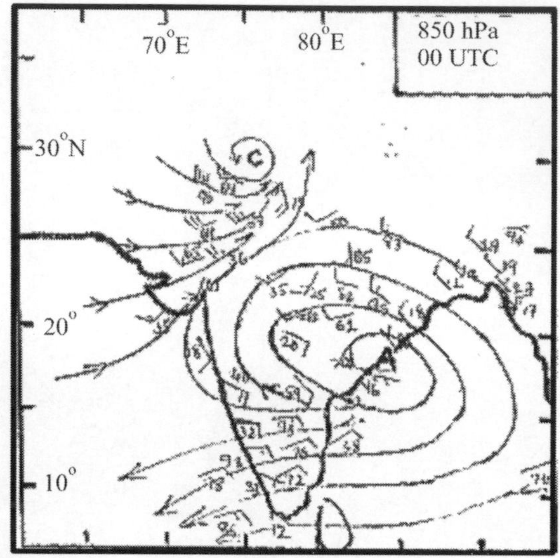

Fig. 7.5 (b) (i) 00UTC wind pattern at 850 hPa.

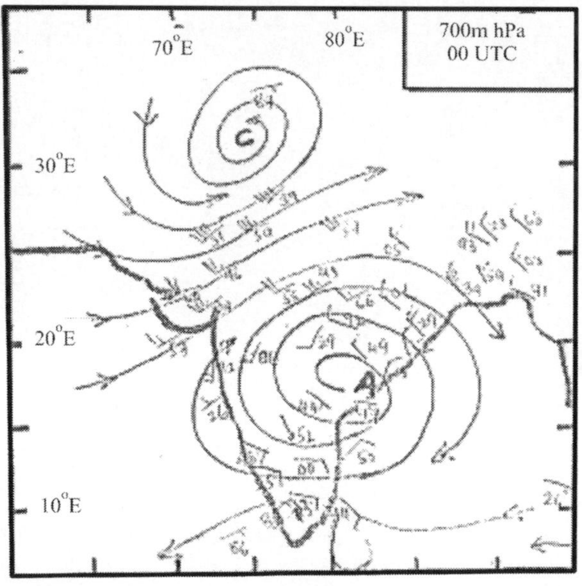

Fig. 7.5 (b) (ii) 00UTC wind pattern at 700 hPa.

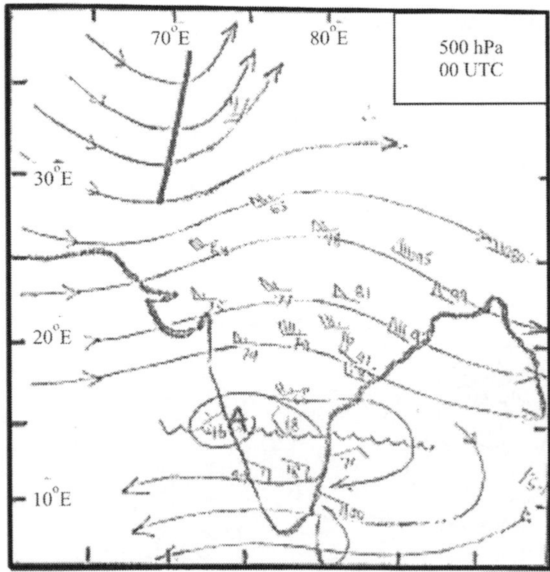

Fig. 7.5 (b) (iii) 00UTC wind pattern at 500 hPa.

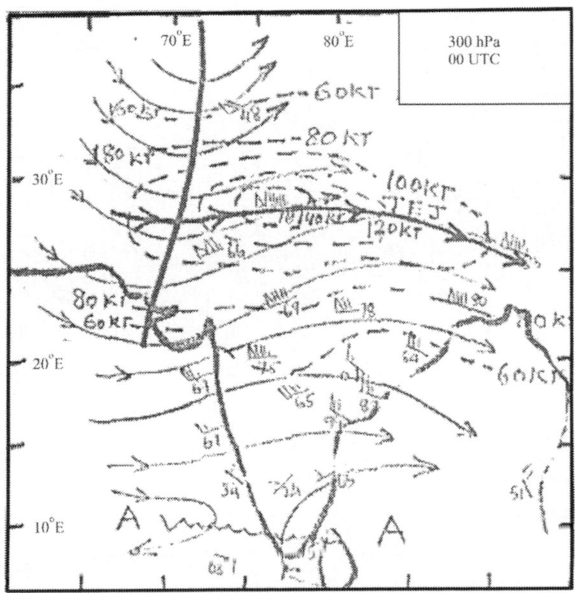

Fig. 7.5 (b) (iv) 00UTC wind pattern at 300 hPa.

Figs. 7.5 (b) (i-iv) Upper wind pattern on 06 January 1989.

(b) Induced Low pressure area or Upper air cyclonic circulation upto 850 hPa.

Towards the second half of January and thereafter western disturbances as surfaces low pressure causing the hailstorm are relatively less common. Most of the documented cases reveal either induced surfaces low pressure or a well marked induced upper air cyclonic circulation over Rajasthan, Punjab, Haryana region as the feature causing hailstorm over northwest India.

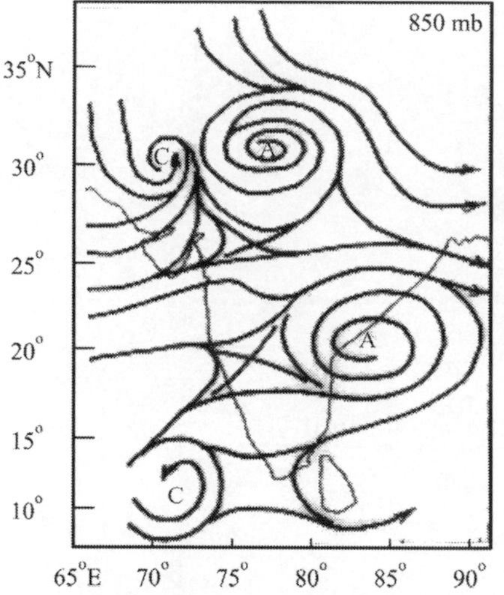

Fig. 7.6 (a) Upper winds at 850hPa, 00UTC, 19 March 81.

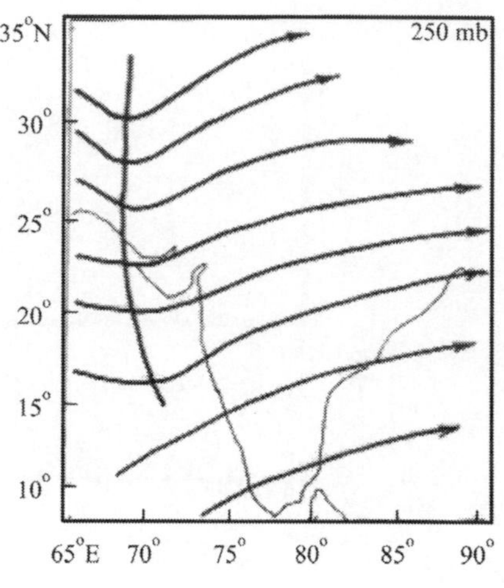

Fig. 7.6 (b) Upper winds at 250hPa, 00UTC, 20 March 81.

Figs. 7.6 (a) and (b) show a situation on 19 Mar 81 (00UTC) when a well marked induced low pressure area on surface was lying over southeast Rajasthan and adjoining parts of Madhya Pradesh. Upper air cyclonic circulation was extending up to 700 hPa in 12 UTC chart of 18 Mar 81. Bikaner experienced hailstorm in late evening of 19 Mar 81. Upper air divergence was provided by a well marked trough aloft along 70^0 E of long. in the westerlies between 300 – 200 hPa, Fig. 7.6 (b). Intermixing of moist air mass from Arabian sea and dry air mass from the north west can be noticed over the region of hailstorm occurrences.

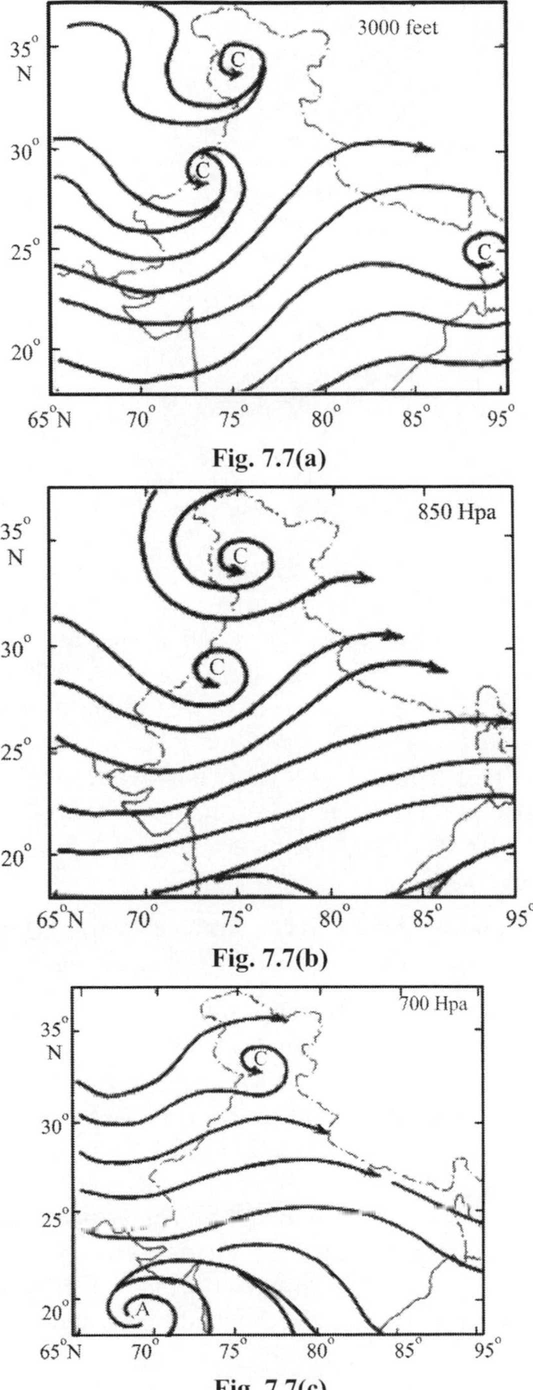

Fig. 7.7(a)

Fig. 7.7(b)

Fig. 7.7(c)

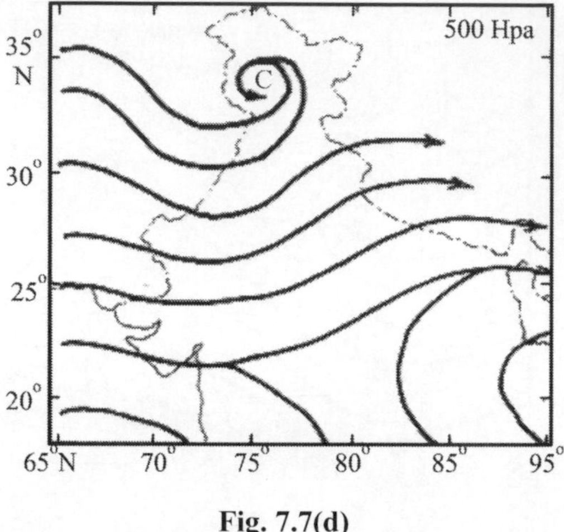

Fig. 7.7(d)

Fig. 7.7 (a-d) 25 February 1990, 12 UTC

Figs. 7.7 (a - d) show another synoptic situation of air mass type hailstorm which occurred over Adampur and Chandigarh region on 26 Feb. in the morning hours. On 25 February 90 there was an upper air cyclonic circulation over J & K up to 500 hPa (Fig. 7. 7 (a - d)). There was one induced system lying over NW Rajasthan as an upper air cyclonic circulation extending up to 850 hPa (Fig. 7.7(a) & (b)). Strong low level convergence ($\approx -7 \times 10^{-5}$ Sec^{-1}) was induced. The upper air divergence ($\approx 2 \times 10^{-5}$ Sec^{-1}) was provided by the trough in the weterlies aloft at 300 hPa along 70^0E long.

(c) Low pressure area over surface with a well marked discontinuity on the surface and upper air chart

Severe hailstorm on the evening of 27 May 1959 near Sikar (Rajasthan) had indicated only a surface low in 18 UTC chart over Punjab and north Rajasthan. Strong vertical wind shear was provided by the jet maxima of 85 kt at 10.5 km (00UTC) which increased to 135 kt at 12.0 km (1200 UTC). See Fig. 7.8 (a), (b) and (c). A well marked discontinuity existed on the surface and upper air. It was particularly pronounced between Jodhpur and Delhi, in the 00 UTC upper air chart, up to 1.5 km - Fig. 7.8 (d). Fig. 7.8 (e) shows the 24 hours wind tendency chart, 1.5 km a.m.s.l. Sikar was located in the double hatched area.

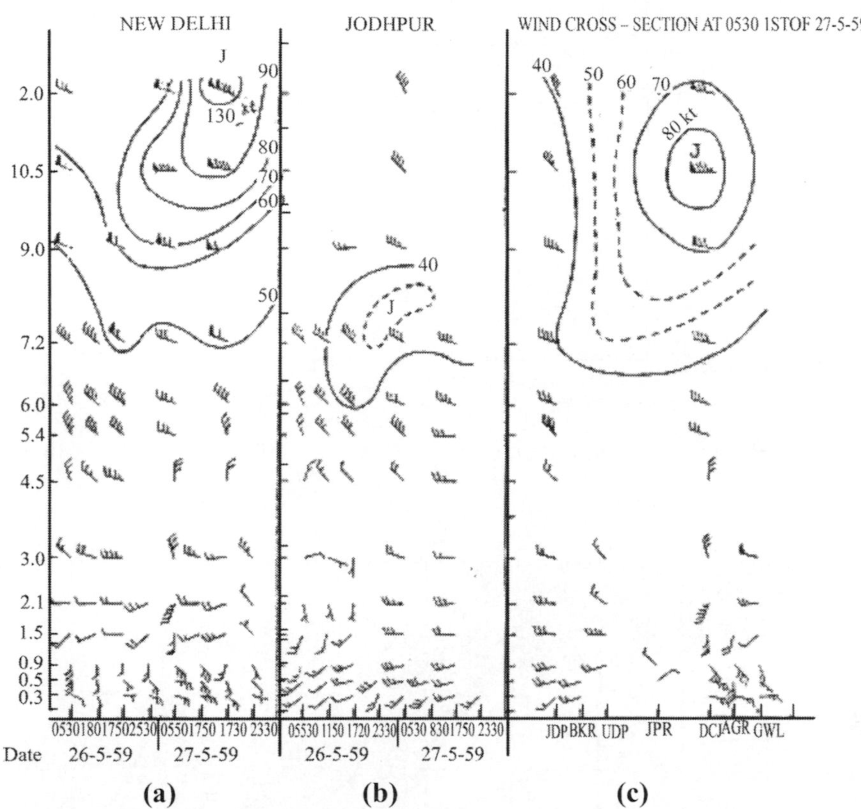

Fig. 7.8 (a), (b) and (c) Vertical time-section of winds at New Delhi and Jodhpur on 26 and 27 May 1959 and wind cross-section at 00 UTC on 27 May 1959.

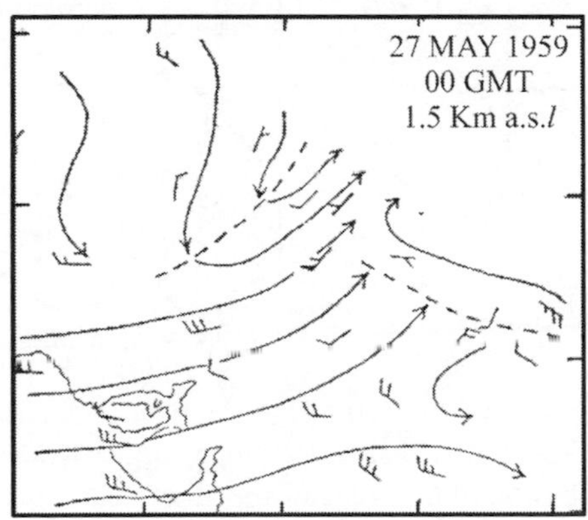

Fig. 7.8 (d) The streamline at 1.5 km a.m.s.l. at 00 UTC of 27 May 1959.

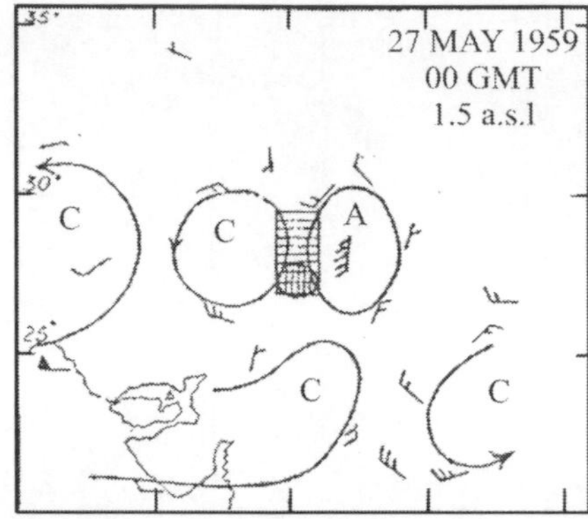

Fig. 7.8 (e) 24 hourly wind change chart.

Fig. 7.8(f) Shows the layer of instability in T-Φ gram of Jodhpur in shaded area.

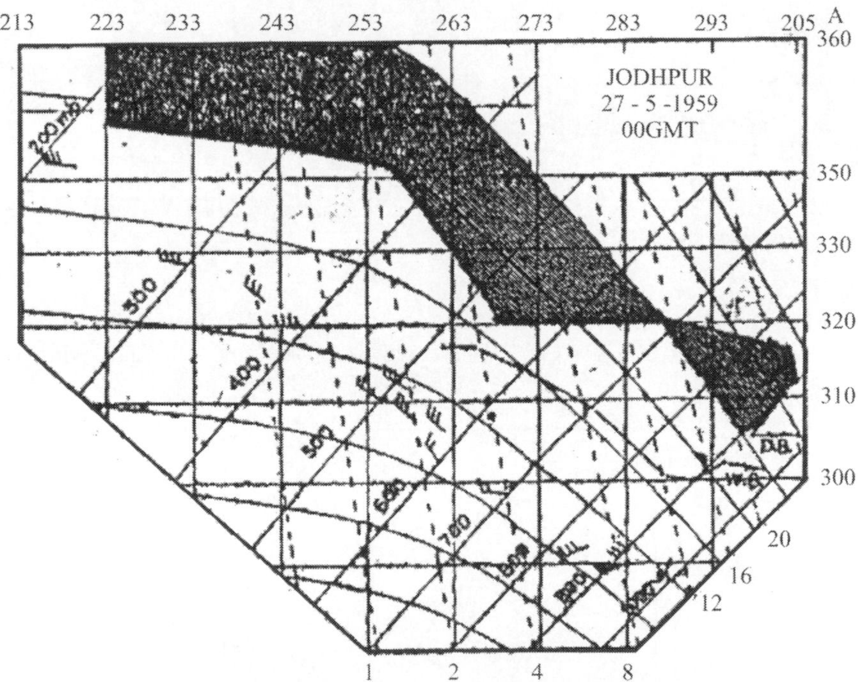

Fig. 7.8(f) T-Φ gram of Jodhpur, 00UTC, 27 May 1959. The hatched upper rectangle shows the large CAPE value. Close proximity of dry and wet bulb near surface (between 900-1000 hPa) indicates moisture accumulation at lower level. Mid level dryness may also be noted.

On the evening on 27 May 1959 by around 1200 UTC IAC Viscount Aircraft suffered with considerable damage while flying 110 to 190 km WSW of Delhi on Jodhpur – Delhi route due to hailstorm. See the photographs 1-8.

Photo-1

Photo-2

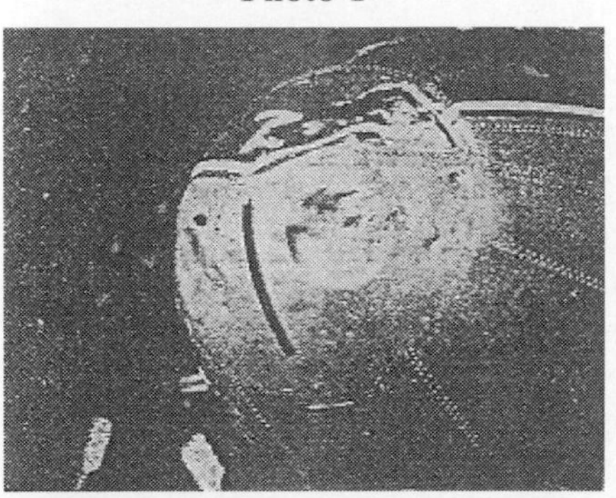

Photo-3 Size of large dent 10"

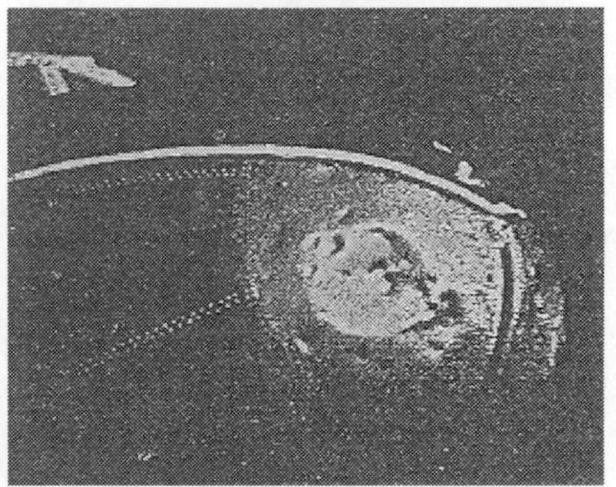

Photo-4 Size of large dent 12"

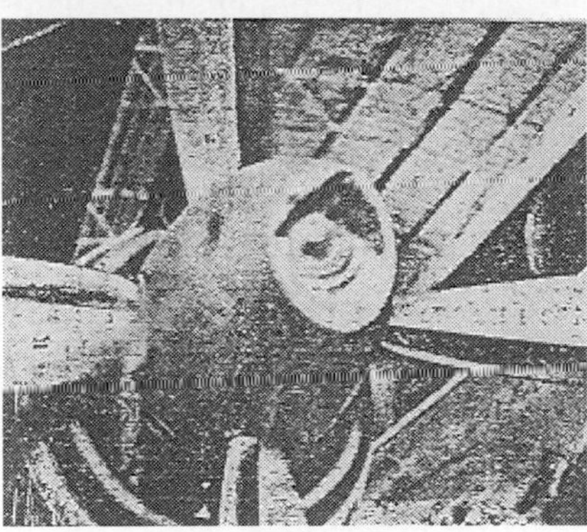

Photo-5

Photo-6 Size of dent 15"x10"

Photo-7 Size of hole 5”

Photo-8 Size of hole 6”x5”

Readers may note the extensive damage to the air craft which caused.

(i) Holes of varying sizes some of them as long as 12.7 cm across Photos (Fig. 1-8).

(ii) The cockpit wind screen had been completely smashed due to impact of hailstones on it. The triplex glass, although it had become opaque due to cracking, was still intact in frames. The glass of one of the window had been burst through by the hailstones and the glass splinters and the hailstones hit the pilot on the head resulting in bleeding.

(iii) Pressurization had failed at 19000 ft level as a result pilot had to immediately descend to 9000 ft level.

(iv) Apart from the holes made on the aircraft surface, by hailstones ripping open the metallic cover, there were innumerable dents both large and small. A few as big as 38×25 cm^2.

(d) Low level confluence of streamlines due to induced low and anticyclone

Pilani in west Rajasthan experienced hailstorm in the early morning of 06 May 81. Cyclonic circulation over Rajasthan at 850 and 700 hPa in association with anticyclone over west Uttar Pradesh provided strong confluence to give low level convergence for convective build up. Upper air trough west of Pilani at 200 hPa provided the upper air divergence.

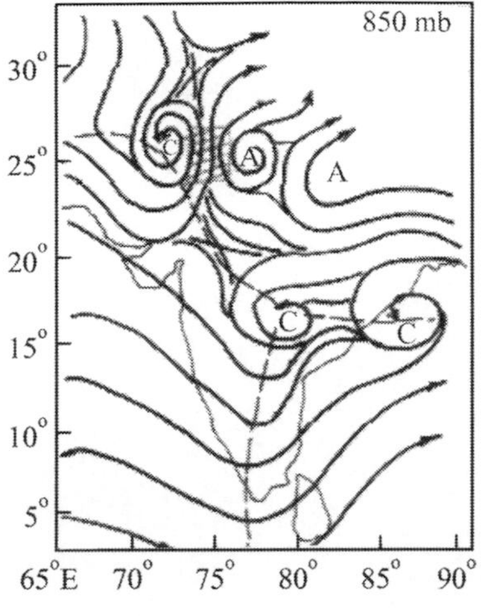

Fig. 7.9 (a) 6 May 1981, 00UTC

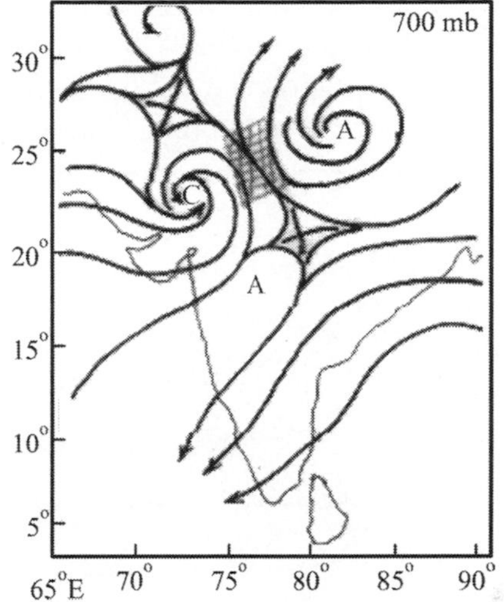

Fig. 7.9 (b) 6 May 1981, 00UTC

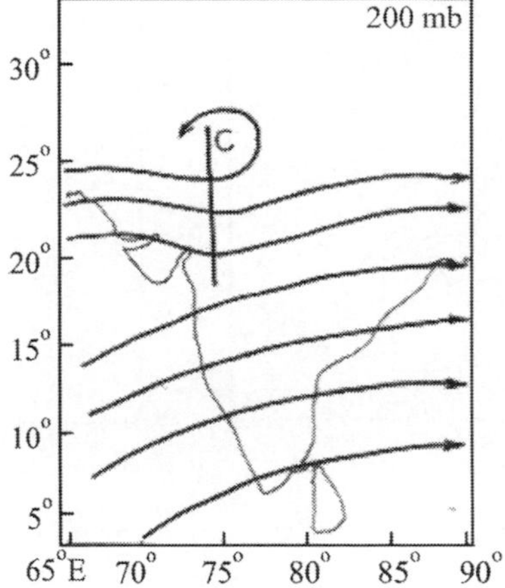

Fig. 7.9 (c) 6 May 1981, 00UTC

(e) Wind Tendency Shear Region

Rao & Mukherjee (1958) suggested regions of strong wind – tendency shear between cyclonic and anticyclonic tendency lines to be regions susceptible to cyclogenesis and convective process. In such a situation hailstorm builds up if

(i) The streamlines at lower level (1.5 km) are inducting moist current in to the region of cyclonic tendency.

(ii) Strong wind shear with height prevails over the region. Rao and Mukherjee (1958) also claimed that the direction of squall in such a situation is the direction of isalobaric winds as shown in all the four cases in Figs. 7.10(a-d).

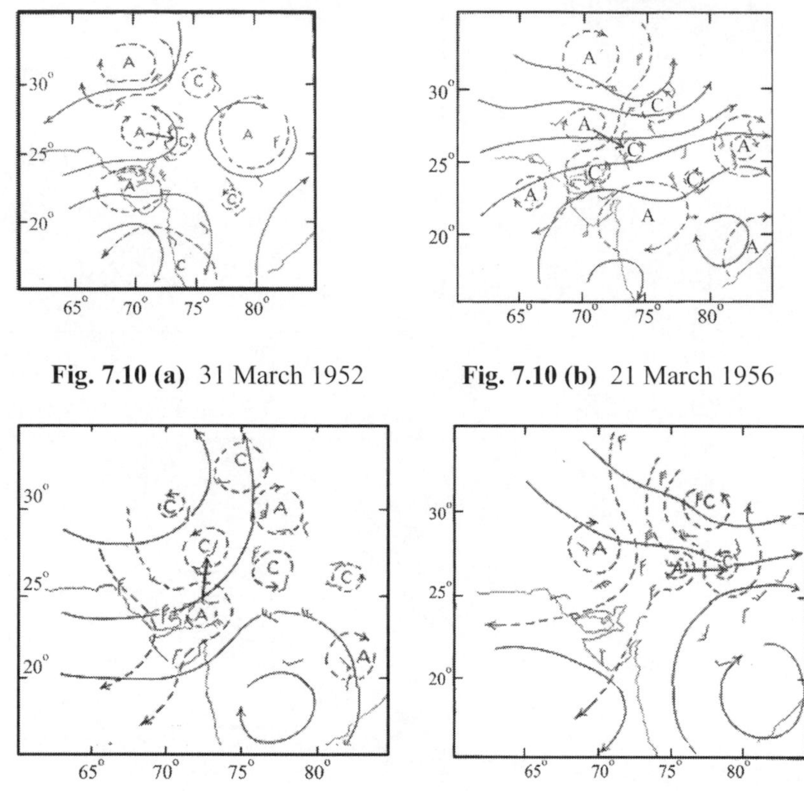

Fig. 7.10 (a) 31 March 1952

Fig. 7.10 (b) 21 March 1956

Fig. 7.10 (c) 18 March 1957

Fig. 7.10 (d) 19 March 1957

Fig. 7.10 Vectorial wind changes at 1.5 Km a.s.l. for 24 hours ending 02 UTC (0730 IST).

(iii) Mull and Kulshrestha (1962) had confirmed the procedure by Rao and Mukherjee (1958) for the hailstorm which occurred near Siker (Rajasthan) on the evening of 27 May 1959. See Fig. 7.8 (e) showing wind tendency charts of 27 May 1959. The large rectangular shaded area in 27 May 1959 (00 UTC) 1.5 km upper air chart shows possibility of hailstorm. The hailstorm actually occurred in the smaller double hatched area shown in the Fig. 7.8(e).

7.2 Vertical Wind Shear

In general westerly jet maxima have been observed over the region of hailstorm. Raju (1982) also noticed that in most of the cases maximum wind level lowers down by 0.2 km. to 1.2 km. Fig. 7.11 shows five cases of hailstorm occurrences and the levels of maximum wind for one day prior to the day of occurrence (D-1), Day of occurrence (D) to One day after the day of occurrence (D + 1).

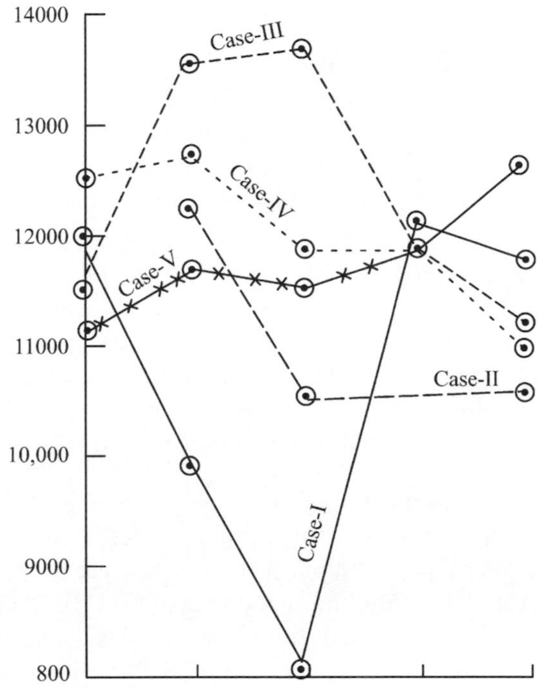

	D-1, 00UTC	D-1, 12UTC	D, 00UTC	D, 12UTC	D+1, 00UTC
Case I: dates	9	9	10	10	11
Case II: dates	12	12	13	13	14
Case III: dates	14	14	15	15	16
Case IV: dates	19	19	20	20	21
Case V: dates	30	30	01	01	02

Fig. 7.11 Maximum Wind Level.

Vertical wind shear between 850 and 300 hPa in all the cases have been noticed to increase significantly from the previous day on the day of hailstorm. Various studies showed it to be ranging between 65 to 112 kt for the winter month of January and February and between 65 to 70 kt during the period March to May.

7.3 Humidity Parameter

Values computed by Singhal (1992), Joshi (1987) and Sathe (19991) indicate that;

(a) Mean Mixing Ratio (MMR), over Delhi, between surface to 850 hPa was found to be 6.0 to 10 gm/kg. (mean 8.0 gm/ kg) on the day of hailstorm during Dec, Jan and Feb months against the mean value 5.8 gm/kg during these months. Mean mixing ratio (MMR) during premonosoon month was noticed to be between 9.2 to 13.4 gm/kg. (mean 11.6 gm/kg) against the mean value of 6.54 gm / kg.

(b) Mean mixing Ratio between 700 – 500 hPa during the winter months (December, January and February) was found to be 1.1 to 3.0 gm / kg (mean 1.5 gm./kg) against the mean of 1.5 gm / kg for these months. For premonsoon month the MMR for hailstorm days was found to be between 1.5 to 3.4 gm/ kg. (mean 2.2 gm/kg.) as against the mean value of 1.7 gm/kg.

With the above figures it may be observed that it is the increase in moisture in the lowest layer only which differentiates the hailstorm favouring atmospheric profile against the normal one. Note the low level isothermal layer in the tephigram of Jodhpur of 00 UTC dated 27 May 59 apparently helping in the horizontal accumulation of moisture for hailstorm formation in lowest layer, (Fig. 7.8(f)). Middle level moisture parameters do not show any change during the hailstorm days. However the middle layer becomes comparatively much drier with the moisture increase in the lowest layer between surface to 850 hPa.

Table 7.3 Values of indices on various days

Days(D)/ Indices	Adampur & Chanigarh 26th Feb.		Amritsar & Adampur 06th Jan	Ambala & Adampur 06th Mar		Ambala & Adampur & Faridkot 18th Mar		Delhi 29th Dec.		
	D – 1	D	D –1	D	D –1	D	D –1	D	D –1	D
Rackliff Index(RI)	28	31	27	37	25	30	29	30	26	31
Modiffed Jefferson Index(MJI)	28	35	23	38	27	35	29	39	25	32

Table (7.3) shows the changes in the values of various indices from D – 1 to D day for the occurrence of hailstorm. It may be observed that in general rise is indicated in both the indices from D –1 to D day. On D day in general RI ≥30 and MJI ≥32.

***Radar Study of Hailstorm Over Northwest India*:** Although the radar echoes, positively identified as associated with hail, have been found to show protruding fingers, hooks, scalloped edges or U – shapes from the edges of the bright, sharp edged, high intensity echoes (convective types), the reverse is not always true (Mull and Kulshrestha – 1962). Thus so far, even with help of radar, it has not been possible to establish a perfect system for identification of hailstorm. Some of the radar photographs taken on 27 May 1959 by high power CPS – 9 radar at Safdarjung, Delhi are reproduced in Fig 7.12 – 7.27. The photographs show very clearly the following features:

(i) Precipitation echoes on the PPI were characterized by oval shapes, sharply defined contour and high echo intensity indicating the strongly convective nature of the clouds present through out the whole day.

(ii) Some of these echoes, in the radar pictures of the forenoon were detached and had fairly clear cut edges indicating that these were associated with cumuliform clouds, while other were either continuous or showed a line structure indicating the presences of cumulonimbus clouds or frontal thunderstorms.

(iii) Precipitation echoes on the RHI were roughly cylindrical in shape indicating existence of extensive cumulonimbus clouds. Some of them exhibited the simplest form, i.e. a single column in some directions (Fig. 7. 13) while others consisted of several columns. (Fig. 7.15 and 7.18).

(iv) Some of these echoes showed a widening at the top indicating the spreading up to the vertical currents. (Fig. 7.16 and 7.17).

(v) One of the RHI picture (Fig. 7.13) showed even the anvil.

(vi) A number of PPI pictures of the forenoon (Fig. 7.12, 7.14) and afternoon (Fig.7.22) showed protruding fingers and hooks, respectively, suggesting the occurrences of hail, but whether they were actually associated with hail or not it is difficult to say, particularly in the absence of any reports.

Different PPI and RHI radar echo photographs taken on 27 May 1959 by the CPS-9 Radar at Safdarjung, New Delhi of the severe Hailstorm near Sikar, Rajastha, India by Mull and Kulshrestha (1962). Below each photo the figures indicate from left to right the Time in

IST, Range in statute miles and elevation/Azimuth in degrees respectively.

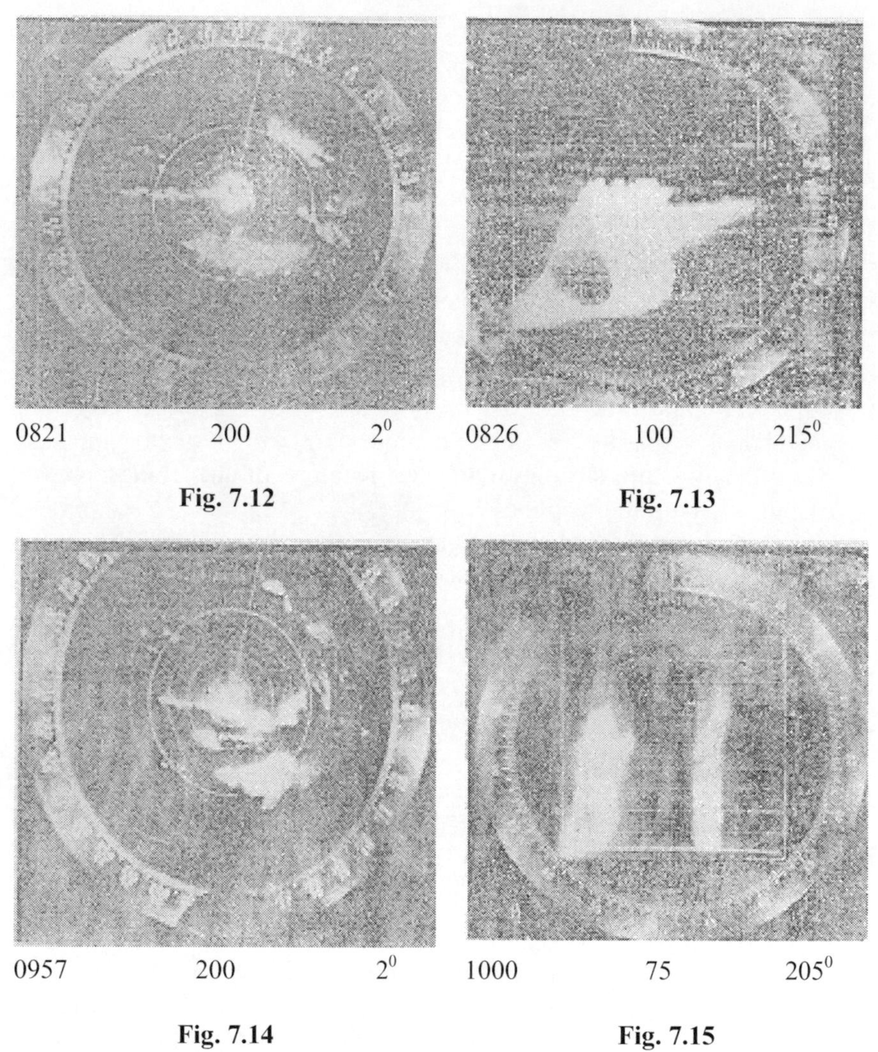

0821 200 2^0

Fig. 7.12

0826 100 215^0

Fig. 7.13

0957 200 2^0

Fig. 7.14

1000 75 205^0

Fig. 7.15

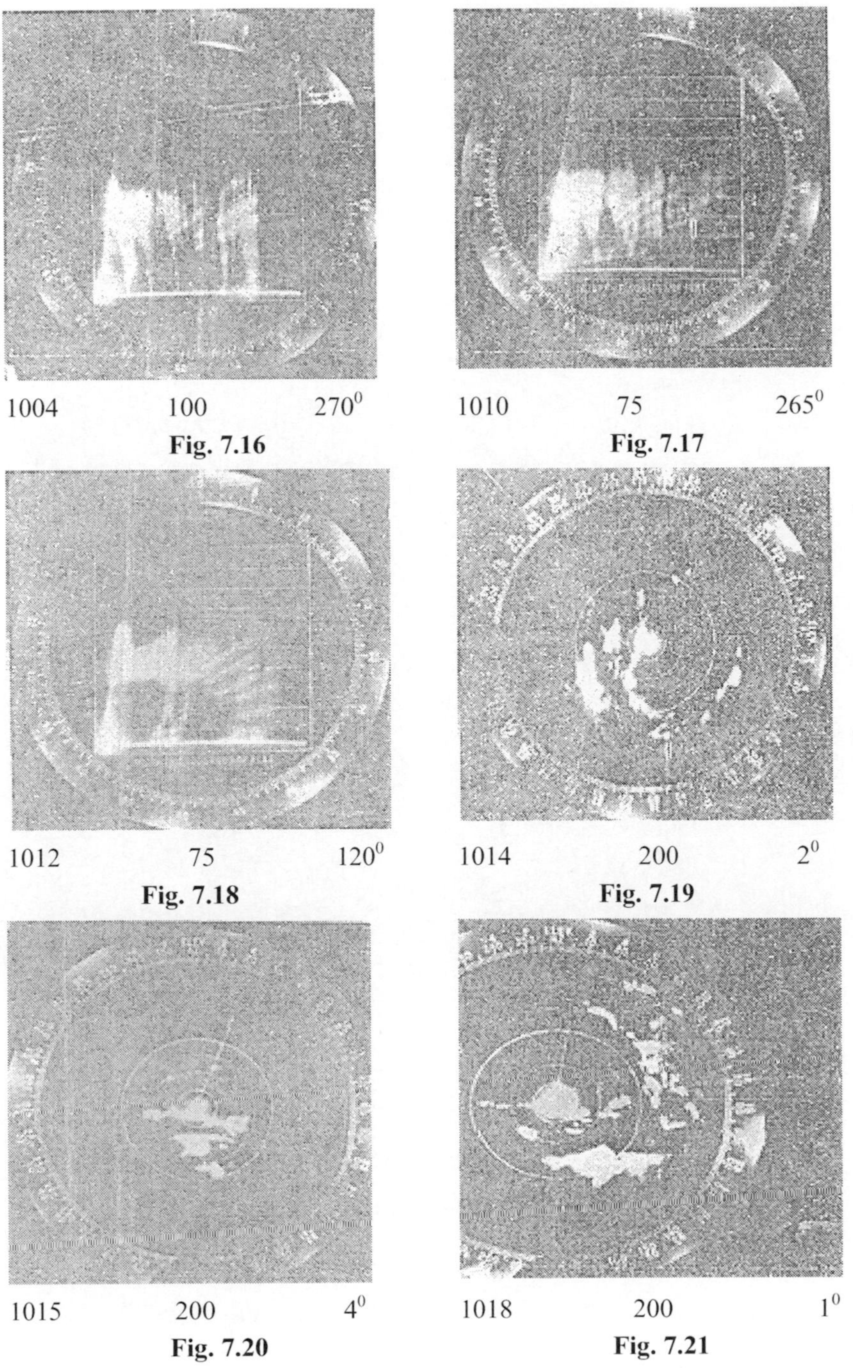

1004 100 270^0

Fig. 7.16

1010 75 265^0

Fig. 7.17

1012 75 120^0

Fig. 7.18

1014 200 2^0

Fig. 7.19

1015 200 4^0

Fig. 7.20

1018 200 1^0

Fig. 7.21

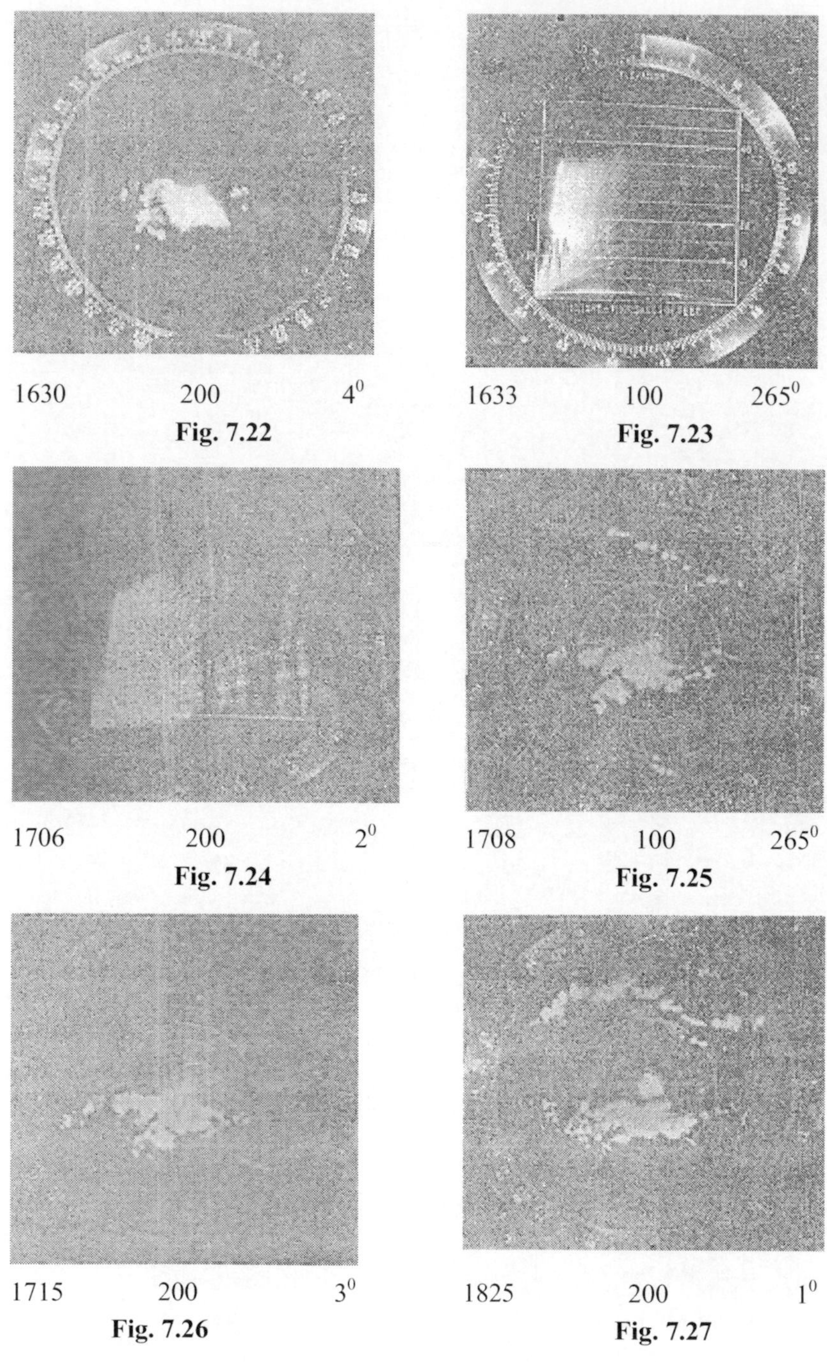

1630 200 4^0
Fig. 7.22

1633 100 265^0
Fig. 7.23

1706 200 2^0
Fig. 7.24

1708 100 265^0
Fig. 7.25

1715 200 3^0
Fig. 7.26

1825 200 1^0
Fig. 7.27

(vii) The PPI picture taken at 1630 IST. i.e. about 45 minutes before the pilot of the Viscount encountered hail (Fig. 7.22) showed a well defined hook shaped protuberance in 265^0 azimuth at about 60 to 80 nmiles range. The RHI picture taken 3 minutes later (Fig. 7.23) showed that the echo extended to about 45, 000ft above ground. Immediate transmission of hail possibility message to the pilot could have avoided the damage to the aircraft.

(viii) The hook shaped protuberances was very prominent in all PPI pictures taken between 1630 and 1730 IST but did not exist in later pictures (Fig. 7.27).

References

1. Fawbush E.J & Miller R.C., (1953), Bull. Am Met Soc. 34 pp 139 – 145.
2. Joshi M.C. (Fit Lt) (1987), on some synoptic and therodynamic aspects of hailstorm occurrence over Northwest India during January – March, JFC thesis, Air forces Asmin College, Coimbatore.
3. Kulahrestha & Jain, (1967), Radar hematology of Delhi and neighourhood occurrence of severe weather,IJMG, vol. 18, pp 105.
4. Misra, P.K. Prased S.K., (1980), Forecasting hailstorm over India, IJMG vol. 31. No. 3.
5. Mull S. & Kulshrestha S. M., (1962), on the formation of hilstorm vol. Spl. NO. Mar. 1962, Ind. J. Met & Geoph, pp. 95 – 103.
6. Raju K.S. (Flt Lt) (1982), on some aspects of hailstorm occurrences over Northwest India, Vatavaran, Vol. 5. No. 1.
7. Rao D. V. & Mukherjee A.K., (1958), India J. Met. Geophys.. 9. pp. 31113 – 3322.
8. Ravikiran & Kumar, P., (1994), Case studies of hailstorm over Gwalior air field J3126, JFC thesis, Air forces Admin College, Conimbatore, India.
9. Sathe P.P. (Flt Lt), (1991), on some aspects of hailstorm occurrences over Punjab, Haryana, and Delhi during January March. JFC thesis, Air force Admin College, Coimbatores, India. : also In Advances in Tropical Meteorology; Meteorology and National Development, Ed. R. K. Datta, 355-364.

10. Singhal H.C. (Sqn Lt), (1992), Mesoscale study of typical; situations leading to unseal hailstorm occurrences during December – January over Delhi and neighbourhood and comparing the same with other non – hailstorm active winter systems SFC, thesis, Air forces Admin College, Coimbatore, India.
11. Tabib, V.D.(1977), Construction of mean model of Tiphigram for hailstorm over North and west Central India and comparison of the same with any Tiphigram as an aid to forecasting hailstorm J. 11/ 4, JFC Thesis, Air force Admin College, Coimbatore, India.
12. Tyagi, A. (1983), Severe convective activity over north India with cut-off low, Vayu Mandal, 13(1 and 2), pp 56.

CHAPTER 8

Hailstorm over Central India

Central India includes the vast plains of India including East Uttar Pradesh, Bihar, West Bengal, Orissa, Madya Pradesh, Gujrat and Maharashtra. It is a highly fertile land. Rabi and Kharif crops of winter and summer respectively includes dal, rice, wheat, oilseeds, sugarcanes, onion and ginger etc. This region also produces bulk of fruits likes mangoes, guavas, bananas and oranges etc. There is great loss of life and property every year due to hailstorm in the region. One of the severe hailstorms on 30th October 1961 is reported to have lashed 46 villages of the Bhind districted of NW Madhya Pradesh causing death of 12 persons & injuring about 100. About 1000 heads of cattle had perished. The hailstones ranged in sizes from a few millimeters to that of a tennis ball. Bigger ones were reported to be weighing more than a quarter kilogram. In the same year hailstorms were reported from Sagar and Uijain on 5th November, from Uijain, Jabalpur and Damoh on 06th November and from Rewa district on 07th November. Some of these hailstorm lasted for 10 to 20 minutes.

Stationwise hail report in the region is mentioned in Tables 8.1.

Table 8.1 Mean Frequency of days with hail-storm over various areas in India
(The Figures represent number of occasions in 100 years)

Name of the Dist	No.of Station	JAN	FEB	MAR	APR	MAY	JUN	JUL	AUG	SEP	OCT	NOV	DEC	YEAR
Balasore	38	5	5	34	34	8	3	-	-	-	-	-	-	89
Cuttack	38	3	-	8	8	8	-	-	-	-	-	-	-	27
Sambapur	38	5	11	3	18	16	13	3	5	13	3	-	-	90
Chalbasa	38	-	3	-	3	13	-	-	-	-	-	-	-	19
Ranchi	38	11	13	13	16	32	-	3	-	3	-	-	-	91
Purea	38	-	5	-	18	11	3	-	-	3	-	-	-	40
Patna	38	11	8	5	13	5	-	-	-	-	-	-	-	42
Gaya	38	3	5	3	5	11	-	-	-	-	-	-	-	27
Naya Dumka	38	-	-	3	11	11	-	-	-	-	-	-	-	25
Gorakhpur	38	11	-	13	3	5	-	-	3	-	-	-	5	40
Varanasi	38	5	21	8	-	3	-	-	3	-	-	-	5	45
Allahabad	38	11	26	11	8	8	-	-	3	-	-	-	3	70
Luoknow	38	5	11	8	5	3	5	-	-	-	-	-	-	37
Barelly	38	8	5	11	11	5	3	-	-	-	5	-	3	51
Roorkee	38	26	24	29	11	-	5	3	-	-	3	3	16	120
Neemuch	38	-	13	8	3	8	3	-	-	-	-	-	-	35
Indore	38	8	13	3	5	24	13	-	3	-	-	-	3	72
Nowgong	38	16	21	8	11	8	-	-	-	3	3	3	11	84
Satna	38	13	8	5	3	3	5	-	-	-	-	-	3	40
Akola	38	3	-	5	3	-	-	-	-	-	-	-	-	11
Amraoti	38	-	3	-	5	3	-	-	3	-	-	-	3	17

Chowdhary et al (1973) carried out detailed study of hailstorms over central India. Some of the important conclusions drawn by their study of over 800 hailstorm reports from Madhya Pradesh and Vidarbha region of central India are mentioned below.

(a) Hailstorms occur with increased frequency from January to March in this area.

(b) Though hailstorm reports have been received from this region during post monsoon period but mainly they occur between December to May.

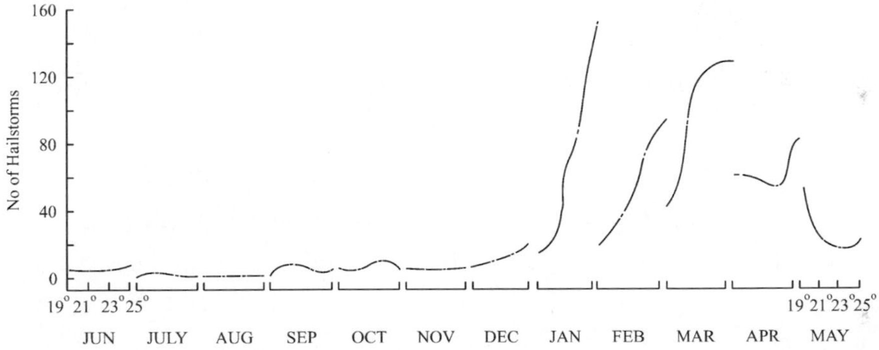

Fig. 8.1 Monthly distribution of hailstorms latitude wise (between 19^0-25^0) in Madhya Pradesh and Vidarbha (1943-70).

(c) They are absent during southwest monsoon period (June to September).

(d) The latitudinal distribution of the hailstorms in this area in each months (giving due weightage to the unequal areas in $18 - 20^0$, $20 - 22^0$, $23 - 24^0$ and $24 - 26^0$ N latitudinal belt) show a peak value in the vicinity of 23^0 N latitude.

(e) The diurnal frequency of occurrences is highest is during 1500 to 1800 hours IST. See Fig 8.2. About 80% of the hailstorms occur between 1400 – 2100 hours IST.

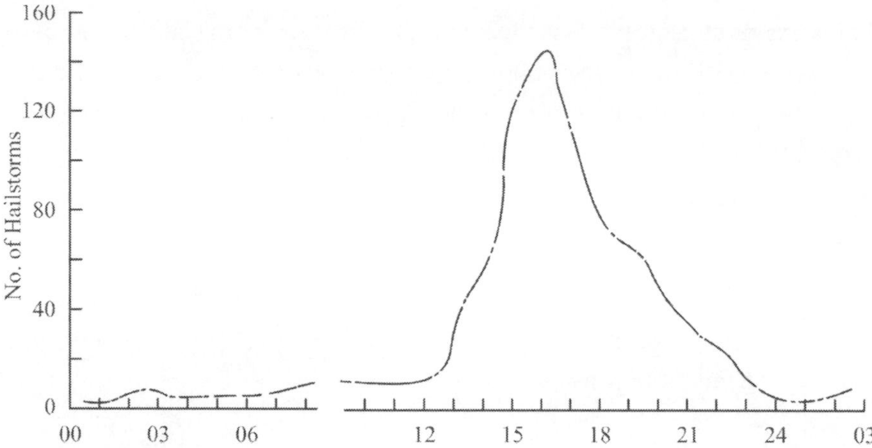

Fig. 8.2 Diurnal frequency of hailstorms in Madhya Pradesh and Vidarbha (Horizontal axis shows time in IST).

(a) Hailstorm of severe types (maximum size of the hailstorm ≥ 2 Cm - Chowdhary et al (1973)) occur in March (Fig. 8.3). Every one out of eight hailstorm is of severe type in this season.

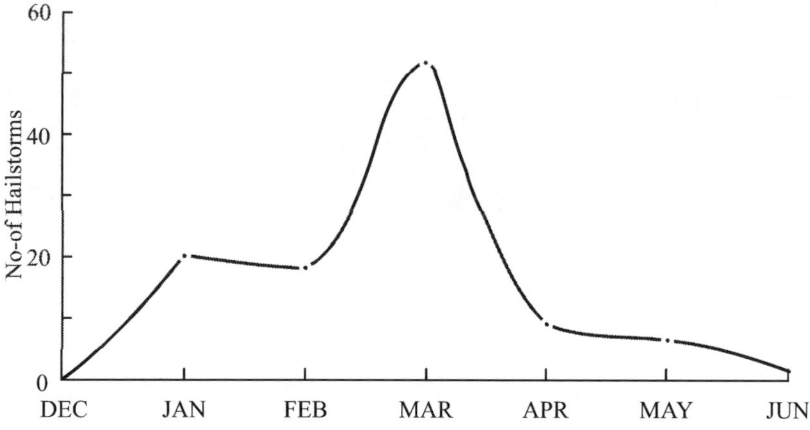

Fig. 8.3 Frequency of severe hailstorms.

Tabib (1977) had produced the mean tephigram of the days of hailstorm occurrences between 1966 to 1969 over Nagpur and Bombay for the months of maximum hailstorm activity i.e. January to March. It was noted by him that freezing level is lower by about 40 hPa in the mean hailstorm days than the mean of the normal days. Tabib (1977), however, did not differentiate between the intensity of hailstorm based on the size of hails. The mean hailstorm sounding produced by Chowdhury et al. (1973) (Skew T – log p or Herlafson diagram) of 12 severe cases at Nagpur is presented in Fig. 8.4 (a). Inversion layer near the ground is significant. No inversion layer may be noted in the case 8.5 (b) which is averaged for 18 light hailstorm cases over Nagpur with diameter less than 2 Cm.

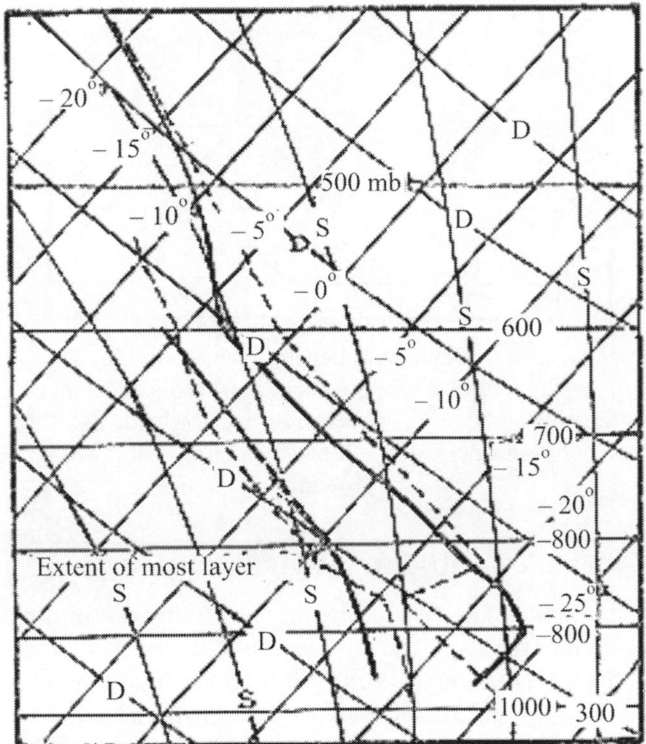

Fig. 8.4(a) Mean hailstorm soundings (plotted on a skew T-log p or Herlafson diagram). Legend: (a) Continuous curves represent the mean Dry bulb and Wet bulb curves for 12 cases of severe hailstorms at Nagpur. (b) The dotted curves represent similar curves as in (a) for airmasses in U.S.A. producing hailstones of 1" diam (25 cases) after Fawbush and Miller (1953).

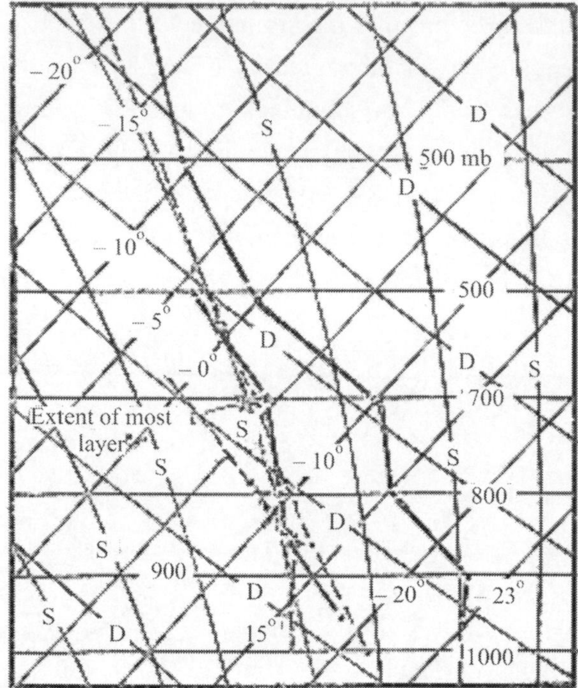

Fig. 8.4(b) Mean hailstorm sounding. Legend: Same as in fig. 8.4(a) except that the mean soundings represent 18 cases of light hailstorms at Nagpur and 68 cases in U.S.A. producing stones of ½" diameter - after Fawbush and Miller (1953). Note no inversion or isothermal layer exists as in Fig. 8.4(a) near or above the ground.

8.1 Favourable Condition for Hailstorm Occurrence

Western disturbances or their secondries moving across north India function as active agents in developing potent zones of convergences associated with severe thunderstorm fields in central India by advecting cold dry air from the north and drawing in warm – moist air from south towards convergences zones. Various synoptic situations which caused hailstorm over central India are being mentioned below.

(a) Western disturbances as an upper air cyclonic circulation/trough in lower levels.

Fig. 8.5(a) shows streamline pattern at 850 hPa (Swaminathan 1962). Hailstorm lashed Gormi (Bhind District) of NW Madhya Pradesh region at 11 UTC on 30 October 1961. Trough from the circulation could be seen extending up to Madhya Pradesh. Upper air divergences was provided by the left exit region of jet maxima at 300 hPa;(Fig. 8.5(b)).

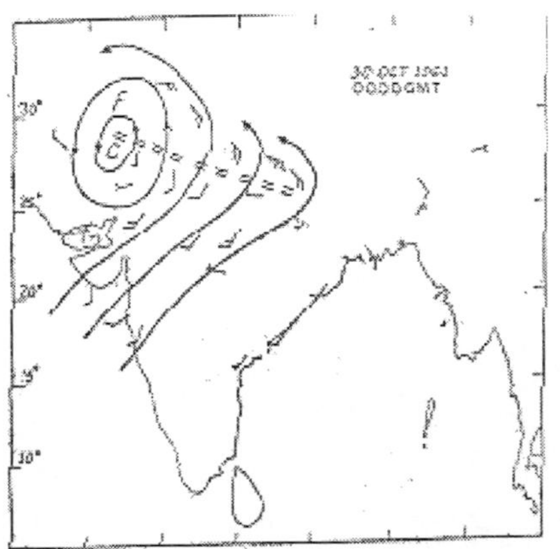

Fig. 8.5(a) Upper winds and Streamlines at 1500 a.m.s.l. Trough line is marked by the double dotted line through the centre of circulation.

Fig. 8.5(b) Upper winds and Streamlines at 9.0 Km a.m.s.l. Trough line is marked by the double dotted line through the jet core.

(b) Trough in easterlies in lower levels.

Fig. 8.6(a) (1 Nov. 1961) shows streamline pattern at 900 hPa. Hailstorm occurred over Ujjain district, between 1030 to 1130 UTC, over west Madhya Pradesh region on 1 Nov. 1961. Swaminathan (1962) had reported that upper air divergence was provided by the right entrance region of the jet maxima.

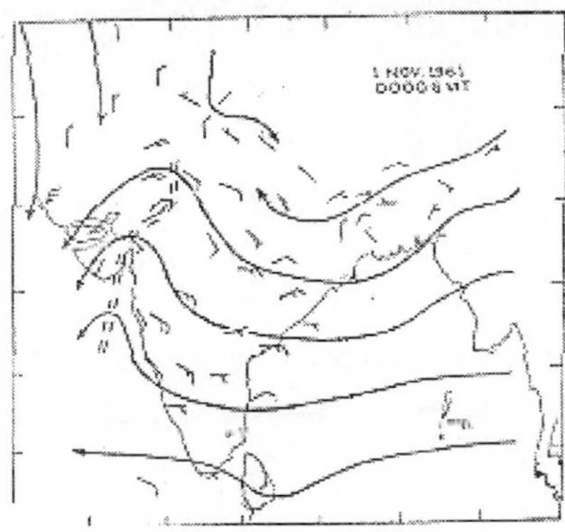

Fig. 8.6(a) Upper winds and Streamlines at 900 a.m.s.l. Trough line is marked by the double dotted line.

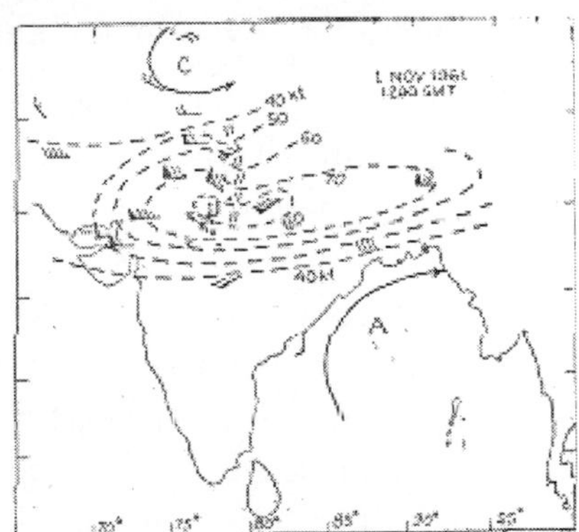

Fig. 8.6(b) Upper winds and Streamlines at 9.0 Km a.m.s.l. Trough line is marked by the double dotted line through the jet core.

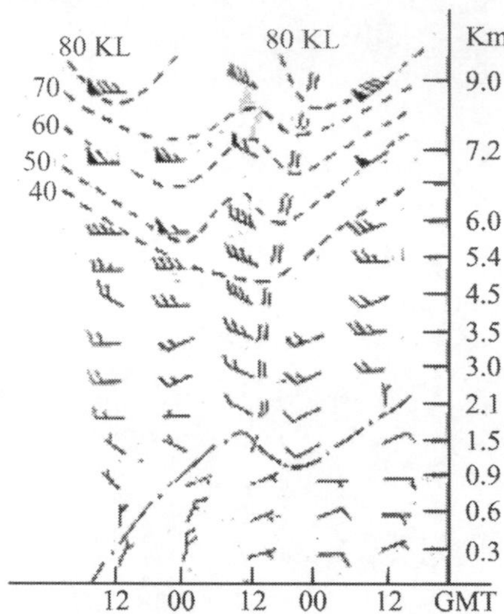

Fig. 8.7 Vertical Time section of winds over Gwalior during 31 October to 2 November 1961. Separation of easterlies and westerlies in lower levels are marked in the figure.

Fig. 8.7 shows the vertical time section over Gwalior during 31 October to 2 November 1961. The passage of upper tropospheric trough in westerlies and lower level trough in easterlies could be clearly marked on the day of hailstorm.

(c) Upper air circulation over north east Arabian sea and coastal Maharashta/Saurashtra region in December at 850 hPa.

Ananad (1987) studied hailstorm occurrences over North Madhya Maharashtra and Marathwada region on the evening on 26th Dec. 1986 and morning of 27th Dec. 1986. Fig. 8.8(a) & 8.9(a) (850 hPa) and Fig. 8.8(b) & 8.9(b) (700 hPa) show the streamline pattern and confluences of moist airmass of Bay Bengal and Pacific origin with the drier air mass from north. Note intermixing of the airmass over Saurashtra, north Maharashtra and adjoining region of northeast Madhya Pradesh.

Fig. 8.8(a) Streamlines at 00UTC, 850hPa, on 26 December 1986.

Fig. 8.8(b) Streamlines at 00UTC, 700hPa, on 26 December 1986.

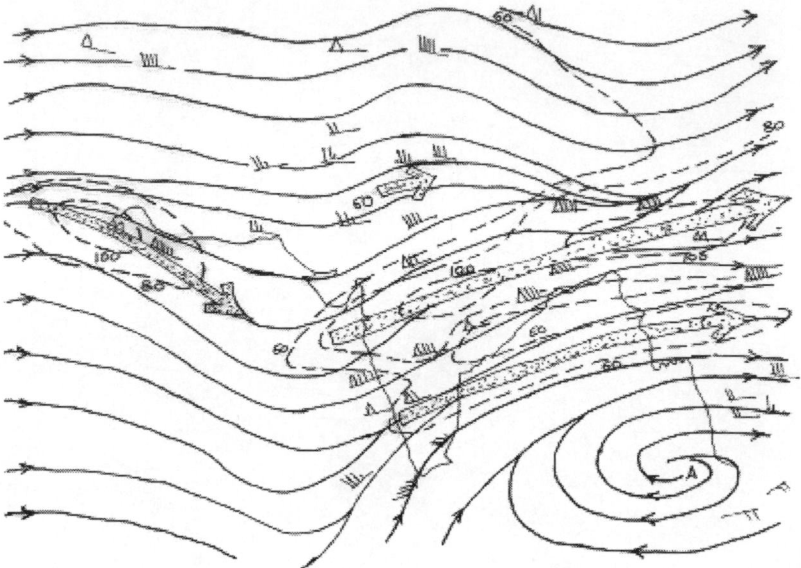

Fig. 8.8(c) Streamlines at 00UTC, 300hPa, on 26 December 1986.

Fig. 8.9(a) Streamlines at 00UTC, 850hPa, on 27 December 1986.

Fig. 8.9(b) Streamlines at 00UTC, 700hPa, on 27 December 1986.

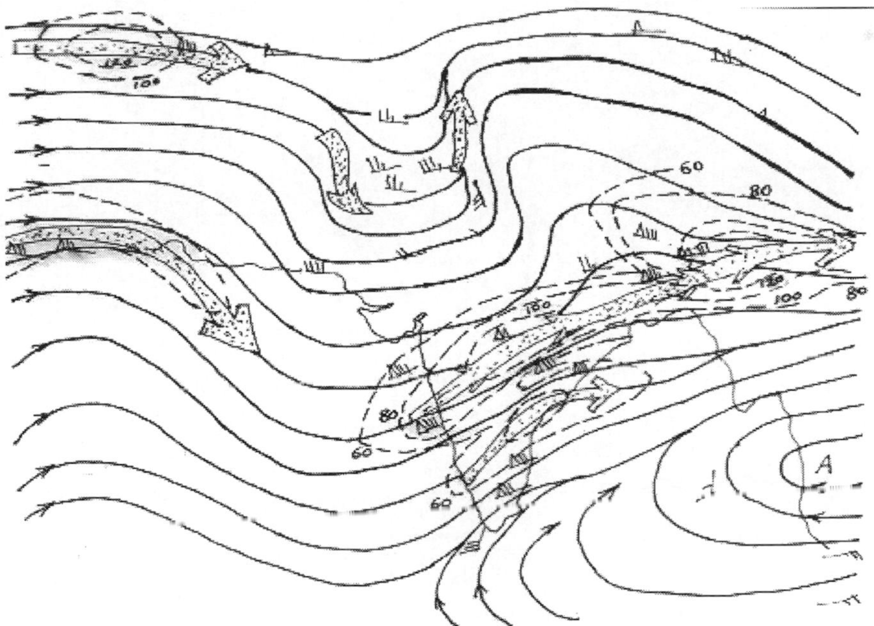

Fig. 8.9(c) Streamlines at 00UTC, 300hPa, on 27 December 1986.

Upper air divergence is provided by a deep north south trough along 72^0 E long at 300 hPa. Convergence provided by left entry region of wind maxima is apparently over ridden by the divergence provided by deep trough in westerlies. See streamline pattern at 300 hPa on 26 and 27 December 1986 in Figs. 8.8(c) and 8.9(c).

(d) Anticyclone in January over Orissa at 900 hPa.

Chowdhury et al (1973) reported that if anticyclones in January is located over Orissa at 900 hPa this may cause intermixing of moist airmass from Bay of Bangal and dry airmass from northwest India, while studying hailstorms which lashed central Madhya Pradesh on 20th January 1959. They produced streamlines of 00 UTC as shown in Fig. 8.10. Area of hailstorm is shown in Fig. 8.11.

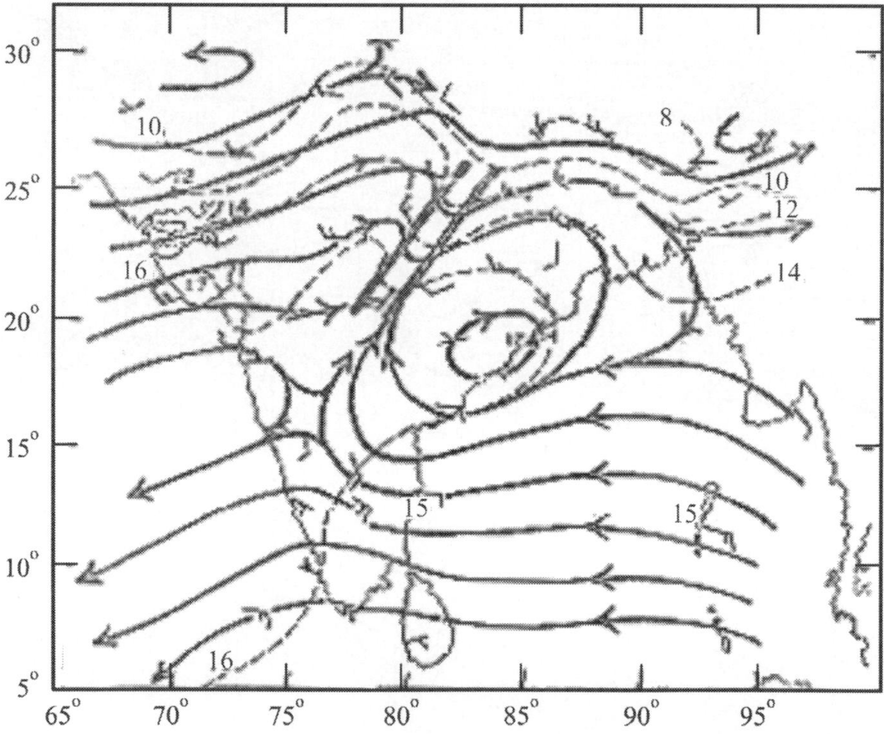

Fig. 8.10 Streamlines and isotherm patterns at 0.9 Km a.m.s.l. on 20 January 1959 (00UTC); Dashed lines represent isotherm in ^{0}C; Boundary zone, as assessed from streamlines, is shown by the two thick lines.

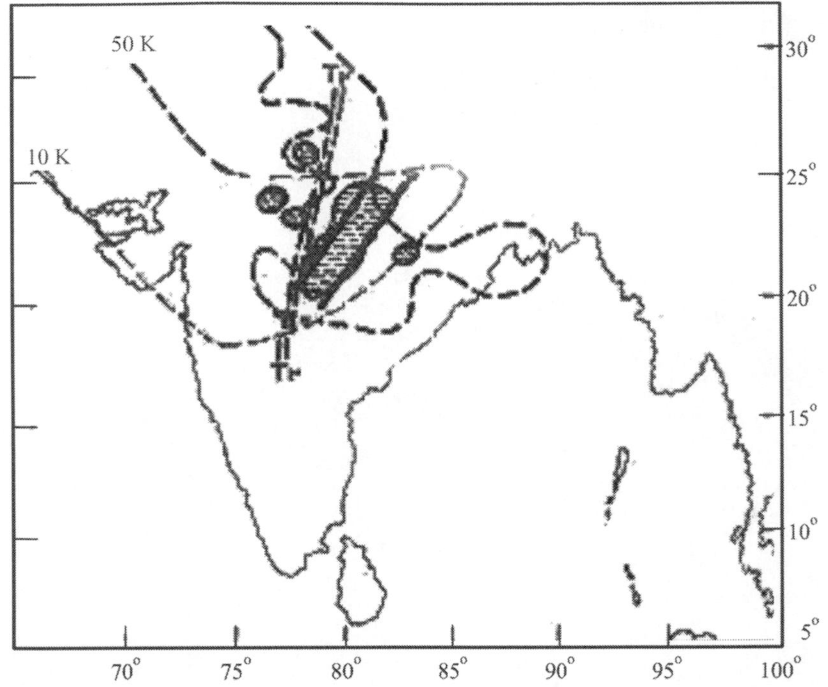

Fig. 8.11 Convective weather pattern during the next 24 hours; Thick lines cover areas of thunderstorm; Area hatched with dots are the areas of hailstorms; 50K isotach & trough and boundary zone of fig. 8.10 are also depicted. Note the active hailstorm area lies with the boundary zone and in the divergence sector of trough. Almost all the hailstorm areas are within the 50 K band of strong winds at 7.2 Km.

(e) Low level cyclonic vortex at 0.9 km embedded in the easterlies and rapidly moving from east to west.

Sharma (1967) studied hailstorm on 26th and 27th February 1964, towards evening, over Kamptee (18.5 km northeast of Nagpur). Rapid movement of the cyclonic vortex could be inferred with the vertical time section of Nagpur as shown Fig. 8.12.

Hailstorm apparently occurred in the forward NE sector of vortex. Upper air divergence was provided by the lowering of layer of maximum wind to 9.0 km. Detailed discussion of the case is presented in section 8.4.

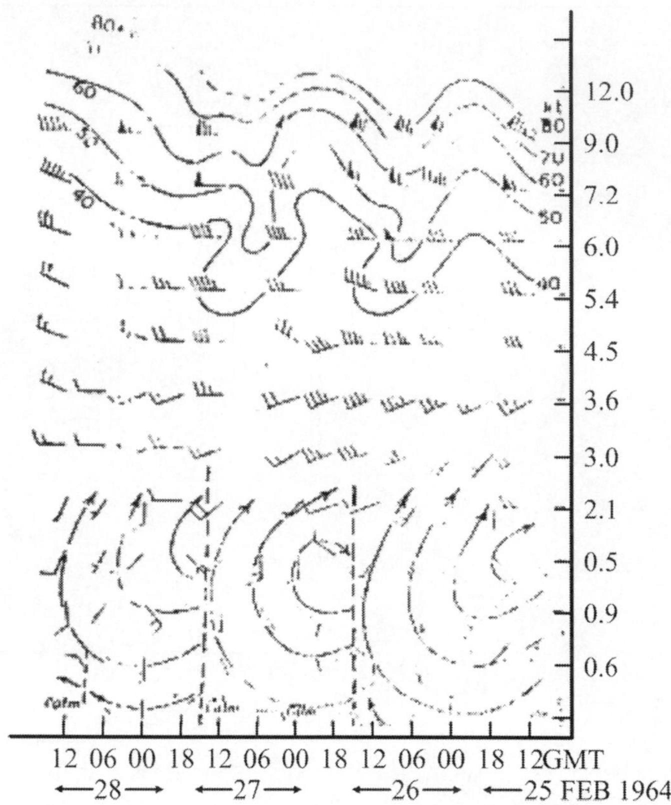

Fig. 8.12 Vertical time-section of wind over Nagpur, 25-28 February 1964.

(f) Anticyclone over central bay of Bengal in lower levels and cyclonic circulation over Orissa and East MP or Coastal Andhra Pradesh.

Severe hailstorm lashed Orissa region on 03rd March 1970. Chowdhury et al (1973) presented the streamlines at 0.9 km. (00 UTC). Intermixing of moist airmass from Bay of Bengal and dry air mass from northwest is clearly discernible in Fig. 8.13.

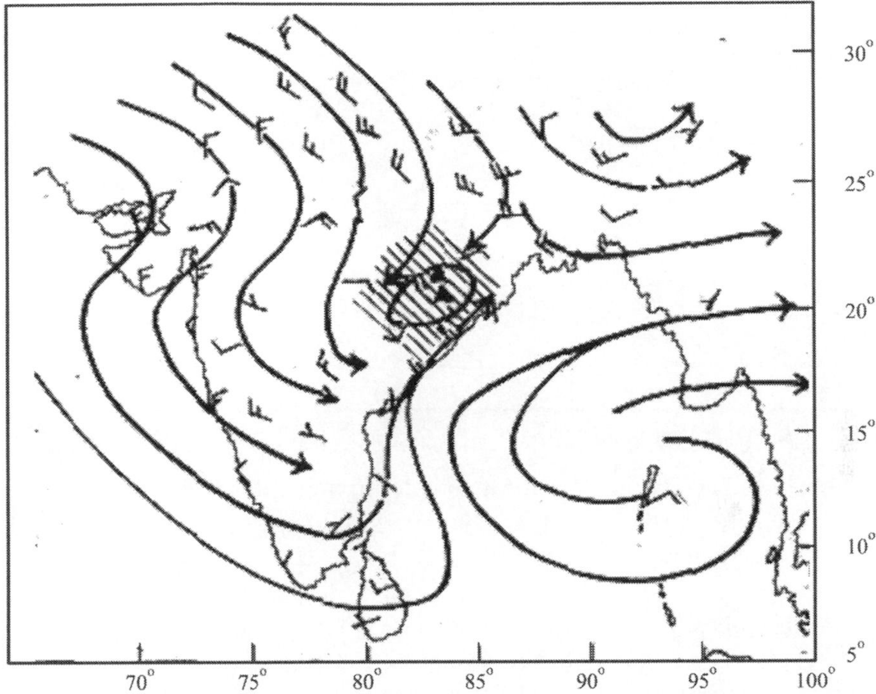

Fig. 8.13 Flow pattern at 0.9 Km a.m.s.l. on 3 March 1970 (00UTC); Triangle represents the places of violent hailstorm. Presence of cyclonic vortex can be noted. Hatched area represents the boundary zone.

8.2 Vertical Wind Shear

A study of average monthwise spatial distribution of hailstorms over Madhya Pradesh and Vidarbha are also closely related to strong vertical wind zones. Wind shear of considerable magnitude persists from December to May. It is well marked from January to March. It is also observed (Chowdhury et a 1973) that in each month from December to May the shear has a tendency to increase towards the northern latitudes. Anand (1967) has noted wind shears of 104 kt and 96 kt between 850-300hPa during the hailstorm days of 26th Dec.86 and 27th Dec. 86,, respectively. Table 8.2 shows the average wind shear over all the latitudes of this region during hailstorm, as computed by Chowdhury et al (1973), with respect to the severity of the storm. For severe hailstorm stronger wind shear may be noted.

Table 8.2 Average wind shear in hailstorm

Category of Hailstorm	Wind Shear (In Knots) (a) Between 1.5 and 9.0 a.s.l. (b) Between 1.5 and 7.2 km. a.s.l.	Actual Wind Speed (In knots) at (a) 9.0 km. a.s.l. (b) 7.2 km. a.s.l
SEVERE	(a) 40 to 45 (b) 30	(a) 47 (b) 35
LIGHT	(a) 30 (b) 23	(a) 35 (b) 30

8.3 Humidity Parameter

Averages mean mixing for the two intermixing air masses at different levels, which give hailstorm, as computed by Chowdhury et al (1973) is given in Table 8.3.

Table 8.3 Mean Distribution of temperature (TT) in ^{0}C and mixing ratio (X) in gm/kg in the moist and dry air masses

Level	TT and X	Surface	900 hPa	850 hPa	800 hPa	700 hPa	600 hPa	500 hPa.	400 hPa	300 hPa	200 hPa	Category
Moist	TT	25.3	21.5	19.1	15.6	7.9	-0.9	-9.2	-20.1	-33.4	-48.1	SEVERE
Air-mass	X	16.0	10.1	7.2	6.2	4.8	3.2					
Dry	TT	20.1	18.3	15.9	11.8	4.1	-5.1	-13.9	-25.5	-39.0	-51.5	
Air-mass	X	8.1	5.9	4.9	4.0	2.1						
Moist	TT	23.7	21.8	19.2	16.0	8.6	0.1	-8.3	-19.1	-33.1	-51.3	LIGHT
Air-mass	X	13.9	8.0	6.4	5.3	3.3	3.2					
Dry	TT	20.2	19.0	15.8	13.7	5.1	-3.9	-13.9	-24.2	-38.9	-49.0	
Air-mass	X	8.6	5.7	4.7	4.1	3.0	3.0					

They also gave the average data for hailstorms which occurred over Nagpur as in Table 8.4. For hailstorm days from January to March average depth of moist layer as computed by Tablib (1977) is shown in Table. 8.5

Table 8.4 Mean distribution of temperature (day bulb and wet bulb) and mixing ratio at 00 UTC at different levels in the air mass over Nagpur for light and severe hailstorm

	Surface	900 hPa	850 hPa	800 hPa	700 hPa	600 hPa	500 hPa	400 hPa	300 hPa	200 hPa	Category of hailstorm
D. B. (^{0}C)	23.2	22.9	20.5	16.8	6.4	-2.9	-11.2	-21.6	-35.6	-53.6	SEVERE
W. B. (^{0}C)	18.1	16.0	13.8	11.5	3.5	-6.2	-	-	-	-	
Mix Ratio (gm/Kg.)	11.1	9.8	9.0	8.1	5.1	3.0	-	-	-	-	
D. B. (^{0}C)	25.1	23.0	19.8	15.6	11.0	-1.1	-10.5	-20.6	-35.7	-53.9	LIGHT
W. B. (^{0}C)	17.5	14.7	12.0	9.4	3.5	-4.9	-	-	-	-	
Mix Ratio (gm/Kg.)	9.8	8.7	7.3	7.4	4.8	3.6	-	-	-	-	

Table 8.5 Depth of moist layer (hPa) during Jan – Mar

Time	Mean depth on Hailstorm days	Mean depth on non-Hailstorm days	Difference
0530 IST	680	830	150
1730 IST	660	830	170

Changes in the Modified Jefferson Index (MJI) for the hailstorm which occurred in the evening of 26th Dec. 1986 and morning by 27th Dec. 1986 have been computed by Anand (1987) for Nagpur as show in Table. 8.6 Note MJI ≥ 30 on D-day

Table 8.6 Modified Jafferson Index (MJI)

	D – 1 (25 Dec.)		D (26 Dec.)		D (27 Dec.)		D + 1 (28 Dec.)	
UTC	00	12	00	12	00	12	00	12
MJI	--	33.5	--	30	31	30.5	28.5	25.5

High value Modified Jefferson Index (33.5 on D – 1 day) could not support the convective build up due to left entry region of strong jet maxima ahead of the trough region which nullified the upper air divergences. See pattern at 300 hPa on 25th Dec. 1986. (Fig. 8.14)

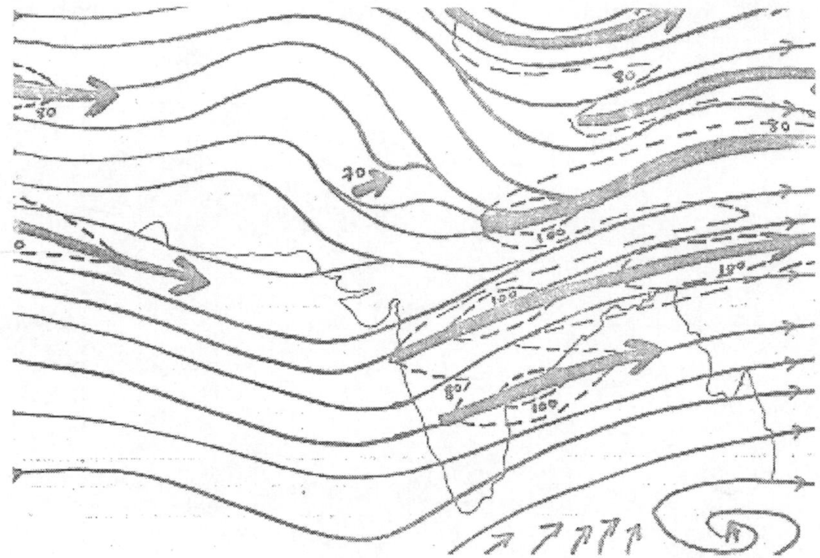

Fig. 8.14 Upper wind pattern at 300 hPa (00UTC) 25th Dec. 1986.

8.4 Radar Study of Hailstorm over Kamptee

Sharma (1967) made detailed study of hailstorm on 26th and 27th February 1964 towards evening over Kamptee which is 18.5 km away in 042^0 azimuthal angles from the radar station at Nagpur. He used Decca types – 41 radar, which has PPI scopes operating on 3 cm wave band and has the peak power output of 30 kw. Its beam width in horizontal planes is 0.75 degree and in the vertical planes 4 degrees.

A solid echo was first observed on radar scope at 1630 IST in the northwest of the station at a distances of about 8 in miles. It was associated with well marked indentation in the west and also a finger like protuberance or a hook having a deep notch towards south of the echo mass. This deep notch could not be seen clearly above 8 degree elevation of the radar antenna. At 4 -5^0 elevations, the hook with the notch was seen quite distinctly. Various observations as shown in Fig. 8.15 (a) to (m) are as follows:

(a) Photographs from (a) to (d) are representing main echo from lower level ($1/2^0$ elevation) to higher level (5^0 elevation) beginning from 1643 IST for full and half gains.

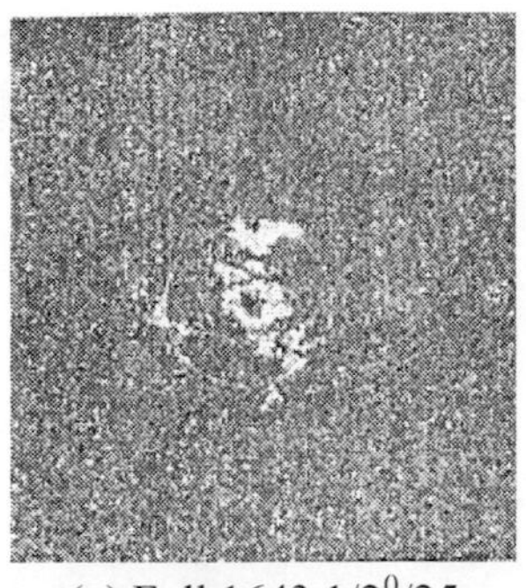

(a) Full 1643 1/2^0/25

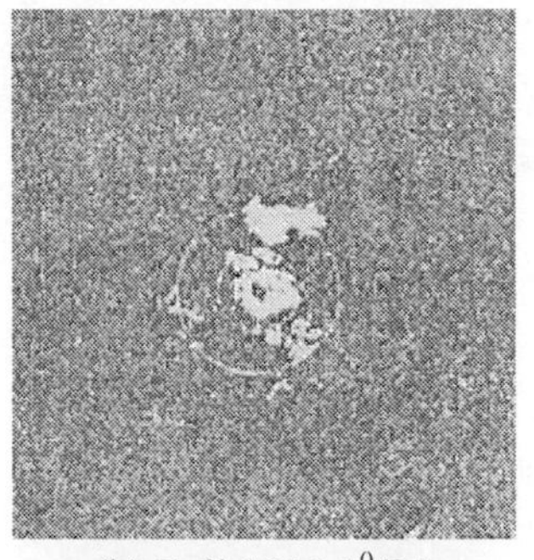

(b) Full 1645 1^0/25

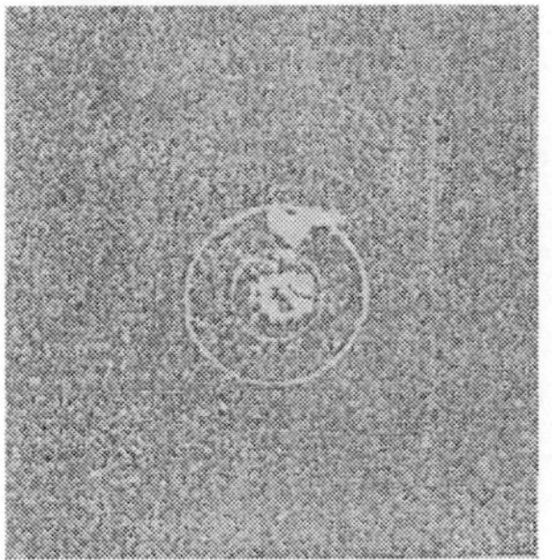

(c) Half 1647 3^0/25

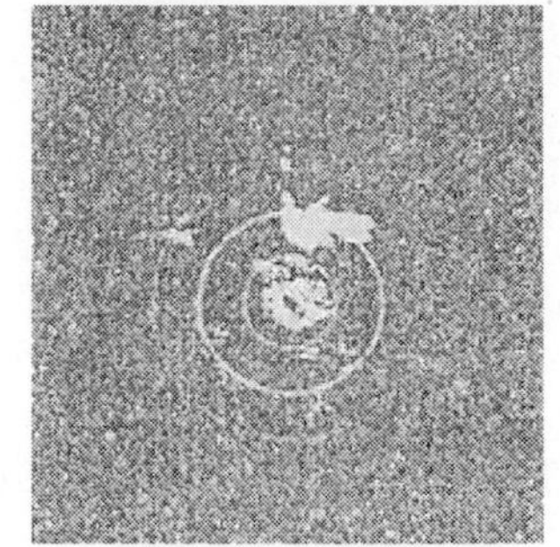

(d) Full 1649 5^0/25

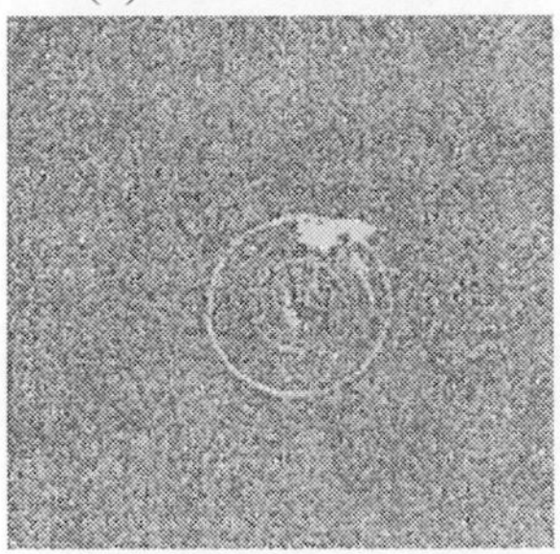

(e) Half 1650 1/2^0/25

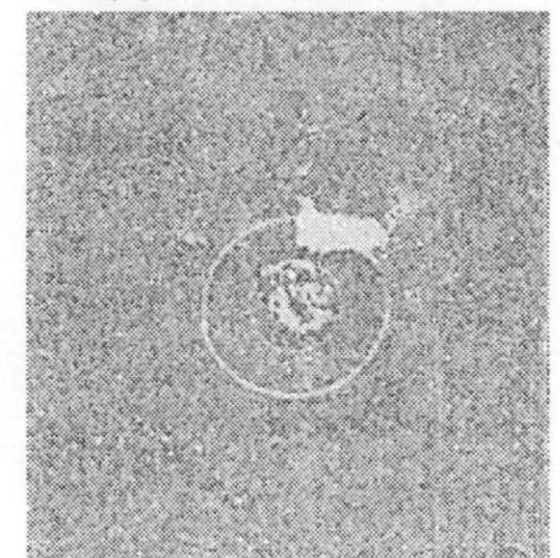

(f) Full 1651 4^0/25

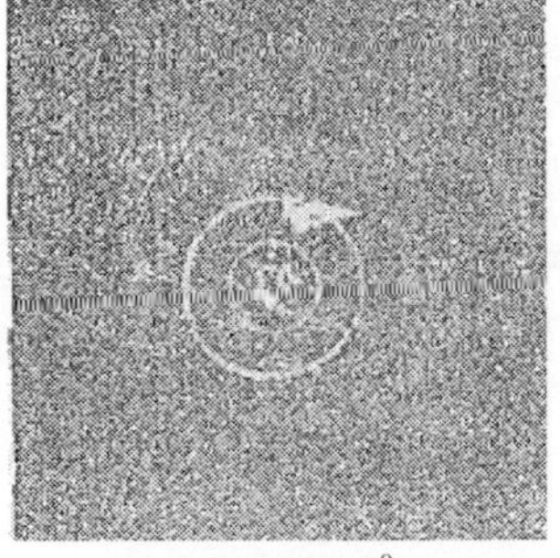

(g) 1/4 1653 4^0/25

(h) Zero 1655 4^0/25

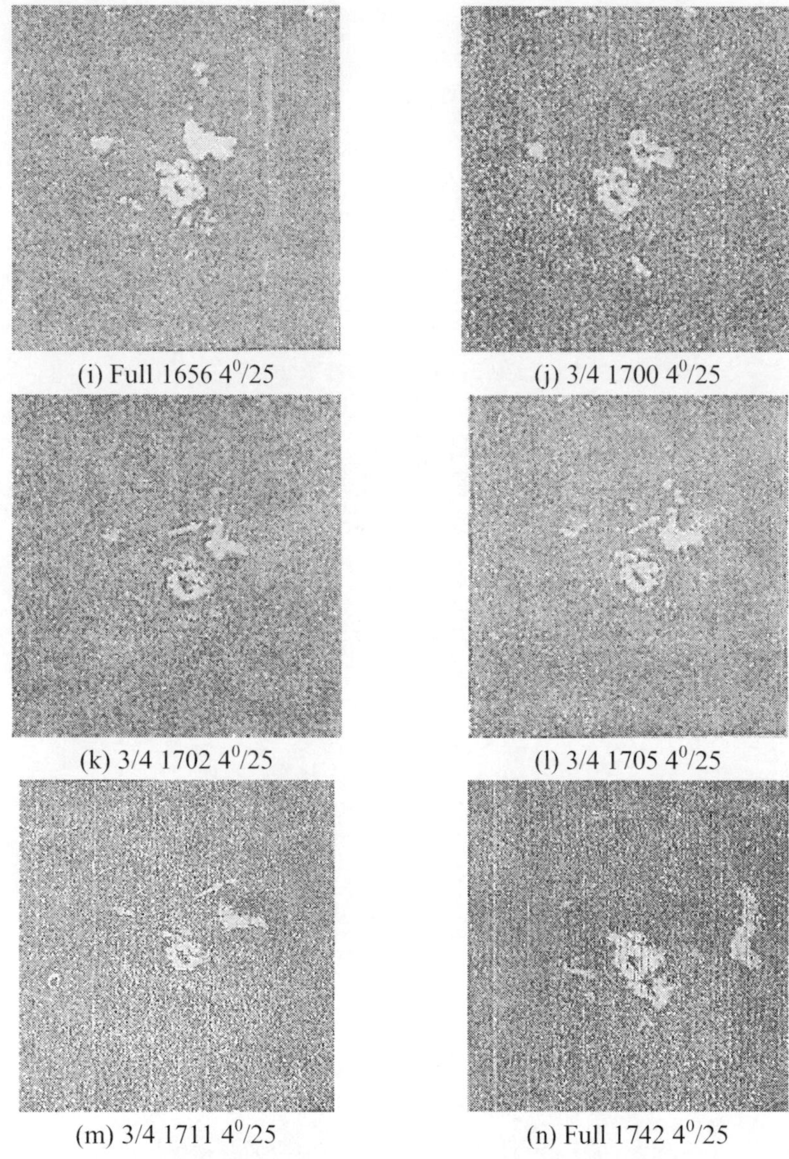

(i) Full 1656 $4^0/25$

(j) 3/4 1700 $4^0/25$

(k) 3/4 1702 $4^0/25$

(l) 3/4 1705 $4^0/25$

(m) 3/4 1711 $4^0/25$

(n) Full 1742 $4^0/25$

Fig. 8.15 a – n 26 January 1964; Figures below the photographs indicate from left Gain of receiver, Time in IST, elevation of radar Antenna and the total range in n. miles.

(b) The last one (d) shows all the special features like
  (i) Forward over – hang towards left hand forward flank
  (ii) An echo – free notch or vault.
  (iii) A hook or a wall.

(c) Notch area and forward overhang could be seen with gain reduced to its half value ; refer photographs (c) and (e).

(d) In (f) the wall or the hook at 1651 IST has been seen to have turned forward towards east and filled up to notch area between the hook and the forward echo mass known to be an echo from the forward over – hang. At this moment the station recorded a rise in pressures of 0.3 hPa.(1650 to 1652 IST). This could be due to microburst.

(e) Hook turning cyclonically forward towards notch could be seen in photo (g) at 1653 IST. Fujita (1958) observed that radius of the curvature of the hook increases with time and then turns cyclonically into miniature eye at the center of the tornado cyclone. This appeared as a hole in the reduced gain picture. In (h) the hole could be seen at 1655 IST. Sferic data also confirmed that this time the number of sferics per minutes were maximum.

(f) A blunt hook or a winder notch at 1656 IST (photo (i) in fives minutes of its being a narrow one suggests that hails might have been released during this time over Kamptee.

(g) Major cell started showing protuberances again in its left and right flanks from 1700 IST again. It kept on moving forward in 110^0 direction till 1711 IST (photo j to m). After that movement could not be made out on the radar scope.

(h) At 1742 IST (photo n) it transformed into bright line types echo. The sferic data showed that their number increased from 100 to 140 from 1700 to 1736 IST and then went on decreasing to 40 only till 1845 IST.

The Kamptee storm was rotational which was moving towards 090 azimuth up to 1649 IST with a speed of 25 KT. As the storm changed the direction of movement between 1649 to 1656 IST forward 110^0 direction its speed slowed down to 17 kt. This may be because at the time of rotation the storm strengthened. The updraft, therefore, became very strong and offered retardation to the speed. Thereafter it again began moving with its original speed of 25 kt. It is quite likely that the energy of the updraft is distributed partly in generating new cells and partly in maintaining the main storm cell. Under such circumstances major cell would lose the vertical momentum temporarily. But after sometimes, it overcomes the retardation offered by the updraft and starts moving again with the original speed. The new echo appearing at 1702 IST at 360^0/ 13 n. m. and the crescent shaped echo (photo (k) to (m)) separating itself from the main echo mass (shown by arrow) had been moving towards 065 degrees azimuth at a speed of 27 kt. Their direction of movement deviated by 45^0 from the main storm cell. This is in conformity with the movement of rotational echo described by Fujita (1965). According to him whenever the hook echo becomes rotational the direction changes abruptly. This intensifies the storms and produces heavy rain and hail; and on left flank non – rotational echo path diverges from the main echo. The crescent shaped echo which diverged fromed the main storm cell by 45^0, which therefore, is non rotational echo. The storm cell is the rotational echo.

References

1. Anand N.K. (Flt Lt), (1987), A study of rare occurrences of hailstorm over Maharashtra during 26-27 December 1986, JFC thesis, Air Forces Administrative college, Coimbatore.
2. Chowdhary A. Banerjee A.K. & Sharma B.D. (1973), Hailstorm in central India, Met monograph, Clim/No.1973.
3. Fawbush E.J & Miller R.C., (1953), Bull. Am Met Soc. 34 pp 139 – 145.
4. Fujita (1958), J. Met 15, pp 288 – 296.
5. Fujita (1965), Mon, Weath. Rev, 93, 2, pp 67 – 78.
6. Swaminathan D.R. (1962), on a destructives Hailstorm on Gormi Area of Bhind district on 30 December 1961. IJMG, Vol. 13, No. 4.
7. Sharma B.L., (1967), Hailstorm near Nagpur IJMG, Vol. 18, No. 2.
8. Tabib V.D. (Flt Lt), (1977), Construction of mean model of tephigram for hailstorm over North and west Central Inda and comparison of same with normal tephigrams as an aid for forecasting hailstorm, J 11/4, JFC thesis, Air Force Administrative College Coimbatore.

CHAPTER 9

Hailstorm over South Peninsular India

Hailstorm frequency is much less over south peninsula than the north India which includes, Andhra Pradesh, Karnataka, Tamilnadu and Kerala. Many cases also go unreported. One such unreported case of hailstorm over Chennai is mentioned hereunder.

"Chennai was in for a mid-afternoon surprise on Thursday 27 September 2007 when a sudden hailstorm, combined with sleet, descended on certain parts of the city. The hailstorm, lasting a few minutes, was witnessed by people in Anna Salai, Cathedral Road and Kodambakkam High Road and in Triplicane and Mylapore". Falling hailstones could be seen in the photograph in Fig. 9.1. However, the Meteorological department said the observatory in Nungambakkam did not record a hailstorm. "We have recorded only 7.5 mm rainfall, between 2-30 p.m. and 5-30 p.m.", was the claim made by Area Cyclone Warning Centre.

Fig. 9.1 Photograph of Falling hailstones over Chennai on 27 September 2007.

They rarely occur in January or February. Tables 9.1 shows the stations with hail fall distribution in this region.

Table 9.1 Mean frequency of days with hail storm over various areas in South Peninsula

Name of the dist	No. of stations	JAN	FEB	MAR	APR	MAY	JUN	JUL	AUG	SEP	OCT	NOV	DEC	YEAR
Auranagabad	38	3	3	3	-	3	-	-	-	-	-	-	-	12
Nizamabad	38	-	-	-	3	-	-	-	-	-	-	-	-	3
Gulbarga	38	-	-	3	3	3	-	-	-	-	-	-	-	9
Raichur	38	-	3	-	-	-	-	-	-	--	-	-	-	3
Hanamkonda	38	-	-	-	-	-	3	-	-	-		-	-	3
Chitaldrug	38	-	3	-	8	8	3	-	-	-	-	-	3	25
Bangalore	38	--	-	3	16	16	-	-	-	-	3	-	-	38
Mysore	38	-	-	5	8	13	5	-	-	-	-	-	-	31
Mangalore	38	-	-	3	13	16	-	3	-	8	5	-	-	48
coonin	38	5		-	5	8	8	-	3	-	3	3	-	35
Calicut	38	-	-	-	-	-	3	-	-	-	-	-	-	3
Thiruvanandh a-puram	38	3	-	-	-	3	3	-	-	-	-	-	--	9
Pamban	38	-	-	-	-	-	-	-	-	--	-	-	-	-
Madura	38		3	-	-	-	-	-	-	-	--	-	--	3
Nagapatnam	38	-	-	-	-	3	-	-	-	-	-	-	-	3

Table 9.1 *Contd...*

Name of the dist	No. of stations	JAN	FEB	MAR	APR	MAY	JUN	JUL	AUG	SEP	OCT	NOV	DEC	YEAR
Trichinopoly	38	--	-	-	-	5	3	-	-	-	-	-	-	8
Coimbatore	38	-	-	3	-	-	-	-	-	-	-	-	-	3
Salem	38	-	-	-	8	10	-	-	-	-	-	-	--	18
Vellore	35-38	-	-	-	-	3	-	-	-	-	-	-	-	3
Cuddapah	38	--	-	-	-	3	-	--	-	-	-	-	-	3
Bellary	38	-	-	5	-	-	3	-	-	-	-	-	-	8
Kurnool	38	-	-	-	-	-	-	-	-	-	-	-	--	-
Nellore	38	-	-	-	-	3	-	-	-	--	-	-	-	3
Masulipatanam	38	-	-	-	3	3	-	-	--	-	-	--	-	6
Cacanada	38	-	-	3	-	8	3	-	-	-	-	-	-	14
Gopalpur	38	-	-	3	5	3	3	-	-	-	-	-	-	14
Kodaikanal	36-37	-	9	33	164	147	22	3	-	8	8	3	-	305

(The figures represent number of occasions in 100 years)

Although hailstorm mainly occur over south peninsula in the coastal or interior Karnataka in the month of April or May, hailstorms have been reported (Lakshminarayana 1988) as south as Trivandrum during May.

Heavy hail shower, however, has been once reported to have lashed 8 villages in Mundargi taluk and two villages of Gadag taluk of Dharwad district killing 11 people and perishing about 400 heads of cattle, sheep and foul etc., perished. The hailswath was reported to be less than a km wide and 6 to 8 km in length. The crops were completely damaged over an area of 6000 acres. Hailstorm reports causing heavy losses of life and property in the region in the months of March to May are innumerable. One such widespread hailstorm activity over Telangana and neighborhood was reported between 11 and 13 March 1981. Due to this over 470 villages of 56 taluks in about 13 districts of Andhra were affected. It caused severe damage to standing crops in about 33,000 acres besides damaging 85,000 houses. It had taken a toll of human lives and about 13,000 livestock.

9.1 Favourable Conditions for Hailstorm Occurrence

Upper air trough in subtropical jet is apparently significant feature to support convective hailstorm activity over south peninsula if there exists, a low level circulation over the region. A few synoptic situations which caused hailstorm south of 18^0 N Lat are being mentioned hereunder.

(a) A Well Marked North – South Trough On Surfaces Embedded With Upper Air Cyclonic Circulations Over South Peninsula.

Hailstorm between 11-13 March 1981 over Telengana was due to the North – south trough (Pandharinath & Bhavnarayana – 1990). This trough was the result of induced low pressure area over northeast Madhya Pradesh on 11 March 1981, which moved eastwards from Rajasthan. A well marked trough could be marked from northeast Madhya Pradesh to Kerala which was embedded with upper air cyclonic circulation over Karnataka and Andhra Pradesh and another over Kerala. See Fig. 9.2.

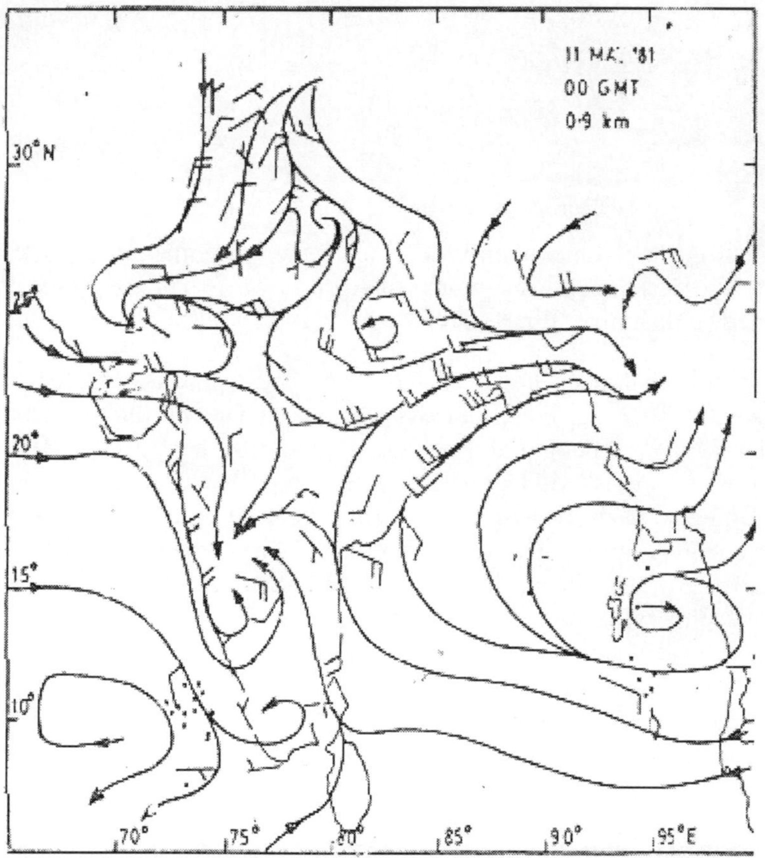

Fig. 9.2 Upper air winds at 0.9Km a.m.s.l. on 11 March 1981 to indicate low level convergence

Contrasting air mass feature could be seen in the cyclonic circulation over Karnataka and Andhra Pradesh. Airmass from northwest India is meeting the moist airmass from Bay of Bengal and Arabian Sea. An upper air divergence is apparent, in figure. 9.3 over Telengana region, at 200 hPa level.

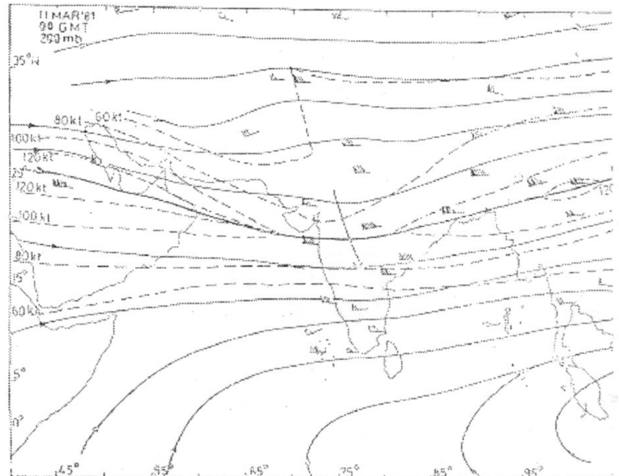

Fig. 9.3 Upper winds at 200hPa on 11 March 1981 indicating trough line in jetstream and location of axis in the jet core break.

It was observed that there was a break in the axis of jet core of 100 kt or more over Telengana and neighbourhood. The axis of trough in the zonal westerlines of 80 kt or more at the same level was passing through the jet core break. Eastward movement of the trough could be seen in Fig. 9.4 on 14th Mar 1981 when the hailstorm activity stopped.

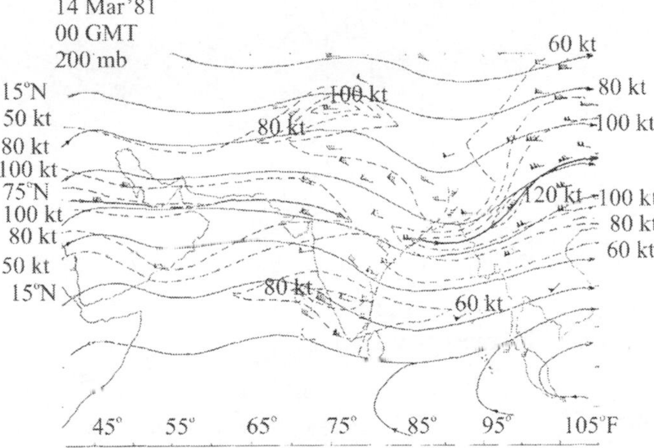

Fig. 9.4 Upper air analysis at 200hPa on 14 March 1981 indicating eastward movement of trough in jetstream.

(b) Shallow Trough of Low Pressure Over Peninsula in January or February.

Such a trough in winter season is likely to cause influx of moisture from Arabian Sea or Bay of Bengal causing confluence of air mass from north India with the moist oceanic air. Venkateshwaran (1967) studied case of hailstorm over Dharwar district on 30th Jan. 1966. Abnormally deep trough in the subtropical jet developed along 73^{0}E long on 30th Jan 1966 (00 UTC). See Fig.s 9.5 to 9.8. Weather showed rapid improvement on 31st January with the eastward movement of the trough in subtropical jet stream.

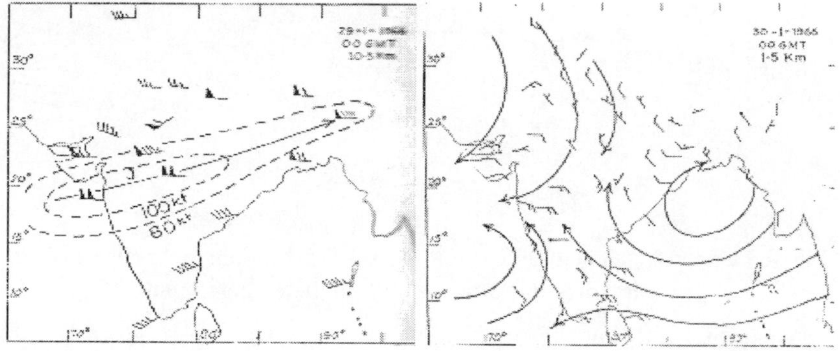

Fig. 9.5 29 Jan 1966(00UTC),10.5 Km

Fig. 9.6 30 Jan 1966(00UTC), 1.5 Km

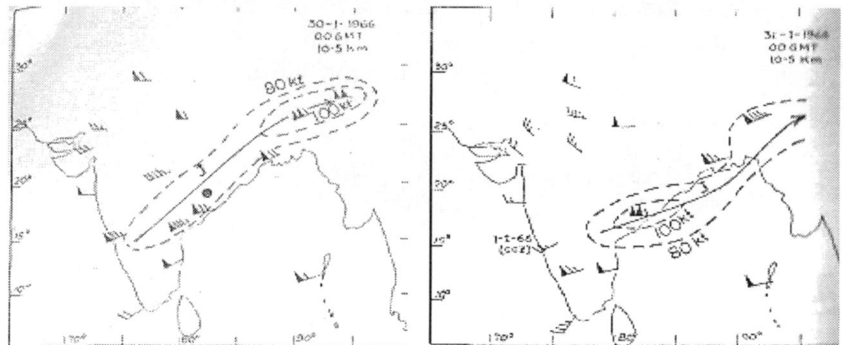

Fig. 9.7 30 Jan 1966(00UTC), 10.5 Km

Fig. 9.8 31 Jan 1966(00UTC),10.5Km

(c) Low Level Cyclonic Circulation Develops Over ITCZ and Moves Northwards or Northwestwards In May.

Lakshminaranya (1988) has mentioned a rare hailstorm over Trivandrum on 17th May 1984. Trivandrum city and airport experienced heavy thunderstorm accompanied by hail between 1320 and 1540 ist on 17th May. Though the size of the hailstones were only 1 cm they fell for a period of twenty minutes from 1420 to 1440 IST, both at airport and in the city – which are 8 km apart. There was a squall from south easterly direction. Maximum speed of the squall was recorded 53 kmph. Neyyattinkara town and its neighbourhood (20 km southeast of Trivandrum) also experienced hailstorm in the afternoon of 17th May 1984 with damages to houses and crops. Alleppy (120 km NNW of Trivandrum) also experienced hailstorm at 1400 IST. On the evening and night of 18th may 1984 also there were reports of hailstorm in several areas of Cannanore district. One day prior to occurrences a lower troposphere cyclonic circulation developed over Sri Lanka and adjoining Comorin area at 850 hPa extending up to 700 hPa. The trough extended up to west central Bay of Bengal. It subsequently moved westward across Comorin, Maldives and Lakshadweep region. Upper air divergences was provided by the trough in easterlies at 300 hPa between central bay of Bengal and Sri Lanka. See Figure 9.9 (a to j) and Fig. 9.10 (a to d). Convective activity stopped with the passage of trough in easterlies at 300 hPa on 18 May 1984 (1200UTC)-Fig. 9.10(d).

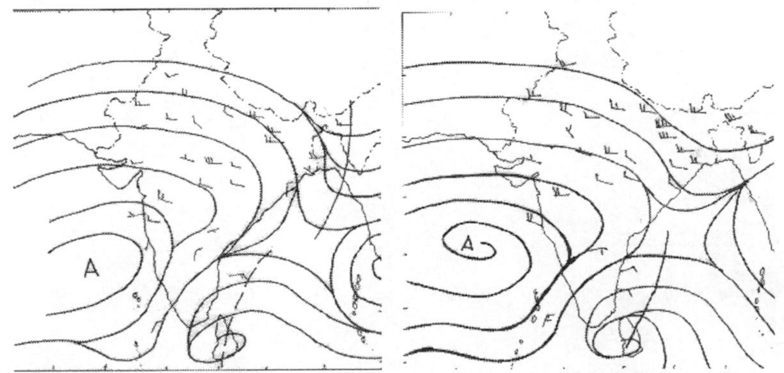

Fig 9.9(a) 16May 84(12UTC),850hPa **Fig 9.9(b)** 16May 84(12UTC),700hPa

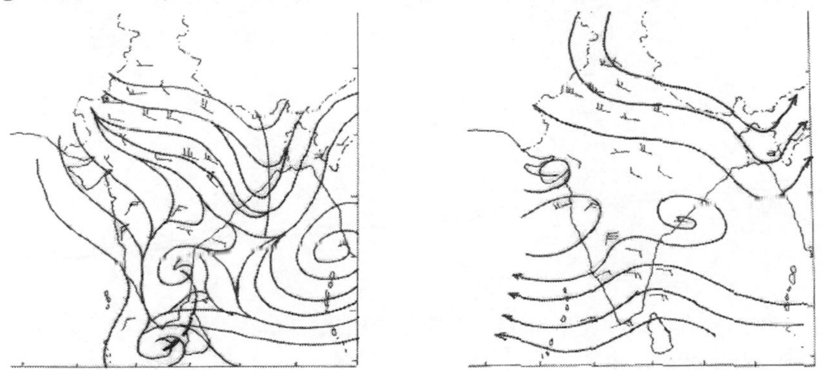

Fig. 9.9(c) 17May 84(00UTC),850hPa **Fig. 9.9(d))** 17May 84(00UTC),700hPa

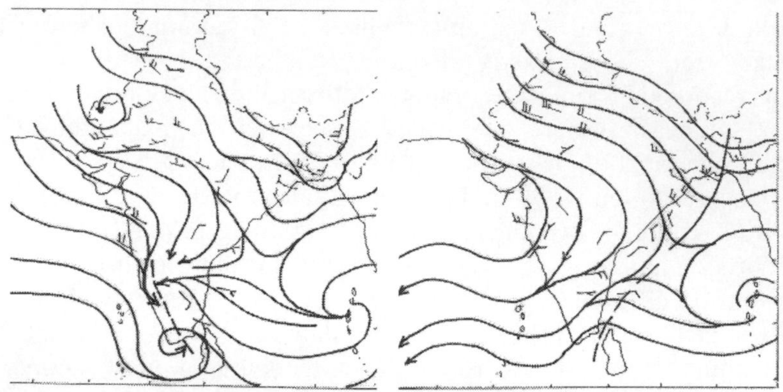

Fig 9.9(e) 17May 84(12UTC),850hPa **Fig 9.9(f)** 17May 84(12UTC),700hPa

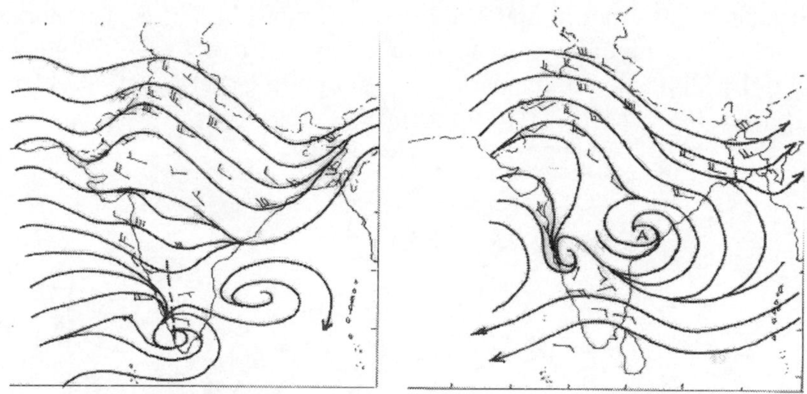

Fig. 9.9(g) 18May 84(00UTC),850hPa **Fig. 9.9(h)** 18May 84(00UTC),700hPa

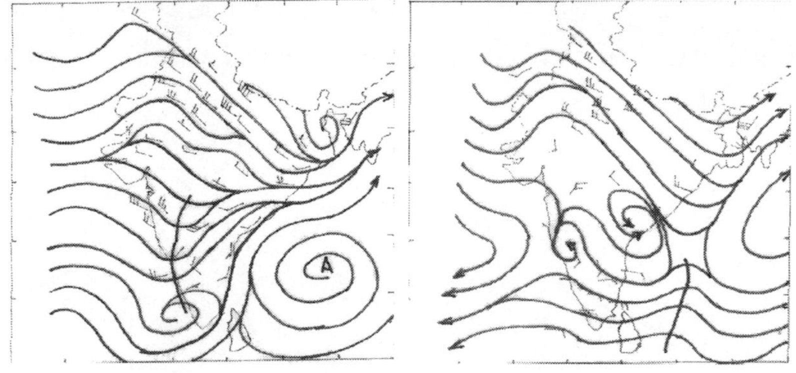

Fig. 9.9(i) 18May 84(12UTC),850hPa **Fig. 9.9(j);** 18May 84(12UTC),700hPa

Fig. 9.9(a – j)

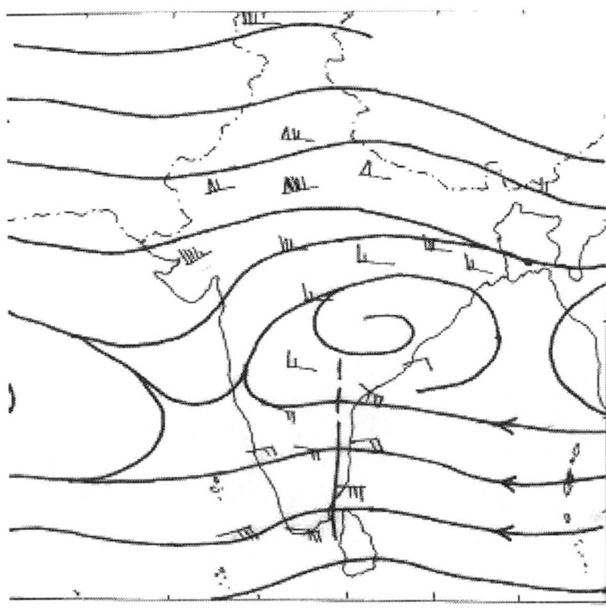

Fig. 9.10(a) 17 May 84(00UTC) 300 hPa

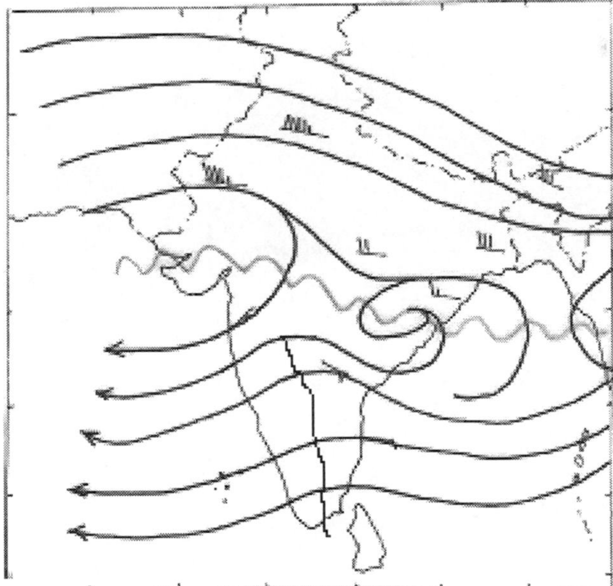

Fig. 9.10(b) 17 May 84(12UTC) 300 hPa

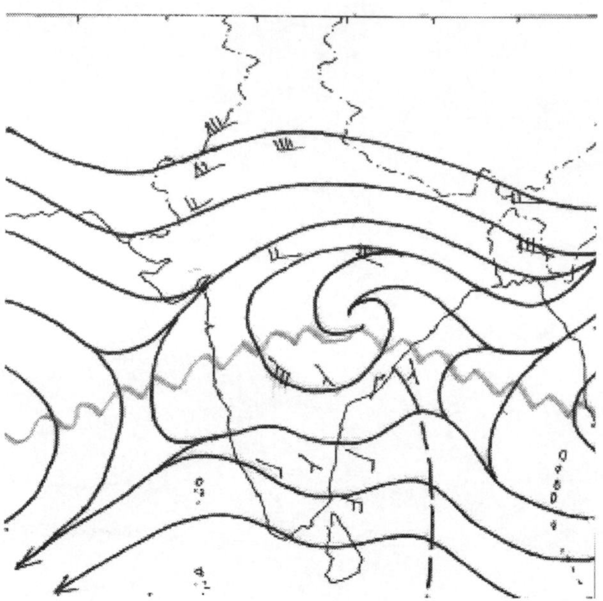

Fig. 9.10(c) 18 May 84(00UTC) 300 hPa.

Fig. 9.10(d) 18 May 84(12UTC) 300 hPa.

Fig. 9.10(a – d)

9.2 Vertical Wind Shear

Windshear for hailstorm for south peninsula presents different picture for the hailstorm occurring in westerly or easterly regimes. While strong vertical wind shear has been reported for hailstorms occurring in westerlies they are not significant in case of those forming in easterlies over Trivandrum. Table 9.2 below shows windshear for the three cases of hailstorm mentioned above. Very low value of wind shear of 22 kt in easterlies may be noted as against relatively much higher values (≈ 70kt or more) in the westerlies.

Table 9.2 Wind shear between levels 850 – 200 hPa in kt

Wind Regime→	Westerly				Easterly
Place (Month), Year→	Telengana (March), 1981			Dharwar (Jan.), 1966	Trivandrum (May) , 1984
Dates→	11	12	13	30	17
Wind Shear between 850 – 200 hPa	82	71	97	75	22

It can be, therefore, said that though strong vertical wind shear is a favorable feature in the hailstorm it is not the necessary condition for its occurrences over south peninsula.

9.3 Humidity Parameter

Contrasting dew point is apparent in the surface chart analysis of 12th March 1981 (Fig. 9.11) when hailstorm occurred over Telengana. Air mass contrast can easily be inferred on seeing the tongue of high dew point jutting into eastern and central parts of Andhra Pradesh. The difference is of the order of 5^0C or more in two different air masses. A rise of dew point of 4 to 12^0C was reported over in 24 hours. On 30th January 1966 when hailstorm was reported over Dharward dew point temperature over interior Mysore and Madhya Maharashtra rose markedly. Dharward and its neighborhood had indicated a rise of 7 to 10^0C over the values on the previous day indicating the encroachment of moist air mass over the region. A contrast of 5^0C or more may in general be taken as contrasting humidity airmasses and confluence zones may be considered to be favorable for the cyclogenesis to help convection in the region. Also a rise of more than 5^0C in dew point temperature within 24 hours, at a station is also favorable for the occurrence of hailstorm.

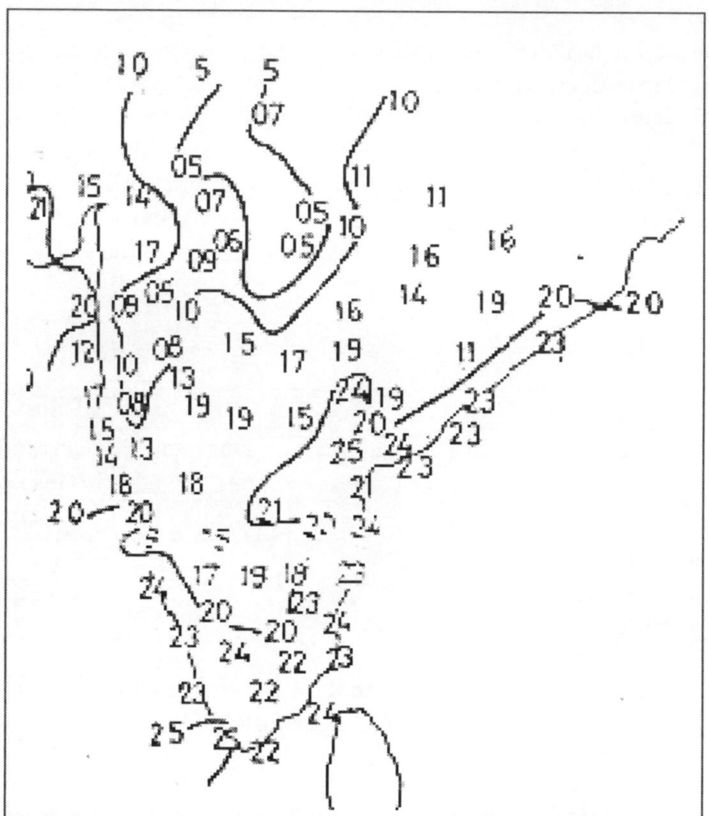

Fig. 9.11 Surface Dew Point Temperature Chart of 12 March 1981(03 UTC). Narrow tongue of high dew points jutting into eastern and central parts of Andhra Pradesh may be observed.

00UTC Tephigram of 17th May 1984 (Fig. 9.12) of Trivandrum shows unique feature of highly moist lowest layer between surface to 850 hPa, and then rapid depletion of moisture between 850 to 700 hPa. The layer between 700 hPa and 400 hPa shows almost uniform dew point depression. Comparatively much drier upper layer, therefore, seems to have served as an evaporation space of the precipitation droplets causing cooling of the middle layer. Though examination of 00 UTC Tephigram of Trivandrum showed insignificant variation in freezing level (only by 20 hPa) it may not give the correct picture of the middle layer at the time of hail fall (0850 UTC to 0910 UTC). Since the size of the hail were quite small ($\leq$ 1 cm) it may be understood that cooling induced by evaporating in the middle layer was temporarily sufficient to produce small size of hailstones only. George K – index in 00 UTC Tephigram also shows quite lower value of only 24.5. Understandably the depth of the moist layer might have extended to higher levels to increase buoyancy at the middle layer between 700 and 500 hPa.

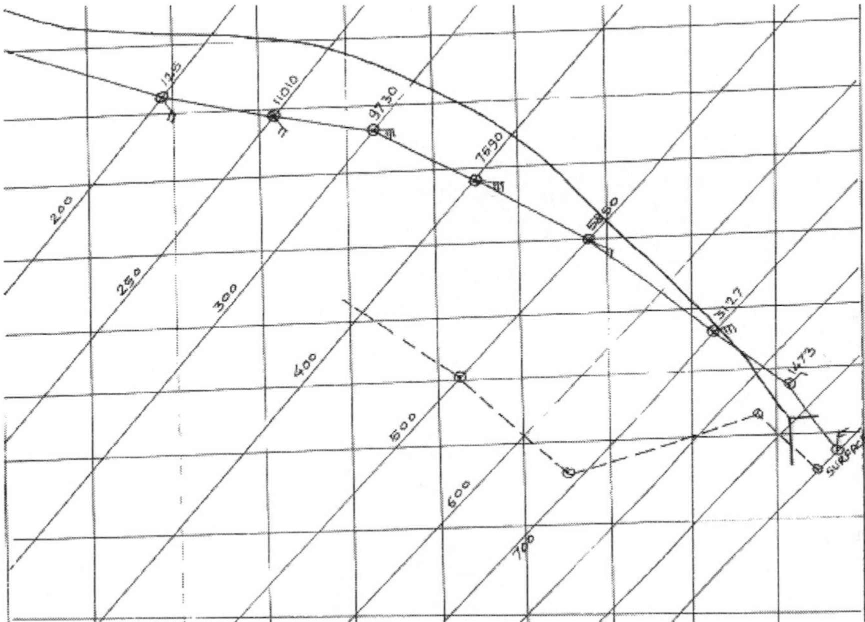

Fig. 9.12 00 UTC Tephigram of 17th May 1984 of Trivandrum.

In case of Telengana, hailstorms occurred between 1030 UTC to 1230 UTC and in the 12 UTC sounding George K – Index showed high value from 33.9 to 37.4. See Table 9.3.

Table 9.3 George K – Index (9 – 14 March 1981); Hailstorm occurred over Telangana from 11 to 13 March 1981

Date	09		10		11		12		13		14	
Time in UTC	00	12	00	12	00	12	00	12	00	12	00	12
K-Value	35.7	34.4	30.1	37.4	29.6	34.2	30.5	33.9	32.1	34.9	29.2	30.9

Although high static instability could be seen on 9 and 10 March 81 also (K > 35) but the corresponding convective build – up could not be realized due to the absence of any upper air divergence feature.

References

1. Lakshminaryana, R. (1988), A rare occurrence of hailstorm over Trivandrum and neighbourhood on 17th May 1984, Vayumandal Vol. 18, no. 3 & 4.
2. Pandharinath N. and Bhavanarayana V. (1990), Hailstorm over Telengana - IJMG – Vol. 41. No. 3.
3. Venkateswaran V. (1967), IJMG, Vol. 18. No. 1.

CHAPTER 10

Hailstorm Modification and Control

Protection of crop from hail damage was in the mind of the researchers as early as the end of 19th century. In1896 Albert Stinger tried out hail cannon to protect the orchards in Styria, Austria. He used gun which comprised of 3 cm mortar with the funnel of steam locomotives mounted above it. The gadget used to make deafening noise on firing and it used to produce ring of smoke which rose in the air with singing noise. Its practical application led to repeated accident. In 1902 the Austro – Hungarian government organized an international conference to evaluate this technique. Later on after scientifically supervised experiments, in Austria and Italy, the method was declared totally useless. Thereafter, there was just no development in hail suppression effort till the end of Second World War when a French General F. L. Rubey developed an inexpensive small rocket of limited altitude range. It carried explosives which a farmer could fire at any threatening cloud. This although, gave psychological relief to the farmers using it, but its actual effect in reducing the hail fall was totally nil. Explosive rockets were also tried extensively by Swiss meteorologists during 1948 to 1952 without any success. All these attempts towards hail suppression were made without any scientific basis of the growth of hailstones in the cumulus cloud.

Hail gun was also developed which transmits shock waves to go vertically up in the air every five seconds. It was claimed that this would smash the hails in the cloud which would eventually fall down on the ground in the form of rain or wet snow. The claim was proved wrong after its installation at several places in India (Gulati 2011).

Scientific weather modification had its beginning in 1946 at General Electric (GE) Laboratory in Schenectady, New York, U.S.A. Irving Langmuir, the leader of the new development was known as USA's one of the foremost Nobel Laureates. Actually he and his laboratory assistant, Vincant J Schaefer discovered that dry ice acts as a good nucleating agent for ice crystals in atmospheres. In the same year Bernard Vonnegut discovered that silver iodide and lead iodide also seemed to possess similar lattice structure as that of ice and they were insoluble in water. Hence they could also act as good nucleating agents for ice. Thus was made, the beginning in the field of weather modification and control.

Inchoate attempts of hail suppression by cloud seeding were made initially by some with optimistic results (MacCready, 1958; Iribarne and dePena, 1962) but some challenged the exercise as economically impractical. Weickmann, (1964), and Schleusener, (1968) stated that seeding-hypothesis has no effect on hail fall and it should be rejected.

As per WMO report (2013) several countries e.g. Argentina, Austria, Bosnia and Herzegovina, Bulgaria, Burkina Faso, Canada, China, Germany, Greece, Macedonia, Romania, Russia, Serbia, Spain and Uzbekistan are engaged in active hail suppression programmes.

In the present chapter a brief discussion on the scientific concept and its implementation on Hail-Control is described in section 10.1. Sections 10.2-10.4 cover the science and technology of atmospheric seeding. Causes of ambiguity in the success during hail mitigation experiments, world over, have been analyzed in section 10.5. Project on Hail-Control under National Initiative on Climate Resilient Agriculture (NICRA), sponsored by Indian Council of Agricultural Research (ICAR) is discussed in 10.6. Future of hail mitigation programmes is presented in 10.7.

10.1 Sulakvelidze Hypothesis of Hail Control

Sulakvelidze (1969) hypothesised that by artificially seeding the hail bearing cloud with the several Cloud Condensation Nuclei (CCN) and

Icing Nuclei (IN) aerosols, the condensed droplets or ice crystals increase within the cloud. They all compete to collect the available water vapour and grow larger. As a result cloud water is distributed in to several small ice crystals or small hails above 0°C isotherm level within the cloud. These ice crystals or small hails would melt into water during their travel below zero degree isotherm and turn into rain or drizzle.

This hypothesis was first applied by Georgia (part of erstwhile Soviet Union) and subsequently laid strong scientific foundation for the control of hailstorm world over. Between mid-1970s and beginning of 1980s two larger experiments were undertaken to evaluate the effectiveness of cloud seeding in Western Europe(Switzerland, France and Germany) known as The Great Experiment (Federeret al, 1986) and another in U.S.A. known as National Hail Research Experiment(Knight et al, 1979).The results showed that statistically there was no significant difference in the occurrence of hail between seeded and not seeded hail bearing cloud. Albeit uncertainty on the effectiveness of hail control by cloud seeding prevailed and World Meteorological Organisation in 2007 decided not to recommend Hail Suppression any more, still several countries continued with their Hail Suppression programmes (WMO 2013).

10.2 Seeding in Atmosphere

Discovery of nucleating agents in laboratory opened scope for its vast application in atmospheric science particularly in weather modification. New questions were borne *e.g.* which are most suitable aerosols which may turn into cloud condensation nuclei? Are they some inorganic or organic compounds; what could be the optimum size of nucleating particles and their appropriate quantity in specific air mass? Particularly because each air mass could carry its own aerosol per volume characteristics based on its place of origin and resident time over the region. Moreover, seeding agents above freezing level and below freezing level have different microphysics related to growth of droplets hence for artificial rain making two different types of seeding agents are used. Above the freezing level, seeding the cloud volume, is known as glaciogenic seeding technique and below the freezing level it is called hygroscopic seeding technique. The terms "cold rain" (developing through a process that involves ice particles) and "warm rain" (developing through a process that is entirely liquid) are often used to describe the two primary cloud microphysical processes that produce precipitation, naturally. Agents used in cold rain are known as Ice Nucleating (IN) agents. The most common agents used in "cold rain" or snowpack augmentation cloud seeding projects are silver iodide and dry ice (frozen carbon dioxide). Other agents like Cupric Sulfide, Kaolinite (Clay mineral), liquefied gases (liquid nitrogen for example) can also be used to seed clouds. Naturally occurring bacteria found over decayed leaves also acts as IN agents at $< -4\ ^{O}C$ temperature. Albeit their effective contribution to atmospheric processes, i.e. their potential to trigger glaciations and precipitation, remains uncertain (Thomas, 2014). Dry ice was later replaced by silver iodide. The flares used for cloud seeding usually contain silver iodide or a salt. Organic compound Carbamide ($CO(NH_2)_2$) or Urea can also induce ice nucleation at cloud temperatures ranging from $+6^{o}C$ to below $-15\ ^{o}C$. Urea is also hygroscopic and soluble in water (78gm per 100 ml at 5 ^{o}C). The process of dissolving one gram of urea in water is accompanied by the absorption of 60.5 calories of heat. Ice nucleating properties of urea depends upon its endothermic heat of solution and high solubility. Hence if milled to a fine size it can act as good nucleating agent in warm cloud, too.

Effective Cold Cloud Ice Nucleating Agents: Silver iodide and lead iodide are most effective nucleus for the crystallization of cooled droplets. AgI as the nucleating agent is well known but the properties of AgI particles depend not only on the generating method but also on the generating environment. AgI is quite sensitive to light. Photons decompose it into silver and iodine. The produced iodine would evaporate and only 'Ag' would remain. AgI-NH_4I aerosols are more photostable than AgI – NaI. This is because surface of AgI particles in former mixture are coated with NH_4I where in later mixture coating is provided by AgI itself. AgI is hydrophobic substances. Its solvability in water is approximately 2 parts per trillion ($2 \times 10^{-9} g\ ml^{-1}$). Solution of AgI with NH_4I or NaI makes it, however, little hygroscopic. But hygroscopicity with NH_4I is found to be more than with NaI. Experiments have also been conducted with lead iodide aerosol. The results agree closely, but lower side, with those for silver iodide. Hydrophobicity of PbI_2 – NaI aerosol is relatively more than those of AgI based aerosol due to larger size of the former. Costs of PbI_2, is however much less than AgI. The reagents are dispersed in the cloud by subliming in the flame of acetone or an electric arc or special pyrotechnical composition. It is also dispersed by explosion. The crystallization temperature threshold of AgI for explosives sublimates is $-6^{o}C$ and for PbI_2 is $-8^{o}C$.

Effective Warm Cloud Condensation Nucleating agents: Warm cloud seeding agents can be produced by hygroscopic flares. These flares contain salts (*e.g.*, ammonium sulphate $(NH_4)_2SO_4$) or sodium chloride (NaCl)).When the flares are burned they produce minute particles of the salt which attract water vapour forming cloud droplets in addition to those already present in the cloud. Hygroscopic seeding can also be done by spraying dry powder of common salt (NaCl) or ammonium sulphate $(NH_4)_2SO_4$) or Urea ($CO(NH_2)_2$) that has been milled to optimum sizes in powder form for the promotion of droplet growth in the clouds.

Physical-chemical processes involved in the activation of particles into CCN are governed by the droplet curvature and solute effects on the water vapour pressure. Theory was developed in the thirties by the Swedish meteorologist HildingKöhler (1936). In those days it was believed that water vapour condensates on CCN are made up only by inorganic material from sea-spray or mineral dust from land. Köhler theory in its original form was inadequate for describing the condensational growth of nano-sized droplets containing organic surfactants. Surfactants are compounds that lower the surface tension (or interfacial tension) between two liquids or between a liquid and a solid. Earlier simplified forms of Köhler theory included completely insoluble and non-hygroscopic species and completely soluble fraction, usually ammonium sulfate ($(NH_4)_2SO_4$) or sodium chloride (NaCl). Problem arises when water soluble organic carbon (WSOC) is present in the atmosphere originating from either biomass burning or Secondary Organic Aerosols (SOA) formation e.g., *cis*-pinonic acid and non-biomass burning (Petters and Kreidenwe is, 2007).As an improvisation to Köhler theory, Nano-Köhler theory was given by Kulmala et al., 2004 to include organic and inorganic compounds in the clusters in a single aqueous solution. κ-Köhler theory (Petters and Kreidenweis, 2007) included solute hygroscopicity in the Köhler equation. Later adsorption activation theory was developed to include adsorption of water on wettable surfaces which can dominate over hygroscopic uptake replacing Köhler theory (Ovadnevaite et al., 2011; Kumar et al., 2011). Nevertheless despite several modifications of Köhler theory from the point of view of weather modification the basic premise is well accepted that large size cloud condensation nuclei LCCN($\approx$5–10 μm) or giant size cloud condensation nuclei GCCN($\geq$ 10 μm) with high hygroscopity act as a good seeding agent. Hence practically for the purpose of artificial rain making, while milling the seeding agent into powder, for direct spray, or for making the pyrotechnic cartridges for smoke generation, we can proceed with this basic premise.

10.3 Seeding Strategy

10.3.1 Sandwich Seeding

Simultaneous seeding of warm cloud (Temperature > $0^{O}C$) with urea below and common salt above the cloud is known as sandwich seeding. The two seeding aircrafts' horizontal fly paths make 45° angle as shown in the Fig.10.1. Lower aircraft flies in the direction of the cloud motion.

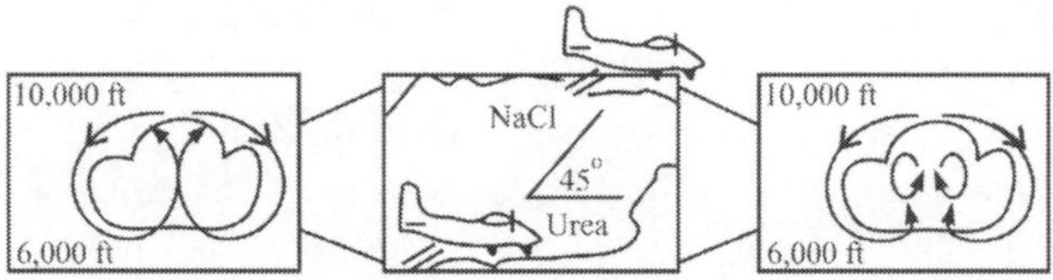

Fig. 10.1 Sandwich seeding for warm cloud

Unseeded growing cumulus is shown in left in Fig. 10.1. Urea seeding (Centre) is below the cloud hence it would immediately enter from the base. After NaCl seeding from the top over a growing cumulus the compensating vertical currents would gradually lead the NaClparticles to reach base of the cloud and enter the cloud from its base.

10.3.2 Super Sandwich Seeding

Seeding of cold cloud (Temperature < $0^{0}C$) is known as glaciations. Silver iodide acts as the best glaciation seeding material. When cold cloud seeding operation is combined with sandwich seeding then it is termed as super sandwich seeding; as shown in the Fig. 10.2.

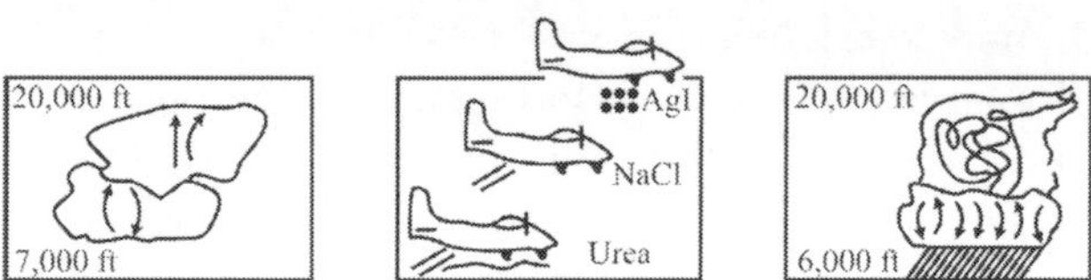

Fig. 10.2 Super Sandwich seeding for cold cloud

Unseeded cloud with base at 7000ft and top reaching up to 20000ft is shown on the left of fig.10.2. Urea seeding is done near the base. NaCl seeding in this case would need entering the cloud from the periphery, so as to avoid the strong updrafts for flight safety. AgI seeding on top could be done either by dropping pyrotechnic cartridges from above the cloud or by entering the cloud from the periphery for wing mounted cartridges. Enhanced turbulence and rain, caused due to seeding is shown in right of Fig.10.2.

10.4 Seeding Technology

Several methods have been adopted to deliver seeding nuclei into cloud. They may be categorized as under:

(i) Ground based dispersion.
(ii) Base dispersion by aircraft.
(iii) Lateral delivery system by firing rockets from aircraft.
(iv) Top-down delivery system by dropping pyrotechnic cartridges from aircraft.
(v) Direct injection by firing rockets from ground from static location.
(vi) Direct injection by firing rocket, transported by the helicopter below the cloud and fired vertically from the ground.

10.4.1 Ground based Dispersion

Two ground based systems are shown in Fig.10.3. They are on a fixed location.

In this propane flame is used to vaporize the seeding solution, which is composed of silver iodide mixed in acetone. The vaporized silver iodide then re crystallizes in the cold air, forming millions of tiny particles which are intended to serve as ice nuclei.

This equipment allows Silver iodide flares to be ignited from remote locations, using customized software and cell phone communications. Each flare burns for approximately 5 minutes.

Fig. 10.3 Photos of ground-based cloud nucleating generators. Courtesy North American Consultants Inc.

Fumes generated on the ground are supposed to be transported into the cloud base by the updraft of the developing cumulus cloud over the region. The generators are positioned in such places so as to maximize the number of silver iodide crystals to reach the base regions of passing storms. The technique suffers with severe restriction of favorable wind and suitable cloud proximity for the maximum dispersion within a cloud.

10.4.2 Base Dispersion by Aircraft

Seeding by aircraft is safer in stratiform clouds but flying below a growing cumulus, near its base is not free from flying hazards as the aircraft may cut across several updraft regimes in its flight path. Refer Fig. 10.4.

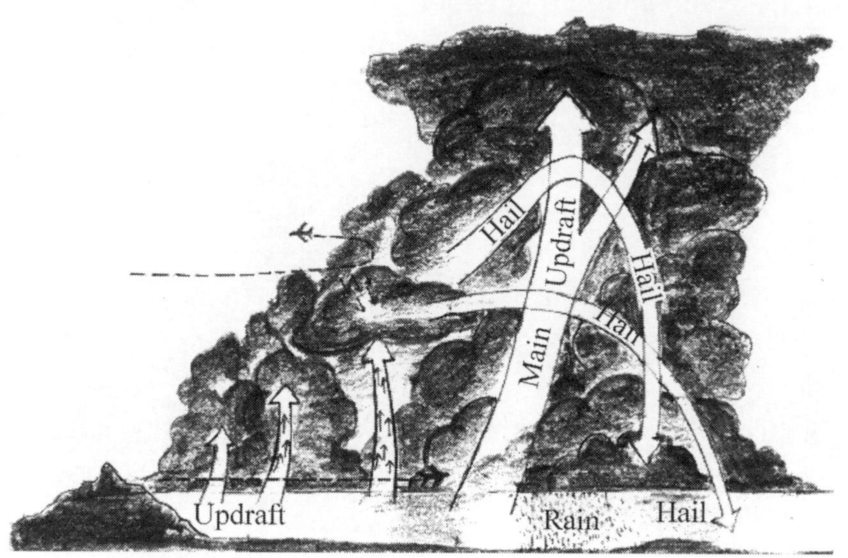

Fig. 10.4 Conceptual model of seeding multicells of growing cumulus. Notice the flight path near the base hazardously cutting across the updraft regimes. Top down seeding is also shown in the figure.

Beside severe flying hazard it is difficult to spot out the effective seeding region in a multicell storm. Seeding in entire region could be wasteful exercise as not all the cell might grow into hailstorm.

Seeding is carried out by Airborne Flame Generator. Airborne Flame Generator is designed to be mounted on the aircraft (Fig 10.5 (a), 10.5 (b) and 10.5(c)). Acetone is used as fuel and solvent for AgI. But since AgI is only slightly soluble in acetone, an iodine salt such as KI, NaI or NH_4I is usually added, in half part, to form a mixture with AgI. Mixture decreases the decomposition rate of AgI in sunlight. It also gives homogeneous solution. Ignited flame temperature is approximately 1000^0C.The exhaust plumes cool to ambient temperature in10 milliseconds. AgI – NH_4I and AgI – NaI are both active nuclei between – 8°C to –16°C; former being little more than the later.

Fig. 10.5 (a) Silver iodide smoke generator for use on aircraft.

Fig. 10.5 (b) Cessna aircraft for airborne flame generator.

AgI Flame generator mounted below the wings of the aircraft

NaCl Flame generator mounted below the wings of the aircraft

Fig. 10.5(c) Flame generators mounted on the wings of the aircraft.

10.4.3 Lateral Delivery System by Firing Rockets from Aircraft

Lateral seeding method needs horizontal firing of rockets carrying Cloud Condensation Nuclei (CCN) or Icing Nuclei (IN) into the cloud from side. Aircraft mounted rocket launching machine is shown in fig. 10.6. This approach not only endangers other flights in the region but also high speed of rocket permits only short interval for dispersing the seeding agents into the cloud.

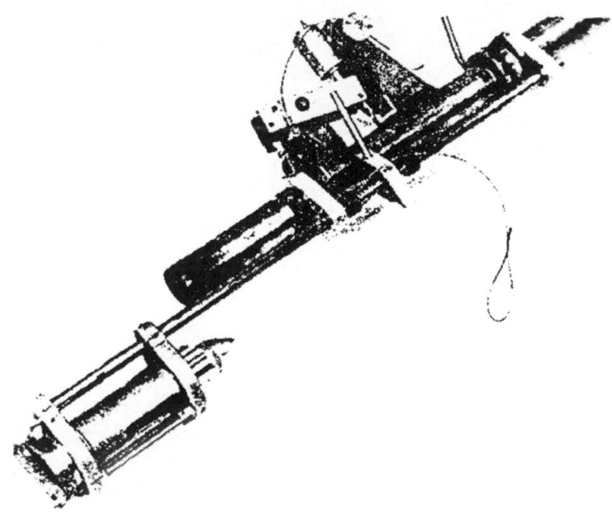

Fig. 10.6 Aircraft-mounted rocket launching mechanism.

10.4.4 Top-down Delivery System by Dropping Pyrotechnic Cartridges from Aircraft

In Top-down delivery system the pyrotechnic cartridges are carried up to the top of the cloud by aircraft and are dropped. Refer Fig. 10.4. They fall by gravity and catch fire. Fumes disperse within the cloud by the in-situ updraft of cumulus cloud. This method is relatively safer and more effective than previous three but this approach also suffers with difficulty of locating the correct spot to drop the cartridges because radar reflectivity may often give diffused information at the growing stage of the multicell cumulus cloud.

For various weather modification objectives different types of pyrotechnic cartridges (Weather Modification, 2009) are available with different chemical proportions of ingredients. They are briefly listed below:

(i) *AgI Ejectable Cloud Seeding Flares*: The flares are mounted on the racks present on the belly of the aircraft fuselage. Refer fig 10.7. Each rack holds 102 cartridges. When triggered, the pyrotechnic is ignited and ejected from the aircraft.

(ii) *AgI Burn-In-Place Cloud Seeding Flares*: This type of flare was designed and engineered to produce the proper size and concentration of Icing Nuclei (IN) while burning at its location e.g. aircraft wings; fig. 10.5(c). Burn-in-Place flare cartridges are available in sizes of 75,100,125, or 150 grams of pyrotechnics per flare.

(iii) *$CaCl_2$ Hygroscopic Burn-In-Place Cloud Seeding Flares*: This pyrotechnic flare is designed for warm clouds. The aerosols produced by the hygroscopic flare are of proper size (≈5mμ) to stimulate cloud to start the rain process quickly and for longer period.

Fig. 10.7 AgI cartridges in the racks mounted on the belly of the aircraft.

Chemical composition of ejectable cartridges (Krauss, et al 2009) is presented in Table 10.1.

Table 10.1 Chemical formulation of pre-existing Pyrotechnic Cartridge

Non perforated AgI cartridges with 2gm formulation	
Material	**Percent by Weight**
Silver iodate	45.0
Potassium perchlorate	27.0
Magnesium	20.0
Nitrocellulose	4.0
Triacetin	4.0

Cross section of the cartridge and igniter are shown in Fig. 10.8(a) and 10.8(b) respectively. Cartridge case is made up of PVC pipe with perforation on the walls of the case. This is wrapped by transparent paper to avoid the contact of atmospheric moisture with the pyrotechnic formulation. Transparent paper quickly burns along with the cartridge.

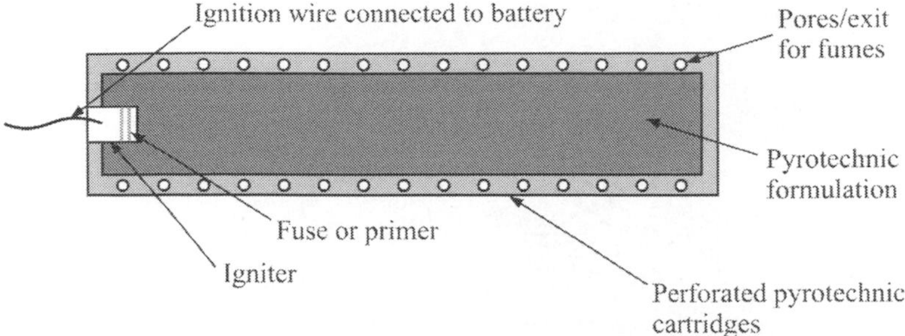

Fig. 10.8(a) Igniter Design of the pyrotechnic cartridge.

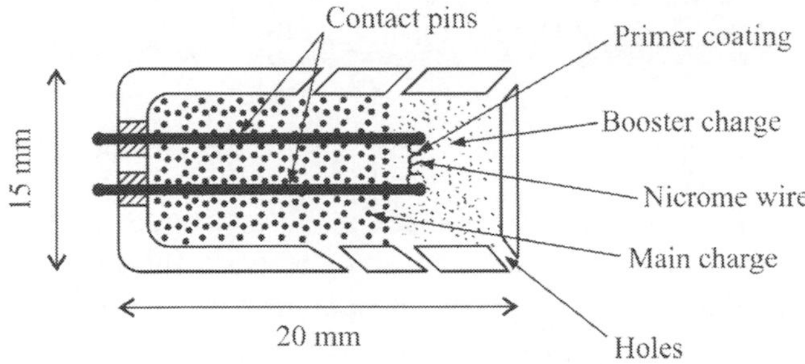

Fig.10.8(b) Cross section of Igniter.

The total weight of the igniter is 25 grams; out of which the main charge is 12 grams. The contact pins are inserted in the case and connected with a Nichrome wire (26 gauge) which is coated with mercuric fulminate primer coating. The main charge is the black powder pellet of 5 to 10 mm size which is a mixture of 75% Potassium Nitrate, 15% charcoal and 10% sulphur. The booster charge is powder of 1 to 3 mm size. As soon as the bridge wire gets red hot, the primer material will explode. This energy is picked up by booster charge and main charge subsequently. Finally hot gases come out which will initiate the burning of AgI mixture, initiating glaciating fumes.

10.4.5 Direct Injection by Firing Rockets from Ground from Static Location

Firing of rockets (Refer Fig.10.9) into cloud is one method of direct seeding (Dennis 1977) and is commonly used method (Radinovic and Curic 2007).

Fig. 10.9 Hail Suppression Project in Serbia. Static base slant firing of rockets.

But it needs large area of air space to be cordoned off to avoid any collateral damage to any other passenger aircraft flying in the path of rockets projectile. Hence it is not an operationally viable strategy in a region of civil aviation routes.

Dual-mode rocket engines are used in weather modification operations. In such rockets the primary mode is without ice-nucleation composition in fuel mixture and in the secondary mode fuel is mixed with Ice Nucleating seeding agents which on emission with the burnt fuel will produce ice-nucleation. Fumes are ejected like rocket exhaust and it will come out automatically after 8th or 9th second of flight duration (Anti-Hail Rockets e.g. A6 and A8, Alan-2, Alazan 5, Alazan 5 and Alazan 6, Alazan 9, KPGR-6, PGR PP-8, PGR PP-6T, Sky Clear 6, WR-98 Series and WR-1D, WR Series). Table 10.2 lists details of some of the ant-hail rockets which may be compared by readers.

Table 10.2 Anti Hail Rockets

	Alazan 2M*, Russia	**Kristall, Russia**	**TG-10, Yugoslavia**	**WR-18, China**	**Alazan-5/6, Russia**	**Alan, Russia**	**Darg, Russia**	**Loza 2/3 (MTT-9M), Bulgaria**	**Sky Clear 6**
Caliber mm	82,5	82,5	80	82	82,5	69	60	60	60
Length, mm	1356	1700	1050	1440	1402	920	1200	1045	1045

Table **10.2** *contd...*

Weight, kg	8.03	12	4.28	8.5	8.5	4.4	4.5	3.85	3.2
Radius of action, km	8	12	10	8	10.5	12	12	7.5	10
Length of a working line, km	5	10	6	5	6	10	7	5	8
Reagent weight, g	630	320	400	725	630	2200	700	400	900
Quantity AgJ, g	12.6	6.4	80	10	12.6/26.4	44	32	50	72
% AgJ	2.0 %	2.0 %	20.0 %	1.3 %	2.0 / 4.2	2.0 %	4.6 %	12.5 %	8.0 %
Ice-nucleation activity per gm at minus 10 °C	3.0E+15	2.0E+15	8.0E+14	1.8E+15	7.0E+15	1.0E+16	1.0E+16	4.8E+15	7.2E+16

Pictures of Anti hail rockets by Serbia and Russia are shown in Fig.10.10.

PGR PP-8 & PGR PP-6TKPGR-6Alazan-5 Russia
Serbia Serbia

Fig. 10.10 Anti hail rockets of Republic of Serbia & Russia.

In the subzero regions within a hail bearing cloud the seeding is optimum in the range of -5° to $-15^{o}C$ temperature. Height of this temperature range of the atmosphere keeps varying on daily basis and also with region and season. Hence if the seeding particles are delivered just after fixed time of flight then delivery may not be in the optimum temperature range to obtain best seeding effects in the cloud.

Vertically launched spin – stabilized airborne cloud seeding rocket system of NHRE (USA) (Fig.10.11) or Russian Oblako and Alazani rockets have been employed in the hail suppression seeding experiments. Oblako rockets have a caliber of 125 mm. It weighs 33 kg and holds 3.2 kg of PbI_2 in pyrotechnique mixture. It could deliver 3×10^{16} nuclei at $-10^{o}C$ temperature. The slant range of the rockets is 12 km and height range is 9.5 km. Oblako rockets are too big and they present flying hazards if parachutes system meant to lower the casing does not open due to the turbulences in the cloud. Alazani rockets are, however, designed to destroy themselves by blowing up in mid-air after discharge of the substances.

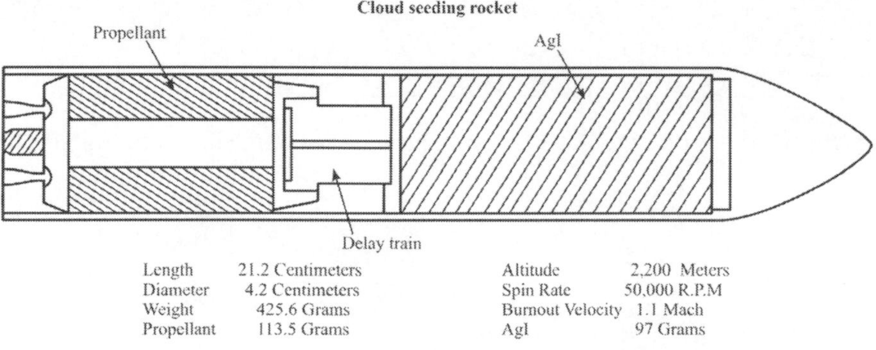

Fig. 10.11 Cloud seeding metallic toy rocket by NHRE(U.S.A.); Configuration and performance specification.

Explosives substances placed in the rocket disintegrate the rocket body into small pieces. Since seeding material can also be delivered in the updrafts of the cloud, in lower levels, hence NHRE (USA) developed non – metallic toy rockets of 21.2 cm long with 4.2 cm diameter and 425.6 gm weight. It could carry about 100 gm of silver iodides and was boosted by 114 gm of propellant. It could attain an altitude of 2.2 km above the launch point.

Artillery guns or antiaircraft guns are also used to disperse the ice nucleating agents in the hail producing zone of the cloud. Russian E1 brus –2 and El brus- 3 shells used 50 to 300 gm of lead iodides. They go up into the cloud and explode spreading the nucleating agent in an area not more than half a kilometer in radius. Their nucleating region is, therefore, limited.

10.4.6 Direct Injection by Firing Rocket, Transported by the Helicopter below the cloud and fired vertically from the Ground

For hail control direct injection by firing rocket, transported by the helicopter below the cloud and fired vertically from the ground has been proposed in NICRA Project(National Initiative on Climate Resistant Agriculture) funded by Indian Council of Agricultural research, Government of India. This approach would be discussed under section 10.6 on NICRA.

10.5 Why Success in Hail Control Mostly Eluded

Despite the Sulakvelidze Hypothesis of Hail Control which triggered momentous effort by several countries on hail suppression field experiments, satisfactory results mostly eluded all over. That led WMO (2007) not to recommend Hail Suppression programme. Nevertheless, without clear evidence it is well accepted that the success in hail suppression are dependent on characteristic of the storm treated, the means of delivery of silver iodide and critical rate of delivery to the storm. Various reasons are categorically discussed below:

Concept of Reaction Time: It has been widely accepted that once hailstones have formed in the cloud it cannot be artificially reduced in size. Hence any exercise to control hailstone size within the cloud must be attempted prior to its embryonic stage. Lower threshold of radar reflectivity for hailstone to appear in the cloud has been well accepted as 45 dBZ (Witt et al, 1998; Singh et al, 2011; Srivastava et al, 2011). This information is important as during this interval only if INs or CCNs are released into the cloud then only we can restrict the larger growth of hail kernels. This period is known as “Reaction Time”. Total Reaction Time (TRT) may be defined as the time taken by any cumulus cloud with reflectivity 20 dBZ to grow till its reflectivity reaches 45dBZ(Kumar and Pati, 2015). Available Reaction Time (ART) is the time actually available within the TRT for action against the threatening cumulus cloud. Seeding of the cloud must be done within the Available Reaction Time. Once the reflectivity of the cloud has grown more than 45dBZ then hails have already formed within the cloud and seeding the cloud thereafter would be of little avail.

Static Location Firing of Rockets: As can be observed from table 10.2 static firing of rocket into the threatening cloud not only has limited range of action i.e., 7.5 to 12 km but also it has limited resident time of the rocket in the towering cloud volume due to rocket speed. Less slant angle will have insufficient seed delivery problem in the cloud. Refer Fig. 10.12. Rocket paths are shown by arrows in small scale cumulus cell of cylindrical towering shape.

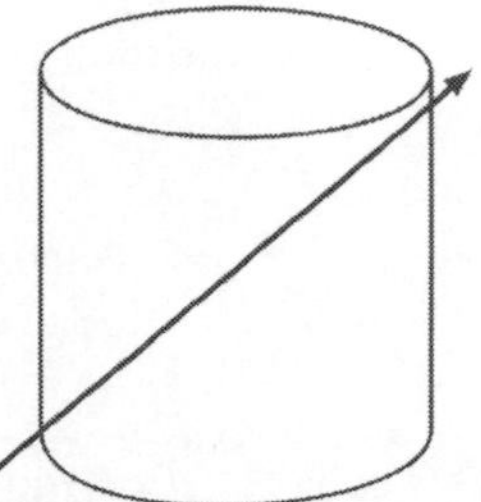

More slant rocket path has more time to remain within cloud vis-à-vis disperse more seeding material therein

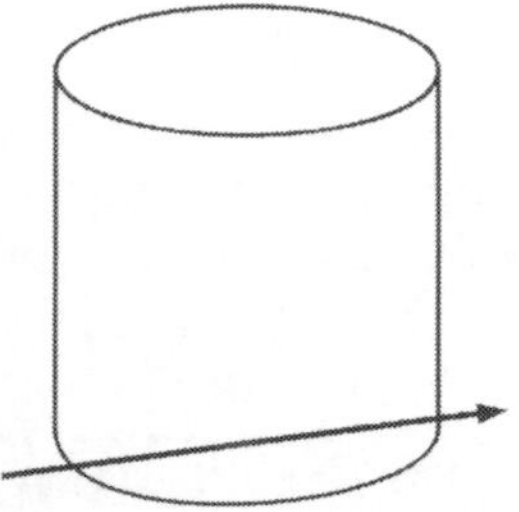

Less slant rocket path has lesser time to remain within cloud vis-à-vis disperse less seeding material therein

Fig. 10.12 Rocket path is shown by arrow. Cylindrical shape represents cumulus cloud.

Seed dispersal by mixing seeding material with rocket exhaust. Schleusener (1968) had hypothesized that seeding of cumulus may increase or decrease the hailfall. Without volume consideration he fixed seeding rate of <1000gm hr^{-1} per storm would increase convection and hailfall and rate of 2000-3000 gm hr^{-1} perstorm would suppress the hailfall. This gave the concept of over seeding as necessary condition for hail suppression. Mixing of IN/CCNs with the rocket propellant results in very short delivery period by the rocket within the cloud. Their CCN/IN concentration of 1.3 to 12.5% is too diluted amount to achieve effective 'over-seeding'. Only TG-10 rockets (Yugoslavia) has AgI as 20%. Even this percentage amount is needed to be improved for effective 'over-seeding'. Under NICRA Project (described in following section) exhaust delivery system was replaced by pyrotechnic cartridge ejection delivery. The pyrotechnic cartridge, having strong concentration of 48% AgI for delivery in the cold region of cloud, was adopted (refer table 10.3) along with NaCl spray in warm region of cloud to ensure 'over-seeding'. Concept of over-seeding is described below.

Table 10.3 AgI Pyrotechnic Cartridge chemical Formulation

Perforated AgI cartridges with 2 gm formulation	
Material	**Percent by Weight**
Silver iodate	48.0
Potassium perchlorate	28.0
Magnesium	12.0
Nitrocellulose	4.0
Triacetin	8.0

It is well known that hailstorms are not forming over deep oceanic surface. Refer the climatology of hailstorm in chapter 1. Their occurrence over some of the coastal or off-shore regions is borne out of microphysical as well as dynamical processes. Dilution of strong sea-salt density of the deep oceanic planetary boundary layer by mixing of the air mass from the land region is the former process. Relatively stronger convections, as compared to those over deep oceans, due to close proximity of land is later process.Without specifying the quantity of seeding material Kuba and Murakami (2010) have noted that seeding can hasten the onset of surface rainfall and increase the accumulated amount of surface rainfall if the amount and radius of seeding particles are appropriate. It necessitates the need for precise definition of exact seeding.

High density of sea salt in air would induce high competition to collect the water from the growing cumulus amongst the in situ aerosols and hence size of resulting liquid droplets or frozen particles would be smaller but their number would be larger. Smaller frozen particles would melt before reaching the oceanic surface (Sulakvelidze, 1969).Concept of over seeding can, therefore, be derived from the salt content in the deep oceanic boundary layer where hailstorms rarely form.Having specified the quantitative definitions of exact seeding and over seeding, definition of under seeding follows by itself. As the salt content in the boundary layer over the ocean is of the order of ≈ ½ to 1.0 kg/km^3 (Foote, G.B., and C.A. Knight, 1977) hence seeding process may be defined as under.

(i) Exact seeding: In this type of operation the amount of CCN/IN dispersed, is exactly ½ to 1.0 kg/km^3 in clouds. This will generate normal rain/shower of size 0.5-6 mm as over oceanic surface.

(ii) Over seeding: In over seeding operation the amount CCN/IN dispersed are more than 1.0 kg/km^3 in clouds. This will result in formation of small graupel or rain or drizzle or virga (smaller than 0.5 mm in diameter) in the cloud. Small graupel would turn in to liquid droplets while travelling below the cloud in above 0^0C temperature of atmosphere.

(iii) Under seeding: In under seeding operation the amount CCN/IN dispersed are less than ½ kg/km^3 in clouds. This will result in a few droplets of rain/shower or hail.

Now with the precise definitions of exact, over and underseeding one can infer that hail suppression can be done only by over seeding of the clouds. Artificial rain making experiments needed exact seeding. Hence we need to determine the aerosol density of the in situ airmass and then supplement the same with the amount of seeding agents which makes it exact seeding. Fixed amount of seeding in all airmasses may allude desired result in rain making experiments. On contrary hail control experiments demand over-seeding which is operationally easy to achieve. Under seeding during hail control is likely to reinforce hail production in convective cloud. Detwiler (2015) has reported a case where storm invigorated, and the volume of the storm containing hail increased dramatically shortly after seeding began albeit under seeding aspect has not been probed in his study.

Woodcock and Spencer (1967) observed that dynamics of warm cloud gets modified due to seeding by NaCl in nearly water saturated atmosphere. They hypothesize that 40 gm of NaCl per kg of air would raise the air temperature by a few tens of degrees. Murthy et al (1975) suggested that the rise in temperature, increase in cloud liquid water content and rise in cloud depth are the dynamic effects of salt seeding. Kuba and Murakami (2010) have noted that seeding by a hygroscopic flare decreases rainfall in the case of large updraft velocity (shallow convective cloud). Posselt and Lohmann (2008), Mahen Konwar et al (2012), and Gregory and Trude (2014) have noted that greater CCN concentrations gave rise to clouds with smaller drops and delayed precipitation development. Warm rain became detectable, i.e. rain water content >0.01 g/Kg, at the top of growing convective clouds when effective radius (radius of sphere of same volume) exceeded 12 mm. Study by Muhammad et al (2016) verified that higher cloud droplet concentration (1000 per cc) in convection results in a 15- 20% low surface accumulated rainfall over both land and ocean than in case of lower concentration (100 per cc). These studies indicate that possibility of enhanced CCN in warm cloud seeding leading to more super cooled water being lifted in the storm and made available for accretion and hailstone growth, will result only in small hailstones if it is over seeding.

Small hailstones are likely to melt before reaching the surface, particularly under tropical conditions.

Delivery of INs in Optimum Temperature Range for Activation: Significance of temperature for achieving the optimum seeding affect for warm cloud was noted by Manohar et al (2001) who suggested altitude range of 4750 – 6250 fta.m.s.l. over Pune, India, as the preferred height for seeding. For cold cloud seeding it is generally accepted that AgI – NH_4I and AgI – NaI are both active nuclei between –8°C to –16°C - former being little more than the later. Cloud chamber experiments conducted by Kumar et al (2014) indicated high efficiency of silver iodide containing pyrotechnic cartridge simulating the International Standard Atmosphere. It was noted that the process of condensation and precipitation attain their optimal value in the temperature range of –19°C and –3.7°C and pressure range of 387 hPa and 533 hPa. The temperature range in International Standard Atmosphere is in conformity with accepted range of –8 to –16°C. Any delivery of seeding agents in cold cloud experiment, therefore, should be made in this temperature range for optimum result.

All the above mentioned concerns were addressed in project on Hail Control, funded by Indian Council of Agricultural Research (ICAR) under its programme on National Initiative on Climate Resilient Agriculture (NICRA) – Kumar et al 2014. Brief description of NICRA project is presented in section 10.6.

10.6 National Initiative on Climate Resilient Agriculture (NICRA) on Hail Control

Hail mitigation operational plans in NICRA project could be divided in following five stages.

(i) **Stage I:** Locating the cloud by radar, which has potential to grow in to hailstorm.

(ii) **Stage II:** Transporting AgI and NaCl rockets by helicopter below the base of the cloud on ground.

(iii) **Stage III:** Firing the AgI and NaCl rockets vertically upward into the cloud from the ground level.

(iv) **Stage IV:** Appropriate level release of the AgI pyrotechnic cartridges in suitable temperature zone in cold region of cloud and spray of common salt's appropriately milled powder, in the warm region of the cloud.

(v) **Stage V:** Recover the helicopter back to base.

10.6.1 Stage I: Locating the Cloud by Radar, which has Potential to grow in to Hailstorm

Pre Hail Detection Algorithm (PHDA) is used after three successive observation. The skill score of occurrence/non-occurrence prediction by PHDA is 0.78. With shorter radar-scan interval of 5 minutes observation the score is expected to be better and Available Reaction Time may further improve to more than an hour (Kumar and Pati 2015).

10.6.2 Stage II: Transporting AgI and NaClRockets by Helicopter below the base of the Cloud on Ground

Hindustan Aeronautics Ltd (India) made CheetahHelicopter can be used as Quick Reaction Platforms for transporting the AgI and NaCl rockets below the cloud. This helicopter can fly very low and can land at any place in case of inclement weather. Fig.10.13 shows the prototype of helicopter with fully extended side attachments for rockets. Picture shows fully extended side attachments with rockets oriented horizontally. Rockets can be reoriented vertically before firing. While flying to destination rockets are oriented horizontally and attachments are in contracted position to ensure smoother aerodynamics. Helicopter is equipped with Navigational Screen and Human Machine Interface (HMI) screen to control rocket fire. Side attachments have capability of carrying 2 AgI and 2 NaCl rockets on each side. Navigational screen displays the area map and also displayed over it is the spot marking geographical location of helicopter as per the Global Navigational Satellite Network (GNSN).

During flying to its destination the height of release of seeding agents are to be fed in the GNSN system based on the latest upper air temperature profile of the day.

Fig. 10.13 Prototype of Quick Reaction Helicopter.

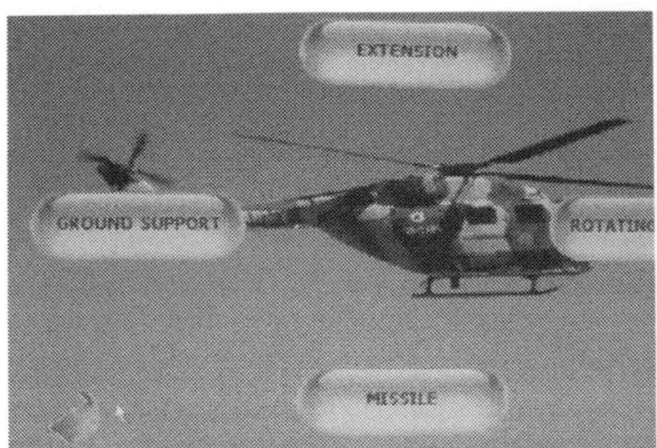

Fig. 10.14(a) HMI front Touch Screen with four options.

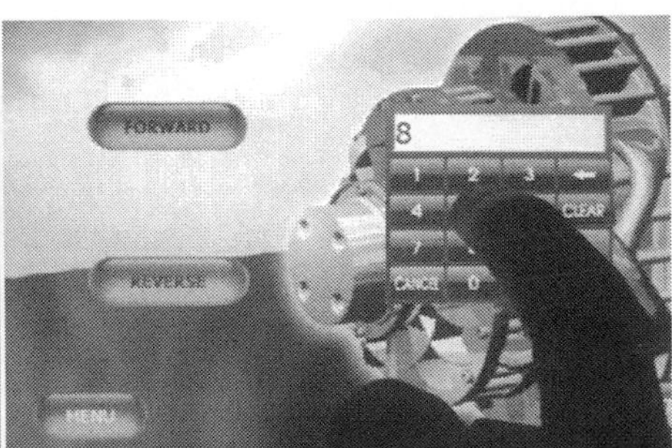

Fig. 10.14(b) HMI Touch Screen with side attachment extension and retraction option options.

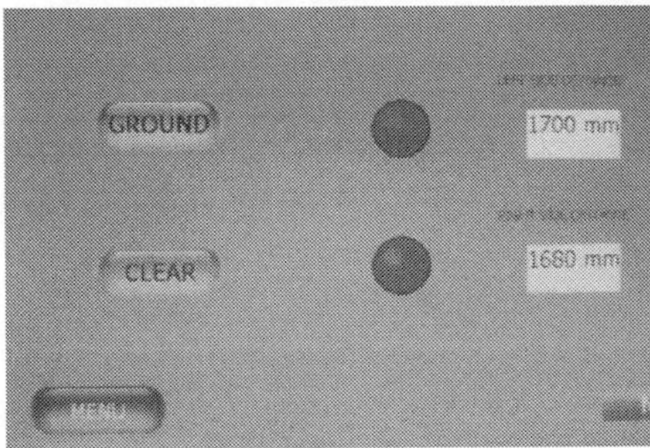

Fig. 10.14(c) HMI Touch Screen with ground support extension and retraction option options.

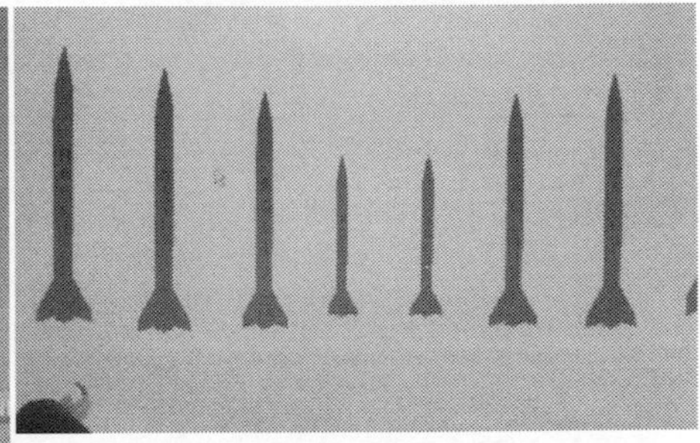

Fig. 10.14(d) HMI Touch Screen with eight rockets – four on each side firing option options.

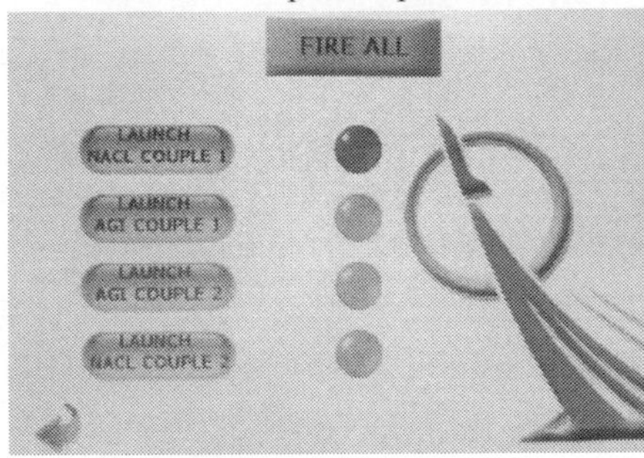

Fig. 10.14(e) HMI Touch Screen with selectively couple firing options. Red spot shows NaCl couple rockets are selected.

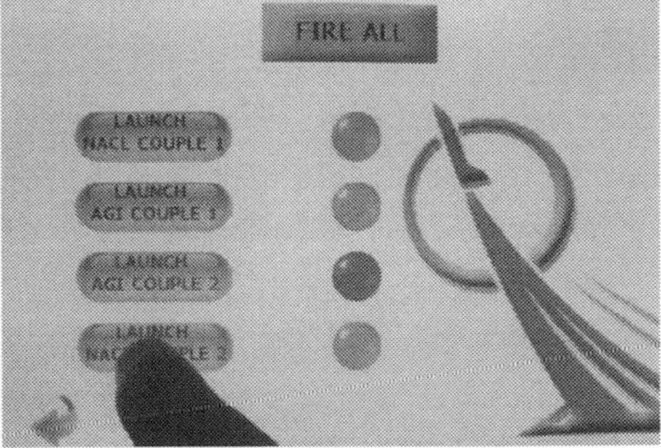

Fig. 10.14(f) HMI Touch Screen with selectively couple firing options. Red spot shows AgI couple rockets are selected.

Fig. 10.14 Human Machine interface(HMI) Screen Installed in front of the Pilot in Cockpit.

10.6.3 Stage III: Firing the AgI and NaClRockets vertically upward into the Cloud from the Ground Level

Immediately after landing below the cloud the attachments are fully extended without stopping the engine, rockets are vertically oriented and fired with the help of Human Machine Interface (HMI) screen, which has touch button system. Fig. 10.14 displays the sequence of side extension and rocket firing through HMI control.

10.6.4 Stage IV: Appropriate level release of the AgI Pyrotechnic cartridges in cold region of cloud and spray common salt's appropriately milled powder in the warm region of the cloud

Two rockets, OLASTRA A1 for AgI pyrotechnic cartridge delivery and OLASTRA N1 for NaCl powder delivery are shown in Fig. 10.15(a) and (b) respectively.

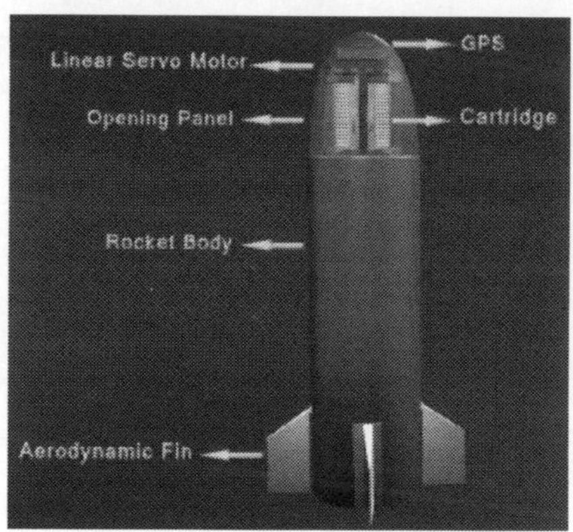

(a) OLASTRA A1 for AgI Rocket: 165 cm length, 10 cm dia.

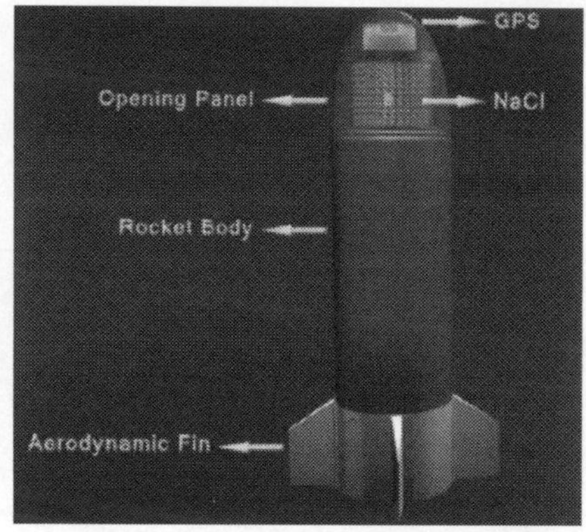

(b) OLASTRA N1 for NaCl powder delivery: 97.5 Cm length10 cm dia.

Fig. 10.15

In the two types of rocket the payload ejection done by Global Navigational Satellite System (GNSS) based control circuit. This circuit will start to work after receiving signals from GNSS Satellite as per the pre fed program. Fig. 10.16(a) shows the GNSS signal received by the AgI rocket after it attains appropriate height and Fig. 10.16(b) shows the ejection of two pyrotechnic cartridges by the servomotor activated by the signal.

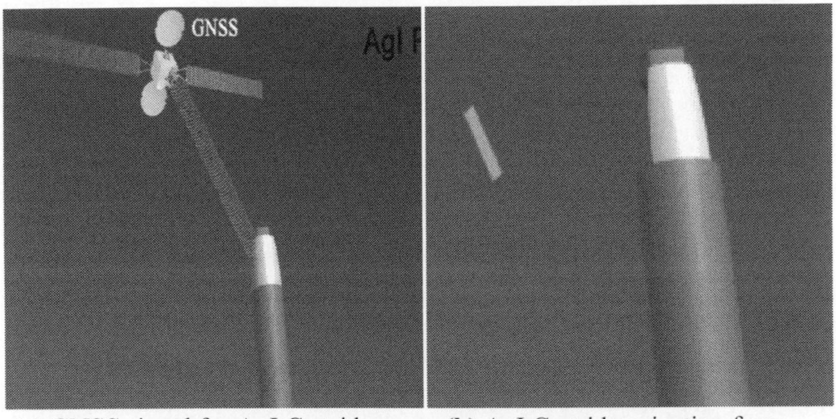

(a) GNSS signal for AgI Cartridge ejection at appropriate height

(b) AgI Cartridge ejection from two sides of the payload section

Fig. 10.16 Height controlled pyrotechnic cartridge ejection system through GNSS in suitable temperature range in the cloud for optimum seeding effect.

10.6.5 Stage V: Recover the Helicopter back to Base

Quick recovery of helicopter back to base is safety requirement. Hence firing of rockets are done with running engine. Immediately after completing the firing of rockets helicopter has to scoot from the place at the earliest.

10.7 The Environmental Impact of Using Silver Iodide as a Cloud Seeding Agent

Environmental impact studies related to silver iodide usage in cloud seeding, were conducted starting in the 1960s and continue to be conducted now.In general all findings to date indicate no adverse environmental and human health impacts (ASCE 2004, 2006; WMA 2005; WMO 2007). Findings indicate no significant effects on plants and animals(Cooper & Jolly 1970; Howell 1977; Klein 1978; Dennis 1980; Harris 1981; Todd & Howell 1985; Berg 1988; Reinking*et al.* 1995; Eliopoulos &Mourelatos 1998; Ouzounidou & Constantinidou 1999; Di Toro *et al.* 2001; Bianchini*et al.* 2002; Tsiouris*et al.* 2002a; Tsiouris*et al.* 2002b; Christodoulou *et al.* 2004; Edwards *et al.* 2005; Keyes *et al.* 2006; Williams & Denholm 2009). The U.S. Public Health Service established a concentration limit of 50 micrograms of silver per liter of water in public water supply to protect human health (Erdreich*et al.* 1985). The concentrations of silver potentially introduced by modern cloud seeding efforts are significantly less than this level. The literature embodies tens of thousands of samples collected from cloud seeding program areas, over athirty-year period showing the average concentration of silver in rainwater, snow and surface water samples, is typically less than 0.01 micrograms per liter. More importantly, these measurements represent the total amount of silver contained in any given sample and are not specific to the form of silver present in a sample. Nevertheless, these measurements show that silver is virtually undetectable in any form in the vast majority of the tens of thousands of samples collected from these areas.

More than 100 Sierra Nevada lakes and rivers have been studied since the 1980's (Stone 1986); no detectable silver above thenatural background was found in seeded target area water bodies, precipitation and lake sediment samples, nor any evidence of silver accumulation after more than fifty years of continuous seeding operations (Stone 1995; Stone 2006). Many of these alpine lakes have virtually no buffering capacity, making them extremely susceptible to the effects of acidification and sensitive to changes in trace metal chemistry. Therefore studies were conductedas part of environmental monitoring efforts to determine if cloud seeding was impacting these lakes. No evidence was found that silver from seeding operations was detectable above the background level. There was also no evidence of an impact on lake water chemistry, which is consistent with the insoluble nature and long times required to mobilize any silver iodide released over these watersheds. Comparisons of silver with other naturally occurring trace metals measured in lake and sediment samples collected from the Mokelumne watershed in the Sierra Nevada indicate that the silver was of natural origin (Stone 2006). Similarly, Sanchez *et al.* (1999) analyzed the chemistry of water bodies and rainwater from samples collected during a summer cloud seeding program in Spain, and determined the silver input from cloud seeding to be indistinguishable from natural inputs. Greek scientists studying the effects on soils, plants and their physiology, atmospheric precipitation, plankton, animals and man, as well as the impact of irrigation and organic matter to AgI leaching from the Greek cloud seeding activities found similar results following the analyses of 2500 soil samples (Tsiouris*et al.* 2002a; Tsiouris *etal.* 2002b).

10.8 Future Prospects

The hail suppression concepts have not changed much over the last 60 yrs. However, the ability of scientists to test the concepts via improved technology, better numerical storm models, better seeding agents, better targeting, and delivery systems make it feasible to design new experiments and make the observations needed to evaluate the steps involved in the concepts.

Knowledge is needed to determine the necessary conditions cloud should meet, to react in a consistent manner with the assumed hypotheses. We need to know more on what criteria are necessary for selecting the target clouds, to know the reaction time to start seeding, to know the places to seed, and the dose of seeding. Natural density of CCN or IN in specific airmass has to be precisely accounted to supplement the dose of exact seeding, under seeding or overseeding as per the objective of the experiment. Under seeding in hail dissipation programme may enhance the hailstones in the cloud instead of dissipating it. Direct measurement of microphysical seeding signatures is still an important step for obtaining knowledge about these required conditions (amount, place, and timing) for seeding. As Kohler's theory is replaced by adsorption theory more innovative research is required to find predictor variables and other evaluation criteria that can reliably be provided through forecasts or direct measurements. Furthermore, development of conceptual models is needed, supported by modeling and physical measurements that give reliable support to the design of a seeding strategy. Young (1996) argues convincingly that early progress in weather modification was due to a healthy interaction between theory and experiment, and that a rejuvenation of this theoretical component is necessary to revitalize the field of weather modification. The early abundant funding and experimentation in hail suppression was partly a result of enthusiasm and unrealistic hopes of quick success. Progress has been slow and research funding has diminished drastically. Risk managers continue to support operational hail suppression programs regardless of the inconclusive scientific results of past experiments. These economic realities warrant scientific experimentation that can be transformed into operational programs, in order to further develop and test specific physical hypotheses. Present day demand is the scientific and political will to merge the research and operational world and design experiments which can further our knowledge and advance the science of hail suppression.

References

1. Berg, N.H. (1988). A Twelve-Year study of environmental aspects of weather modification in the Central Sierra Nevada and Carson Range. The Sierra Ecology Project, Unpublished report on file at Pacific SW Res. Station, Forest Service USDA, Albany CA.
2. Bianchini, A., M. Grosell, S.M. Gregory, C.M. Wood (2002), Acute silver toxicity in aquatic animals is a function of sodium uptake rate, Env. Sci. &Techn., **36**, 1763-1766.
3. Browning, K.A., J-P Chalon, P.J. Eccles, R.C. Strauch, F.H. Merrem, D.J. Musil, E.L. May, and W.R. Sand, 1976, Structure of an evolving hailstorm. Part V: Synthesis and implications for hail growth and hail suppression. Mon. Wea. Rev., 104, 603-610.
4. Browning, K.A, and G.B. Foote, 1976, Airflow and hail growth in supercell storms and some implications for hail suppression. Quart. J. Roy. Met. Soc., 102, 499-533.
5. Charlton, R. B., B. M. Kachman, and L. Wojtiw, 1995, Urban hailstorms: A view from Alberta. Natural Hazards, 12, 29-75.
6. Chisholm, A. J., and J. H. Renick, 1972,The kinematics of multicell and supercell Alberta hailstorms, Alberta Hail Studies, 1972. Rep. 72-2, 24-31.
7. Christodoulou, F.V. Samanidou, A.G. Zachariadis, H.-I.A. Constantinidou (2002α). Soil Silver Content of Agricultural Areas Subjected to Cloud Seeding with Silver Iodide. FreseniusEnvironmental Bulletin, **11**, 697-702.
8. Cooper, C.F., W.C. Jolly (1970). Ecological Effects of Silver Iodide and other Weather Modification Agents: A review. Water Resources Research,**6**, AGU, 88-98.
9. Dennis, AS, Hail: Areview of Hail Science and Hail Suppression, Edited by G. Brant Foote and Charles A. Knight, AMS, Vol.16, No. 38 Dec. 1977, pp181-193.
10. Dennis, A. S. (1980), "Weather Modification by Cloud Seeding". International Geophysics, Series, **24,** Academic Press, New York, NY, 21-22.
11. Detwiler Andrew, 2015, A case study of cloud seeding to reduce hail damage, Journal Of Weather Modification, April, Volume 47, pp 26-41.
12. Di Toro, D.M., H.E. Allen, H.L. Bergman, J.S. Meyer, P.R. Paquin, R.C. Santore (2001), Biotic Ligand Model of the acute toxicity of metals. 1. Technical basis. Environmental Toxicologyand Chemistry, **20**, 2383-2396.

13. Dixon, Michael, and Gerry Wiener, 1993, TITAN: Thunderstorm Identification, Tracking, Analysis, and Nowcasting - A Radar-based Methodology. J. Atmos. and Oceanic Technol., 10, 6, 785-797.

14. Edwards, R., A. Huggins, J. McConnell (2005), "Trace Chemistry Evaluation of the Idaho PowerCompany Operational Cloud Seeding Program 2003 to 2005". DRI Publication #41223.

15. Eliopoulos P., D. Mourelatos (1998), Lack of genetoxicity of Silver Iodide in the SCE Assay in vitro, in vivo, and in the Ames/Microsome Test. Terratogenesis, Carcinogenesis andMutagenesis, **18**, 303-308.

16. Erdreich, L., R. Bruins, J. Withey (1985), Drinking Water Criteria Document for Silver (Final Draft). U.S. EPA, Washington, D.C., EPA/600/X-85/040 (NTIS PB86118288).

17. Farley, R.D., 1987, Numerical modeling of hailstorms and hailstone growth. Part III: Simulation of an Alberta hailstorm - natural and seeded cases. J. Climate Appl. Meteor., 26, 789-812.

18. Farley, R.D., P. Nguyen, and H.D. Orville, 1994, Numerical simulation of cloud seeding using a three-dimensional cloud model. J. Wea. Modif, 26, 113-124.

19. Farley, R.D, H. Chen, H. D. Orville, and M.R. Hjelmfelt, 1996: The numerical simulation of the effects of cloud seeding on hailstorms. Preprints,13th Conf. On Planned and Inadvertent Weather Modification, Atlanta, GA, Amer. Meteor. Soc., 23-30.

20. Federer, B., A. Waldvogel, W. Schmid, H.H. Schiesser, F. Hampel, M. Schweingruber, W. Stahel, J. Bader, J.F. Mezeix, N. Doras, G. D'Aubigny, G. DerMegreditchian, and D. Vento, 1986, Main results of Grossversuch IV. J. Climate Appl. Meteor., 25, 917-957.

21. Foote, G.B., and C.A. Knight, 1977,*Hail: A Review of Hail Science and Hail Suppression*, Meteorological Monographs, Vol 16, American Meteorological Society, 277 pp.

22. Foote, G.B., 1984, The study of hail growth utilizing observed storm conditions. J. Climate. Appl. Meteor., 23, 84-101.

23. Gregory Thompson and Trude Eidhammer, A Study Of Aerosol Impacts On Clouds And Precipitation Development In A Large Winter Cyclone, 2014, JAS, Oct., Vol., 71, 3636-3658.

24. Gulati, Vishal, Jun 26, 2011, "Himachal's anti-hail guns fail to deliver", IANS – Sun.

25. Harris, E. (1981), "Environmental Assessment and Finding of No Significant Impact". Sierra, Cooperative Pilot Project, Bureau of Reclamation, Denver, Co.

26. Howell, W. E. (1977), Environmental impacts of precipitation management: Results and inferences from Project Skywater. Bull. Amer. Meteor. Soc., 58, 488-501.
27. Humphries, R.G., M. English, and J. Renick, 1987, Weather modification in Alberta. J. Weather Modification, 19, 13-24.
28. Iribarne, J. V., and R.G. de Pena, 1962, The influence of particle concentration on the evolution of halstones, Nubila, 5, 7-30.
29. Keyes, Conrad G. (2006), “Guidelines for Cloud Seeding to Augment Precipitation”. 2nd Edition. American Society Civil Engineers, Reston, VA.
30. Klein, D.A. (1978). "Environmental Impacts of Artificial Ice Nucleating Agents" Colorado State University, Dowden, Hutchinson & Ross, Inc. Library of Congress Catalog Card Number, 78-7985.
31. Knight, C. A., and N. C. Knight, 1970, Hailstone embryos. J. Atmos. Sci., 29, 1060-1065.
32. Knight, C. A., and N. C. Knight, 1979, Results of a randomized hail suppression experiment in northeast Colorado, Part V: Hailstone embryo types. J. Appl. Meteor., 18, 1583-1588.
33. Knight, C. A., and N. C. Knight, 1999, Hailstorms. Met. Monographs, Amer. Meteor. Soc., Severe Storms Monograph (In Press).
34. Knight, C. A, L.J. Miller, N.C. Knight, and D. Breed, 1982, The 22 June 1976 case study: Precipitation formation. In Hailstorms of the Central High Plains, Vol. II: Case Studies of the National Hail Research Experiment. C.A. Knight and P. Squires, eds. Colorado Assoc. Univ. Press, Boulder, 61-89.
35. Köhler, H.: The nucleus in and the growth of hygroscopic droplets, Trans. Far. Soc. 32, 1152–1161, 1936.
36. Krauss, T.W., and J.D. Marwitz, 1984, Precipitation processes within an Alberta supercell hailstorm. J. Atmos. Sci., 41, 1025-1034.
37. Krauss, T. W., Sin’kevich, A. A., Veremey, N. E., Yu, A. Dovgalyuk and Stepanenko, V.D., 2009, Estimation of results of cumulonimbus cloud modification aiming at hailstorm mitigation in Alberta(Canada) on the radar and numerical modelling data, Vol.34,No.4, pp 218-227, Allerton Press.
38. Kuba, N. and M. Murakami 2010, Effect of hygroscopic seeding on warm rain clouds – numerical study using a hybrid cloud microphysical model, Atmos. Chem. Phys., 10, 3335–3351, 2010
39. Kumar P., Jaykumar D., DebprasadPati and Bhardwaj Shweta, Project Report of NICRA Project on ‘Hailstorm Management in Agriculture’, Indian Council of Agriculture Research, April 2014, Vol.1 and 2.

40. Kumar P, and Pati Deb Prasad (2015),Radar imageries information extraction and its usein pre-hailestimation algorithm, 66, 4, Oct 2015, 695-712.

41. Kulmala, M., Kerminen, V.-M., Anttila, T., Laaksonen, A., O´Dowd, C. D.: Organic aerosol formation via sulphate cluster activation, J. Geophys. Res., 109, D04205, doi:10.1029/2003JD003961, 2004.

42. Kumar, P., Sokolik, I. N., Nenes A.: Measurements of cloud condensation nuclei activity and droplet activation kinetics of fresh unprocessed regional dust samples and minerals, Atmos. Chem. Phys., 11(7), 3527– 3541, doi:10.5194/acp-11-3527-2011, 2011.

43. List, R., 1960, Growth and structure of graupels and hailstones. Physics of Precipitation, Geophys. Monograph, 5, Amer. Geophys. Union, 317-324.

44. List, R.,, 1985, Properties and growth of hailstones, in Thunderstorm Morphology and Dynamics (E. Kessler, ed.), Norman, OK, Univ. of Oklahoma Press, 411pp.

45. Posselt R. and U. Lohmann, Influence of Giant CCN on warm rain processes in the ECHAM5 GCM Atmos. Chem. Phys., 8, 3769–3788, 2008

46. MacCready, P. B., Jr., 1958, The lightening mechanism and its relation to natural and artificial freezing nuclei. Recent Advances in Atmospheric Electricity, New York, Pergamon Press, 369-381.

47. Macklin, W.C., 1977, The characteristics of natural hailstones and their interpretation. Meteor. Monograph, 38, 65-88.

48. MahenKonwar, R. S. Maheskumar, J. R. Kulkarni, E. Freud, B. N. Goswami,and D. Rosenfeld, 2012, Aerosol Control on Depth Of Warm Rain In Convective CloudsJournal Of Geophysical Research, Vol. 117,D13204

49. Manohar, G.K. Kandalgaonkar, S.S. and Tinmaker M.I.R., 2001, Cloud Liquid Water content responses to hygroscopic seeding of warm clouds.Current Science, Vol. 80, 4. 25 February, p – 555

50. Marwitz, J.D., 1972a,The structure and motion of severe hailstorms. Part I: Supercell storms. J. Appl. Meteor., 11, 166-179.

51. Marwitz, J.D., 1972b,The structure and motion of severe hailstorms. Part II: Multicell storms. J. Appl. Meteor., 11, 180-188.

52. Mather, G. K., M. J. Dixon, J. M. DeJager, 1996, Assessing the potential for rain augmentation - The Nelspruit randomized convective cloud seeding experiment. J. Appl. Meteor., 35, 1465-1482.

53. Mikhailov, E., Vlasenko, S., Martin, S. T., Koop, T., Pöschl, U.2009: Amorphous and crystalline aerosol particles interacting with water

vapour: conceptual framework and experimental evidence for restructuring, phase transitions and kinetic limitations, Atmos. Chem. Phys., 9, 9491–9522.

54. Miller, L.J., and J.C. Fankhauser, 1983, Radar echo structure, air motion and hail formation in a large stationary multicellular thunderstorm. J. Atmos. Sci., 40, 2399-2418.
55. Muhammad E. E. Hassim1, W. W. Grabowski, and T. P. Lane, 2016, Impact of aerosols on precipitation over the Maritime Continent simulated by a convection-permitting model, Atmos. Chem. Phys.,Published 25 May, 2016
56. Murthy P.S.R., A.M. Selvam and Bh. Ramamurty, 1975, Summary of Observation indicating dynamic effects of salt seeding in warm cumulus cloud, J.Appl. Meteo., 14, 829 – 637.
57. Orville, H. D., 1996,A review of cloud modeling in weather modification. Bull. Amer. Meteor. Soc., 77, 7, 1535-1555.
58. Ovadnevaite, J., Cerburnis, D., Martucci, G., Bialek, J., Monahan, C., Rinaldi, M., Facchini, M. C., Berresheim, H., Worsnop, D. R., O´Dowd, C.:2011, Primary marine organic aerosol: a dichotomy of low hygroscopocity, 60 and high CCN activity, Geophys. Res. Lett., 38, L21806, doi:10.1029/2011GL048869.
59. Ouzounidou G., H.-I.A. Constantinidou (1999). Changes in growth and physiology of tobacco and cotton under Ag exposure and recovery. Are they of direct or indirect nature? Arch.Environ. Contam. Toxicol.,**37**, 480-487.
60. Petters, M. D., Kreidenweis, S. M. A single parameter representation of hygroscopic growth and cloud condensation nucleus activity, Atmos. Chem. Phys., 7, 1961–1971, 2007. Cloud Seeding to Augment Precipitation, ASCE, Reston, VA, 9-47.
61. Radinovic´ DJ, C ´ uric´ M. A.2007,specific evidence of hail suppression effectiveness in Serbia. J Wea Mod, 21:75–84.
62. Reinking, R. F., and B. E. Martner, 1996: Feeder-cell ingestion of seeding aerosol from cloud base determined by tracking radar chaff. J. Appl. Meteor., 35, 1402-1415.
63. Reinking, R.F., N.H. Berg, B.C. Farhar, O.H. Foehner (1995). “Economic, Environmental, andSocietal Aspects of Precipitation Enhancement by Cloud Seeding,” Manual 81, Guidelines for cloud seeding to Augment Precipitation, ASCE Peston, VA, 9-47.
64. Rudolph, R.C., C.M. Sackiw, and G.T. Riley, 1994. Statistical evaluation of the 1984-1988 seeding experiment in northern Greece. J. Weather Mod., 26, 53-60.

65. Sánchez, J. L., J. Dessens, J.L. Marcos, J.T. Fernández (1999), Comparison of rain-water silver concentrations from seeded and non-seeded days in Leon (Spain). J. Weather Mod., **31**, 87-90.

66. Scheckedaz P.T. Chagnon S. A. (1970), A study of crop hail insurance records for northeastern Colorado with respects to the design of the national hail experiment, Illinois State water survey projects report, pp 47.

67. Schleusener, R. A.,1968, Hailfall damage suppression by cloud seeding – A review of the evidence, J. Appl. Met., Dec., Vol. 7, pp 1004-1011.

68. Smith, P. L., L. R. Johnson, and D. L. Priegnitz, 1997,An exploratory analysis of crop hail insurance data for evidence of cloud seeding effects in North Dakota. J. Appl. Meteor., 36.

69. Stone, R.H. (1986), “Sierra Lakes Chemistry Study.” Final Report to Southern California Edison Co., Contract No. C2755903.

70. Stone, R.H., K. Smith-Miller, P. Neeley (1995), Mokelumne Watershed Lake Water and Sediment Silver Survey. Final Report to the Pacific Gas and Electric Company, Technical and Ecological Services, San Ramon, Ca.

71. Stone, R.H. (2006), “2006 Mokelumne Watershed Lake Water and Sediment Survey.” Final Report to the Pacific Gas and Electric Company, Technical and Ecological Services, San Ramon, Ca.

72. Srivastava K, Lau S, Yeung HY, Bhardwaj R, Kannan A M, Singh H(2011)"Use of SWIRLS Nowcasting systems for quantitative precipitation forecasting using Indian DWR data" ,VOLUME 63, Number 1,Mausam, (P 1-16).

73. Sulakvelidze G.K., 1969, Rainstorm and Hail, IPST Press, Jerusalam.

74. Thomas C. J. Hill' Bruce F. Moffett, Paul J. DeMott, Dimitrios G. Georgakopoulos, William L. Stump and Gary D. Franc, Measurement of Ice Nucleation-Active Bacteria on Plants and in Precipitation by Quantitative PCR, Accepted manuscript posted online 6 December 2013, Appl. Environ. Microbiol. February 2014 vol. 80 no. 4 1256-1267.

75. Carbamide, Urea as an Ice Nucleant for Supercooled Clouds, JAS,AMS, pp 197-201.

76. Tsiouris E.S., A.F. Aravanopoulos, N.L. Papadoyiannis, K.M. Sofoniou, N. Polyzopoulos, M.M., Christodoulou, F.V. Samanidou, A.G. Zachariadis, H.-I.A. Constantinidou (2002α), Soil Silver Content of Agricultural Areas Subjected to Cloud Seeding with Silver Iodide. Fresenius Environmental Bulletin, **11**, 697-702.

77. Tsiouris E.S., A.F. Aravanopoulos, N.L. Papadoyiannis, K.M. Sofoniou, F.V. Samanidou, A.G. Zachariadis, H.-I.A. Constantinidou (2002b), Soil Silver Mobility in Areas Subjected to Cloud Seeding with AgI. Fresenius Environmental Bulletin, **12**, 1059-1063.

78. Todd, C.J., W.E. Howell (1985), "Weather Modification". In Handbook of Applied Meteorology, David D. Houghton, Editor, John Wiley and Sons, Chapter 38, 1065-1139.

79. Wakimoto, R.M., W-C Lee, H.B. Bluestein, C-H Liu, and P.H. Hildebrand, 1996, ELDORA observations during VORTEX 95. Bull. Amer. Meteor. Soc., 77, 7, 1465-1481.

80. Warburton, J.A., S.K. Chai, R.H. Stone, and L.G. Young, 1996, The assessment of snowpack enhancement by silver-iodide cloud-seeding using the physics and chemistry of the snowfall. J. Wea. Mod., 28, 19-28.

81. Weather Modification, 2009, INC, Atmospheric Resources Management Technologies.

82. Weickmann, H. K., 1964, The language of hailstorms and hailstones, Nubila, 6, 7 – 51.

83. Williams, B.D., J.A. Denholm (2009), Assessment of the Environmental Toxicity of Silver Iodide-With Reference to a Cloud Seeding Trial in the Snowy Mountains of Australia. . J.Weather Mod., 41, 75-96.

84. World Meteorological Organization, 1996: Meeting of Experts to Review the Present Status of Hail Suppression. Tech. Doc. No. 764.39pp.

85. World Meteorological Organization - 2007, CAS, Management Group, Second Session Oslo, Norway, 24-26 September 2007, Cas-Mg2/Doc. 4.4.1 (31.Viil.2007) Item 4.4.1.

86. World Meteorological Organization - 2013, CAS, 6th Joint Science Committee of The World, Weather Research Programme,

CAS/WWRP/JSC6/Doc 3.6, (Xx June 2013), Item: 3.6, Report From Expert Team On Weather Modification Research For 2012/2013ung, K. C., 1993, Microphysical Processes in Clouds. Oxford Univ. Press. New York, 427.

87. Woodcock, A. H., and A.T. Spencer(1967), Latent heat released experimentally by adding Sodium Chloride particles to the atmosphere. J. Appl. Meteo., 6, 95-101.
88. Young, K. C., 1996, Weather modification - A theoretician viewpoint. Bul.ofAm.Met. Soc. 77, 11, 2701-2710.
89. Zrnic, D.S., 1996, Weather radar polarimetry – trends towards operational applications. Bull. Amer. Meteor. Soc. 77, 7, 1529-1534.

CHAPTER 11

Damage Due to Hail and Risk Cover

Hailstones can batter all kinds of crops, fruits, flowers to a pulp, with dire consequences for farmers. Betel vine and even the root crops like potatoes, onion, groundnut etc. cannot escape the terrible damaging effect of hailstones. Freidman (1976) developed for the first time reliable estimates of national property loss from hail. Hail damage during 1990's has been calculated to be $ 1.2 billion per year for both property and crop loss (Chagnon 1999). This is comparable to actual damage caused by tornadoes (Kunkel et al- 1999 and NCAR-2001). The area of America from Texas to Montana and from the foot hills of the Rockies to Mississippi River, is known as ''Hail Alley'', and in this belt farmers find substantial outlay on hail insurance an absolute necessity. Hail causes about $1 billion dollars in damage to crops and property each year, according to the National Oceanic Atmospheric Administration (NOAA), U.S.A. In Kolkhozes in the region of Telavi in Georgia, the areas damaged by hails in 1959 were only 284 hectare, but in 1966, despite anti hail work carried out by the Geophysics Institute of Georgian

Academy of Sciences, it exceeded to 3000 hectare. Over the Kabardino-Balkarian region, these variations were larger. In a study during 1951 to 1966, the areas damaged by hail amounted in 1959 to 266 hectare, which varied to 19000 hectare in 1963. In India mainly from the beginning of February up to the end of May, hailstorms accompanied by rain and sometimes thundersqualls cause colossal loss of crops, damage of houses and other properties amounting to crores of rupees. This is not only a matter of concern to the farming communities but its effects are far reaching in our every day life and agro-economy as well.

A hail cloud of average thickness has an average reserve equivalent to that of a uranium atomic bomb. However analytical assessment of the damage caused by hail during any year or over any region poses enormous problems. Annual variability of the hail falls, further adds to these difficulties. Hailstones not only damage property including roofs and automobiles as well as agricultural crops but also pose a significant danger to the public.

11.1 Hail Impact Energy

Laurie (1960) derived hail sizes and correlating kinetic (impact) energies of hail in the 1960's. Laurie graphed the relationship between terminal velocity, hail diameter, and the approximate kinetic (impact) energy, as in Table 11.1. Laurie developed this information from data collected by Bilham and Relf (1937).

Table 11.1. Terminal velocities and energies of hailstones

Diameter		Terminal Velocity			Approximate Impact Energy	
inches	cm	ft/s	mi/hr	(m/sec)	ft lbs	Joules
1	2.5	73	50	22.3	< 1	< 1.36
1-1/4	3.2	82	56	25.0	4	5.42
1-1/2	3.8	90	61	27.4	8	10.85
1-3/4	4.5	97	66	29.6	14	18.96
2	5.1	105	72	32.0	22	29.80
2-1/2	6.4	117	80	35.7	53	71.9
2-3/4	7.0	124	85	37.8	81	109.8
3	7.6	130	88	39.6	120	162.7

Terminal velocity for hail diameter 10 mm was measured by Matson and Huggins (1980) as 15-20 m s^{-1}. Theoretically computed Hail Terminal velocity for hail diameter 40mm, 60mm and 80 mm assuming C_d = 0.5 and density = 900 kg/m^3 are 19.8, 24.2 and 28.0 m/s respectively. For 14 cm diameter giant 1970 Coffeyville, Kansas, hailstone Roos (1972) estimated fall speed 47 m s^{-1} .

Changnon (1970) had noted wide variability in the energy values of hailstreaks from season to season. Average energy of 27 hail-producing systems in 1967 was 3757.9erg/cm^2 whereas in 1968 average energy of 20 hail-producing systems was 2559.8erg/cm^2. Hailstreak mean energy values also exhibited wide ranges with minimum value of 1.45erg/cm^2 to a maximum of 184698.3erg/cm^2. The median energy values for 136 hailstreaks was 89.02erg/cm^2 and the median areal extent 7.9mi^2 (20.46 km^2). The median of the maximum energy values per system of 323.98erg/cm^2 was 74 time greater than the median of the minimum energy values (4.378erg/cm^2). The minimum energy values sampled in one hail-producing system was as high as 596.89erg/cm^2 and was as low as 1.45erg/cm^2 in another. This indicated great variability in hailstreak energy values in any given hail-producing system.

11.2 Damage to Property

The amount of damage depends on the number of hailstones as much as on their size. Over many parts of the world severe hailstorms frequently cause considerable damage to buildings, crops, and automobiles, resulting in large economic and insured losses. For example Munich hailstorm on 12 July 1984 (EUR 1.5 billion losses, converted to 2015 values; Heimann and Kurz, 1985; Höller and Reinhardt, 1986; Kaspar et al., 2009) and the two supercells on 27/28 July 2013 triggered in the vicinity of the low pressure system Andreas caused an insured damage of 2.8 billion EUR in Germany, representing the costliest insured loss event worldwide in the year of 2013 (SwissRe, 2014). One year later, hailstorms associated to the event Ela on 8–10 June 2014 were responsible for insured damage of 2.3 billion EUR in France, Belgium, and Germany. In Russia, hail affects around 5000 km^2 of agricultural area each year (Abshaev and Malkarova, 2006). In Hungary hail is the second most important damage source next to drought, and accounts to a damage of 48 million EUR per year in this sector (Kemény et al., 2012). The hail damage in China was reported to be valued as high as billions of U.S.dollars in 2004 alone (Dong et al. 2006).

Hail size of 5 mm is often assumed as threshold for agricultural losses. Vulnerability of crops, however, depends on the plant type. Changnon (1971), found that damage to wheat, corn and soybeans to be closely related to hailstones with a diameter of 6.35 mm. Larger thresholds generally apply for other insurance types such as automobiles or buildings. Field and laboratory investigations (Stucki and Egli, 2007; Gessler and Petty, 2013), have proven that buildings may become damaged by falling hailstones larger than approximately 25 mm (~1 in.).

Fig 11.1 gives a glimpse of the varied types of catastrophe caused by hailstorm.

A view of road after apocalyptic hailstorm over Sydney, Australia on April 25, 2015.

Hailstorm on December 11, 2015 over the southern city of Cordoba, Argentina. Fields on the way to San Antonio, San Carlos Road, the Road to 60 Cuadras, Bouwer, and La Carbonada, Argentina

Carcass of livestock perished during lashing of Hailstones at Hyderabad, Andhra Pradesh, India on 29 Jan 2013.

More than 5000 birds lie dead at Moukori village in Shailakupa upazila under Jhenidah district of Bangladesh due to hailstorm on 6 April 2015

Fig. 11.1 A glimpse of the various types of destructions caused by hailstorm.

Table in Appendix 'A' shows the losses due to hail damages over different parts of India during 1986-1991. It may be roughly estimated that total loss in respect to property, cattle and crop in the country comes to annually around 200-350 million rupees at the rate of 1990. States which are venerable to hail damage are Assam, Tripura, Uttaranchal, Andhra Pradesh, Bihar, Jharkhand, Karnataka, Maharashtra, Madhya Pradesh, Orissa, Rajasthan, Uttar Pradesh and Gangetic West Bengal.

11.3 Loss of Life

Losses of life due to Hailstones are not uncommon in different parts of the world. One of the most notorious regions for large hail is northern India and Bangladesh, which have reported more hail-related deaths than anywhere else in the world and also some of the largest hailstones ever measured. The greatest loss of life from single hailstorm occurred on April 30, 1888 in Moradabad and its neighboring districts of North India when 246 persons perished from hailstone injuries. Mainland China is also notorious for killer hailstorms. Second largest loss of life has been recorded from China where on June, 19, 1932 in Honan province of China, 200 persons were killed and thousand injured in a severe hailstorm (Flora-1956). Certain locations in North America (such as the area around Calgary, Alberta) have gained the nickname "Hailstorm Alley" among meteorologists for the frequency of hailstorms and their severity. A five years report released by Insurance Information Institute, U.S.A.for 2011-2015 has reported 112 injuries due to hail hit but no fatalities there.

Table 11.2(a) shows year wise total deaths or injuries caused in India from 1991 to 2000. It may be noted that on an average during the decade 28 persons died each year (standard deviation = 38). 1997 was the worst year, which recorded 130 deaths by hailstone hit in any single year. A large number of injuries were also reported from various parts of the country. Many of them remain, however, unreported. Based on some cases which are reported in newspapers, varying from a few to as many as 320 suffered injuries, each year, due to hailstones in India only.

Table 11.2(a)

Year	Deaths due to hit by hailstone	Suffered injuries due to hit by hailstone
1991	14	Many
1992	9	Many
1993	8	3
1994	23	320
1995	13	320
1996	3	140
1997	130	63

Table **11.2 (a)** *contd...*

Year	Deaths due to hit by hailstone	Suffered injuries due to hit by hailstone
1998	61	200
1999	5	50
2000	19	Many
1991-2000	285	>1200(Approx.)

State wise loss of life during 1986-2000 is mentioned in Table 11.2(b). Data is based on only reported cases in news papers. During 15 years period, it can be observed that deaths due to hailstone injuries are very high in the state of Bihar, Uttar Pradesh, Maharashtra and Rajasthan in general. 8 persons had died in Kanpur (Uttar Pradesh) on 28 February 1992. Prior to 1986, too, there were several sporadic cases of hailstone-deaths of more than 10 persons at a place from various other states. From the Dharwar district of Karnataka there had been a report of deaths of 11 persons due to hailstones on 30 Jan 1966. In one of the hailstorm, on 30 Oct 1961 in Bhind district of Madhya Pradesh 12 persons had died and 97 were injured due to hailstones. 9 persons died near Nagpur (Maharashtra) in February 1979. There had been a report of death of 18 persons on 13 March 1981 from Telengana in Andhra Pradesh.

Table 11.2(b) State wise loss of life or injuries to humans between 1986-2000

States	Deaths	Injuries*
Andhra Pradesh	08	
Assam and other adjoining north eastern states	40	More than 100
Bihar	97	50
Karnataka	15	Many
Madhya Pradesh	17	10
Maharashtra	67	Many
Orissa	08	32
Rajasthan	48	263
Uttar Pradesh	64	More than 100
West Bengal	15	30
Total	379	>2000(Approx.)

*Figures are as per the news paper reports and not official

11.4 Damage to Automobiles

Figure 11.2(a) shows the broken windshield of a car. The size of stones which caused the damage to windshield were about an inch larger than the stones shown in picture. It was, however, noticed by the storm chasers that even golf ball sized stones do not always break the glass. It depends on the hardness of the stones and the angle of the impact. Occasionally a motorist may be lucky and get hit with a large soft stone when temperatures aloft are warmer. The soft stones crack apart absorbing the impact. In this case shown in Fig. 11.2(a) the stones were rock hard.

Shattered windscreen of car

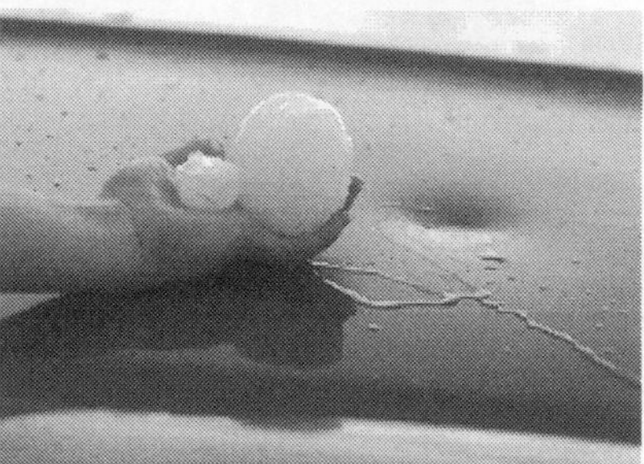

Dented hood of car

Fig. 11.2(a)

Angle of impact is affected by the relative velocity of the falling stone. Higher is the relative fall velocity the higher is the angle of impact. On vehicle generally impct speeds are less that 50 m/s. It was observed that 40 mm hail impacting a bonnet of Proton Satria car (Fig.11.2 (b)) gave a dent diameter range of 9 mm to 28 mm and a dent depth range of 0.2mm to 1.49 mm.

Fig. 11.2(b) Proton Satria car.

The simulated results showed that a 40mm hail impact on mild steel plate of size 10x10 cm^2, 15x15 cm^2 and 25x25 cm^2 yielded dent depth between 1.79mm to 1.59 mm and dent diameter between 25.6 mm to 24. 9 mm (Somasundaram-2012).

11.5 Damage of Aircrafts

Aircrafts flying through hailstorms are subjected to short but intense bombardment which leaves the aircrafts alarmingly pitted with high impact speed in excess of 80 m/s with larger deflection after impact than over lower speed cars on ground. Almost every year in spite of utmost care and vigilance exercised by the aircrew to avoid flying through hailstorms several reports are received of considerable damage to aircrafts, caused by the hailstones. Most of them go unreported in scientific literature. To quote a few are following,

"On 27 May 1959, a Viscount aircraft flying on Karachi (Pakistan) – New Delhi route at 5700 meter amsl encountered a severe hailstorms which damaged the aircraft to such an extend that it had to be sent back to its manufacturer for repair. Almost all the leading edges of exposed surfaces were perforated with holes, some of which were as large as 12-15cm in diameter. One opening on the starboard side of the nose of the aircraft was perhaps, even larger in size and another hailstone pierced directly through the wind screen, completely smashing it and seriously injuring the pilot. Apart from holes made on the aircraft surface by the hailstone ripping open the metallic cover, there were innumerable dents both large and small, some of them as large as 37 × 25 cm". Refer Fig. 11.3.

Fig. 11.3 Photograph of Viscount Aircraft Damaged by Hailstones on 27 May 1959 while flying on Karachi (Pakistan) and New Delhi (India) route.

On 28 January 1959, while flying near Jaipur (India) at 6 km height, a jet aircraft got caught in a severe hailstorm. The damage recorded in the case were; Engine swirl vanes dented, tail plane leading edges dented and fabrics on the leading edges of the rudder fin damaged. Large hails have been observed at the height of even 9000 meter by a US Air Force jet Bomber which was forced to land after being critically damage by the hailstones which exceeded 5 cm in diameter. Attempt were, however, made by Harrison (1956) of American Airlines and by subsequent workers to demonstrate that radar could be used to help indicate hail bearing cloud but techniques are not 100% fool proof.

11.5.1 Effect of Hailstone on Aircraft in Flight

The angle at which large stones would impinge on approaching aircraft could be calculated considering the relative velocity of fall and speed of the aircraft at the given height. The relative fall velocity of a hailstone is terminal velocity, diminished by the speed of updraft. If Q be the relative velocity of a hailstone impinging on an aircraft of speed I, the angle of impact (angle above horizon) is given by Tan^{-1} (Q/I).Table 1.3 and 1.4 by Saha (1961) gives the angle of the impact of hailstones of diameter on two aircraft, one flying at 6 km ASL and at airspeed of 200 kt and another at 9 km at the airspeed 300 kt, assuming average updraft (u) of 16 m/sec and extremely strong updraft of 30 m/sec at both heights. The negative values of relative fall velocity are indicative of upward movement of small size hailstones under strong updraft. Likewise negative angles of impact would mean that hailstones impinging on aircraft at angles slightly below the horizon. Therefore it infers that the angle of impact of hailstone on aircraft in general decreases with the speed of aircraft and the speed of updraft inside the cloud. Also higher is the relative fall speed the higher is the angle of impact. Jet aircraft flying at the speed through Cb cloud are, therefore liable to be hit at low angles of elevation and it is mostly the leading edges of exposed surface like the nose, windscreen, leading edges of wings, rubber, fine and engine swirl vanes which are likely to be affected. The following conclusions may be derived regarding size of hailstones likely to be encountered by aircraft at different heights:

(a) The region above 10.5 km will be free from hailstones.

(b) Aircraft flying in the layer 10.5-9.0 km stands the risk of being hit by hailstones of size up to 3cm in diameter. But the real danger will probably lie in the layer 9.0-4.5 km where giant size hailstones of diameter ranging from 3-15 cm may hit the aircraft. The danger continues even below the freezing level right upto the ground.

Table 11.3 Relative fall velocities of hailstones and angles of impact on aircraft, flying at level 6 km a.s.l., aircraft speed 200kt and updraft(u)

Diameter (cm)	2	3	4	6	8	10	12
Relative fall velocity Q (m/s) (u=16 m/s)	-2	0	5	20	36	60	78
Angles of impact, (degree)	-1	0	2	11	20	31	38
Relative fall velocity. Q (m/s) (u=30 m/s)	-16	-14	-9	6	22	46	64
Angles of impact, (degree)	-9	-8	-5	2	12	26	37

Table 11.4 Relative fall velocity of hailstones and angles of impact on aircraft flying level at 9 km a.m.s.l., aircraft speed 300 kt and updraft (u)

Diameter (cm)	2	3	4	6	8	10	12
Relative fall velocity, Q (m/s) (u=16 m/s)	0	4	9	24	43	71	94
Angles of impact (degree)	0	1	3	9	16	25	32
Relative fall velocity, Q (m/s) (u= 30 m/s)	-14	-10	-5	10	29	57	80
Angle of Impact(degree)	-5	-4	-2	4	11	21	27

11.5.2 Hail Outside Thunderstorms

Although hail is created in the strong thunderstorm updrafts, it can and frequently does fall through the clear air or the masking clouds surrounding the cumulonimbus cell. The presence of a strong vertical wind shear spreads the cirrus top downwind or upwind (refer section 5.7) and perhaps gives a tilt to the thunderstorm, thus increasing the likelihood of hail fall out of the active cell. Consequently the possibility exists of encountering hail for a distance of 2 or 3 km in the direction of the intense radar echoes and upto 10 km on the downwind sides. There have been numerous reports in commercial publications and service periodicals of hail encounters well away from radar echoes of the thunder clouds. A study by Beck (1959) limited to 103 inflight hail encounters revealed the following,

(a) About 30% of all encounters below 3.0 km occurred in clear air outside the thunder cloud.

(b) Approximately 40% of the encounters in the range of 3.0 km to 6.0 occurred in clear air and 82% of these clear air encounters were under an overhanging cloud of thunderstorm.

(c) 20% of the hail encounters above 6.0 km were found in clear air. All encounters were beneath the anvil cloud or other overhanging clouds of the storm.

A major difficulty in avoiding hailstorm during flight is because of the fact that thunderstorm are frequently masked by altostratus. Inadvertent penetration of hail columns of thunderstorms masked by cirrus clouds have been reported at levels as high as 10.5 km. Such penetration apparently occurs because masking cirrus shield end to resemble cirrus deck not associated with thunderstorms. Large isolated thunderstorms particularly those occurring in the great plains of United States as well as frontal squall lines, frequently generate large cirrus shields that have been observed to extend as far as 100 miles downwind from the storm. Large hail fall may also be encountered while flying through the Rear Flank Downdraft (RFD) which is the region of dry air wrapping around the back of a supercell thunderstorm in otherwise clear air.

11.6 Roof Top Damage by Hail Fall

Roof damage by hailstone damage is common feature all over world. This hail roof damage can happen anywhere in the country, usually in conjunction with severe thunderstorms or tornadoes. While many factors affect the extent of hail damage, a roof will sustain the following list contains the most common factors that effect the extent of roof hail damage:

- Age of existing roof system.
- Surfacing of existing roof system.
- Type of substrate or insulation under roof membrane.
- Size of hail stone.
- Density of hail stone.
- Ambient temperature.
- Angle of impact.
- Type of roofing system.

Recently there had been attempts to correlate the damage with the size of hailstone (Crenshaw and Koontz - 1991, Timothy et al 1999).

Table 11.5

Hail size (in)	Hail size (cm)	Typical damage threshold
½	1.3	FEW, IF ANY, ROOFS DAMAGED. Bushes and trees- leaves stripped; crops- damaged. Oxidation coatings on paint, wood, metal- spattered. Thin elastomeric coatings on polyurethane foam roofs- cracked or broken. Thin aluminum vents, fins on air conditioning units, lead sleeves on soil stacks, window screens, aluminum awnings- dented
¾	1.9	THRESHOLD SIZE FOR DAMAGE TO ROLL ROOFING AND DETERIORATED ASPHALT COMPOSITION SHINGLES, ESPECIALLY WHERE UNSUPPORTED. Painted wood surfaces, deteriorated gray-black slates (especially at corners)- chipped. Most aluminum vents, flashings, valleys, siding- dented.
1	2.54	THRESHOLD SIZE FOR DAMAGE TO MOST LIGHTWEIGHT ASPHALT COMPOSITIONSHINGLES. Thin and/or deteriorated wood shingles, shakes- occasionally punctured or cracked. Single-pane windows, thin skylight shells- cracked or broken.
1-1/4	3.2	THRESHOLD SIZE FOR DAMAGE TO MOST HEAVY WEIGHT ASPHALT COMPOSITION SHINGLES, WOOD SHINGLES, AND OLDER MEDIUM SHAKES. Automobile body metal- dented; galvanized metal vents- dented. Older plastic skylights- cracked or broken.
1-1/2	3.8	THRESHOLD SIZE FOR DAMAGE TO CLAY TILE, SLATE, MEDIUM SHAKES, AND MODIFIED BITUMEN SINGLE-PLY MEMBRANES. Automobile body metal- extensive denting
1-3/4	4.4	THRESHOLD SIZE FOR DAMAGE TO HEAVY SHAKES AND CONCRETE TILES. Metal vents- caved in. Bare spots and blisters on deteriorated built-up roofs-bruised or punctured.
2	5.1	THRESHOLD SIZE FOR DAMAGE TO JUMBO SHAKES, WELL-SUPPORTED UNBALLASTED BUILT-UP ROOFING, AND UNBALLASTED EPDM.
2-1/2	6.4	THRESHOLD SIZE FOR DAMAGE TO WELL-SUPPORTED BALLASTED BUILT-UP ROOFING, BALLASTED EPDM, AND METAL PANELS.

Notes:

1. Threshold is the smallest hail size at which damage can occur.
2. These guidelines are for hard hail ice hailstones that strike the impacted material, in relatively good weathered condition, perpendicularly to it surface.
3. These general guidelines apply in most circumstances, but there are exceptions. Determining factors include material properties, deteriorated condition, and underlying support.
4. This information is based on testing and field experience

The higher the size of the hailstone, the higher the terminal velocity. In turn, higher terminal velocity means higher kinetic energy at the time of impact to cause the damage. In Fig. 11.4 terminal velocity computations are made under no wind conditions. If wind is also accounted for then the energy of impact increases causing more damage.

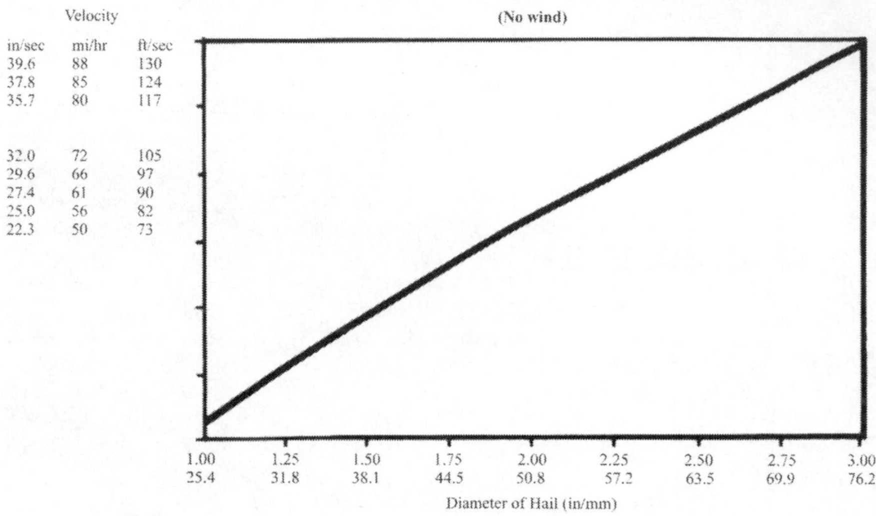

Fig. 11.4 Terminal velocity under no wind conditions.

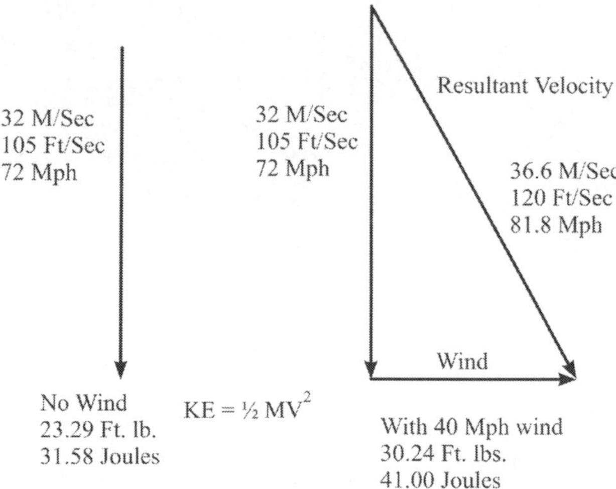

Fig. 11.5 Note the increase in kinetic energy when wind of 40 mph (64.4kmph) is added.

In Fig.11.5 wind of 40 mph is vectorially added to the terminal velocity of falling hail under no wind condition. The total kinetic energy increases from 31.58 Joules to 41.0 Joules. As the terminal velocity increases, with the size of the hailstones, resulting kinetic energy of the hailstone also rapidly increases. Fig. 11.6 shows the relative increase of kinetic energy under different wind speeds of 0, 20, 40 and 60 mph or (0, 32.2, 64.4, 96.6 kmph).

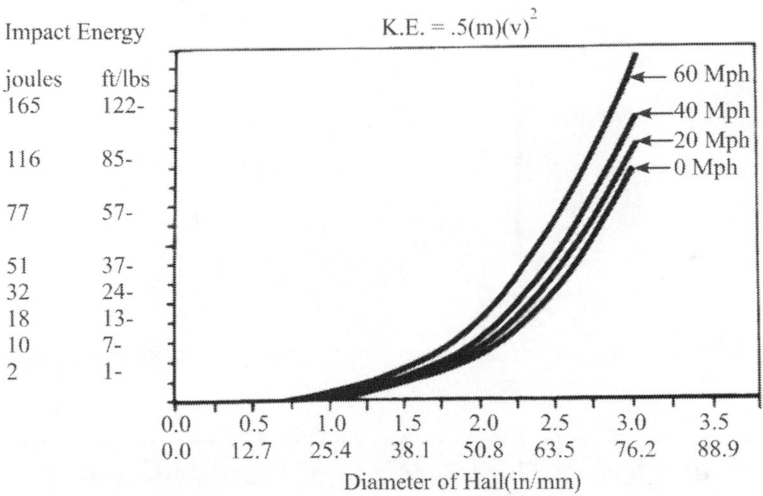

Fig. 11.6 The relative increase of kinetic energy under different wind speeds of 0, 20, 40 and 60 mph (0, 32.2, 64.4, 96.6 kmph).

Damage to the roof is also caused due to the angle of impact. Hail hitting at right angle would deliver maximum impact energy. Refer Fig. 11.7. Higher the perpendicular component of the force to the roof, higher is the damage to the roof.

While hail suppression continues to elude scientists, sophisticated radar has been developed as explained in chapter - V that can detect the presence of hail before it falls to the ground. Eventually, warnings may be issued as much as 15 minutes to even 60 minutes or more prior to hail strike, allowing pilots to avoid threatening air space, people to seek shelter, and property to be protected.

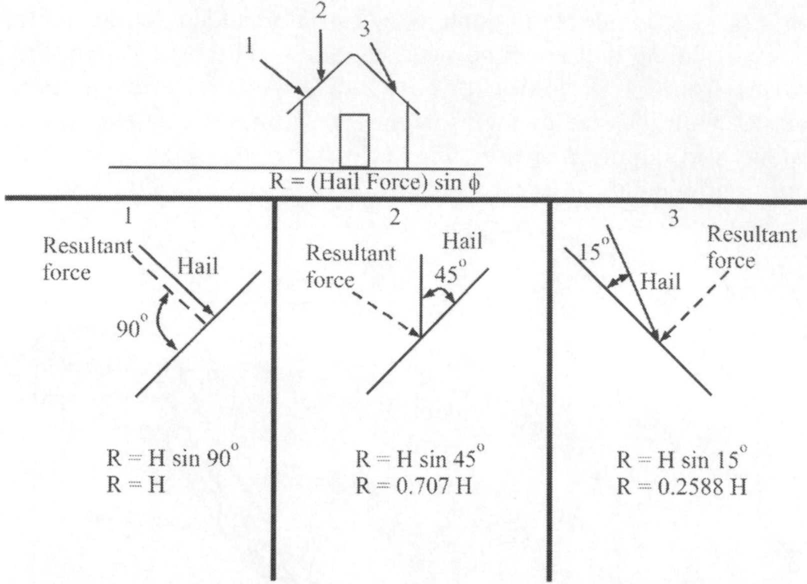

Fig. 11.7 Higher the perpendicular component of the force to the roof, higher is the damage to the roof .

11.7 Protection against Hail Damage to the Crops

Farmers seek monetary compensation for hail damage, but in the past, farmers had no recourse when their crops were destroyed. They were left to their own ingenuity to try to suppress hail. In the 14th century, people in Europe attempted to ward off hail by ringing church bells and firing cannons. Hail cannons were especially famous in the wine-producing regions of Europe during the 19th century, and modern versions of them are still used in parts of Italy. After World War II, scientists across the world experimented with cloud "seeding" as a means of reducing hail size. In Soviet Georgia, scientists fired silver iodide into thunderclouds from the ground. Such methods supposedly stimulated the formation of large numbers of small hailstones, which would melt before they reached the ground, but subsequent comparable experiments performed in Switzerland and the United States did not confirm Soviet theory. Fig. 11.8 shows maze field severely battered by hailstones.

Fig. 11.8 Crop Damage due to hailstones.

11.8 Assessing the risk of Hail Damage and Insurance Cover

All over the world agriculture is synonymous with risk and uncertainty. Agriculture contributes to 24% of the GDP and any change has a multiplier effect on the economy as a whole. Economic growth and agricultural growth are inextricably linked to each other. Crop insurance helps in stabilization of farm production and income of the farming community. It helps in optimal allocation of resources in the production process. Indian Government has been concerned about the risk and uncertainty prevalent in agriculture. Unfortunate serial suicides of farmers in Maharashtra who got caught in a debt trap due to crop failures and the devastating effect it had on their families had been the tragic reality of agriculture in India. Rashtriya Krishi Bima Yojana (RKBY), launched in India in 1999 under the National Agricultural Insurance Scheme (NAIS), insures farmers against the crop damage. NAIS is being implemented by 23 states and two Union territories. During the 12 crop seasons (from Rabi 1999-2000 to Kharif 2005), 7.51 crore farmers were covered over an area of 2.2 crore hectares insuring a sum of Rs.70,696 crore. Claims paid was Rs.7207 crore against premium income of

Rs.2226 crore; benefiting more than two crore farmers in the implementation of NAIS so far. Worlwide crop insurance is similarly prone to losses. By and large the claims ratio has been around 500 per cent. Insurance companies may feel that crop insurance is a liability not a profitable proposition at all. Estimating crop loss due to an unexpected weather event is difficult so also estimation of potential yield and actual yield.

Risk assessment is key factor for the insurance companies to launch any scheme effectively. Correct regional and seasonal risk assessment can replace the present approach of uniform premium for all types of crop damage to linking the risks with the type of crops. Risk assessment at any point can be computed if the climatological isolines chart of the event are available. Insurer has to compute the value of the event at the place where crop insurance is to be provided. If the place falls over the isoline in the map the value is directly read from the isoline. If the place falls in between two isolines then following method may be used to interpolate the value of the event at the place.

Method of Computation of Risk at a Place

Chapter – I has figures of climatology of hailstorms in India and neighbourhood for all months. Hence to assess the risk of hail damage at any place the monthly climatology can be usefully applied. If the place falls over the drawn isolines then direct reading of probability of occurrence of hailstorm could be made. If the value is to be interpolated at any point or place 'P' in between two isohailstorm lines then following method could be used. Refer figure 11.9 and zoomed picture below it.

1. Draw a circle of a radius such that minimum two isolines are cut by the circle.
2. Draw shortest distances from the point 'P' on all the isolines.
3. Arrange values of isolines (h) in descending (or ascending) order. If any two values are equal then any one may be ordered first.
4. Arrange values of shortest distances (l) in ascending (or descending) order. If any two values are equal then any one may be ordered first.
5. Take the sum of the products of h and l ordered sequentially.
6. Divide the sum in (4) by the sum of all the lengths e.g. $(l_1 + l_2 + l_3)$

Detailed computation is shown in Table 11.6.

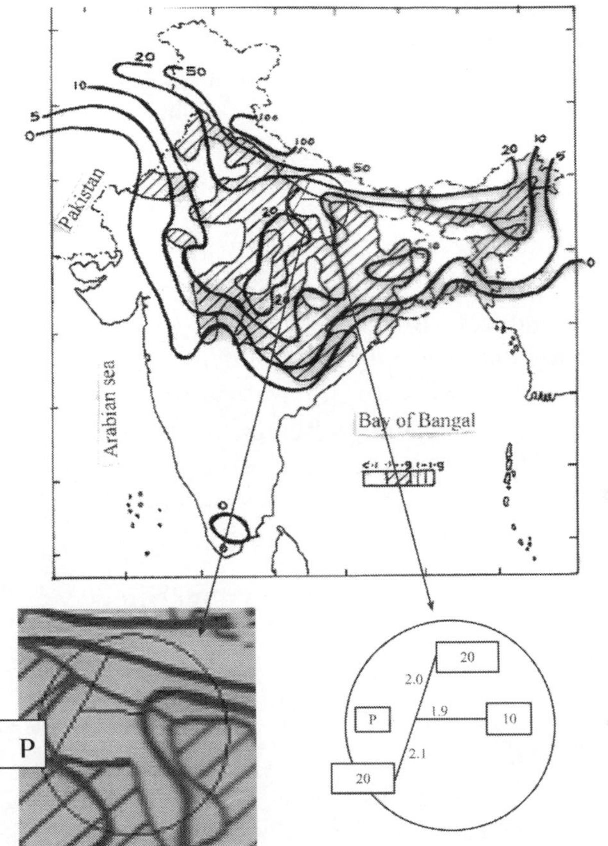

A neat diagram of computation

Fig. 11.9 (Top) Average number of hailstorms in February. Hatched scheme by Philips and Daniel (1976) and isolines of frequency of days of hailstorms in 100 years by Ramdas et al (1938); (Lower-Left) Enlarged circle and point 'P' where Average value of hailstorm in February is to be computed. (Lower-Right) Distance from 'P' to isohail lines in the southwest (h_3=20), northeast (h_2=20) and east (h_1=10) are 2.1 cm (l_3), 2.0 cm (l_2), and 1.9cm (l_1), respectively. Hence value of h at point 'P' is 16.5 as per the details explained in Table 11.6.

Table 11.6

Values of Isolines in descending order	h_2=20	h_3=20	h_1=10
Values of lengths of perpendicular distances drawn from point 'P' to isolines in ascending order(Cm)	l_1=1.9	l_2=2.0	l_3=2.1
Value of h at point P	$h=(l_1 h_2 + l_2 h_3 + l_3 h_1)/(l_1 + l_2 + l_3)$ h = (1.9x20 + 2.0x20 + 2.1x10)/(1.9+2.0+2.1) = 16.5		

For assessing the hail risk Storm Prediction Centre, NOAA, National Weather Service of U.S.A. also has designed maps. For that the probability values were estimated from a 30-year period of severe weather reports over U.S.A.from 1982-2011. The procedure to create the maps is as follows:

1. Reports for each day are put onto a grid 80 km × 80 km.
2. If one or more reports occur in a grid box, that box is assigned the value "1" for the day. If no reports occur, it's a zero.
3. The raw frequency for each day at each grid location is found for the period (number of "1" values divided by number of years) to get a raw annual cycle.
4. The raw annual cycle at each point is smoothed in time, using a Gaussian filter with a standard deviation of 15 days.
5. The smoothed time series are then smoothed in space with a 2-D Gaussian filter (SD = 120 km in each direction).

Readers are encouraged to visit National Weather Service of U.S.A. website for viewing the maps.

References

1. Abshaev, M.T., Malkarova, A.M., Buranova, I., 2006. On the climatology of hail in the North Caucasus. Mater. VI Conf. Young Scientists KBSC. Russian Academy of Sciences, Nalchik, pp. 5–12 (in Russian). URL http://vgistikhiya.ru/climatinfo_page.html.
2. Beck, R.E. 1959, Hail aloft summary, MATs Flyer, 6 No. 2, 13 pp.
3. Bilham, E.G. and Relf, E.F., "The Dynamics of Large Hailstones," *Royal Meteorological Society, Vol. 63,* 1937, pg. 149.
4. Changnon Stanley A, Halstreak, Journal. of Atm. Sc.,Vol. 27, Jan 1970, pp109-125
5. Changnon, S.A., 1971. Note on hailstone size distributions. J. Appl. Meteorol. 10, 168–170.
6. Changnon Stanley A. (1999), Data and Approaches for Determining Hail Risk in the Contiguous United States, Journal of Applied Meteorology 1999; 38: 1730-1739.
7. Crenshaw, Vickie A. and Koontz, Jim D., "Simulated Hail Damage and Impact Resistance Test Procedures for Roof Coverings and Membranes," *RCI Interface*, May 2001, pg.4.

8. Compu-Weather, www.compu-weather.com.
9. De U. S., Dube R. K. and Prakash Rao G. S., Extreme weather events over India in the last 100 years, Journal of the Indian Geophysical Union, 9, 173-187 (2005).
10. Dong, W. J., Q. Zhang, J. X. Guo, and Y. Chen, 2006: *Annals for Chinese Nature Disasters in 2004* (in Chinese). China Meteorological Press, 193 pp.
11. Flora, S.D. 1956, Hailstorm of United states, University of Oklahama on, Norrmann, Oklahamam, 201 pp.
12. Friedman, D. G., 1976: Hail suppression's impact on property insurance. TASH Rep. 11, Illinois State Water Survey, Urbana, IL, 32 pp. [Available from Illinois State Water Survey, Champaign, IL 61820.].
13. Gessler, S., Petty, S., 2013. Forensic Engineering: Hail Fundamentals and General Hail strike Damage Assessment Methodology. CRC Press, pp. 25–65 Ch. 2–4.
14. Greenfeld, Sidney H., "Hail Resistance of Roofing Products," *Building Science Series (BSS) 23, NationalBureau of Standards*, August 1969.
15. HailTrax.com,www.hailtrax.com/faq
16. "Impact Resistance of Prepared Roof Coverings," *Standard UL 2218 Underwriters Laboratories Inc.,* May 31, 1996.
17. Harrison, H.T., 1956, The display of weather echoes on 5.5 cm radar, United Airlines, Met Circular, No. 39, Denever, Colorado (Abridged version Aeronautical Engineering Review, 15, pp 102-109, 1956).
18. Hairston, J. "Report of Test on Hail Impact of Roof Panels from General Testing and Inspection Agency, Inc.", *National Bureau of Standards, Test No. 73-510,Washington, D.C.,* November 1972.
19. Heimann, D., Kurz, M., 1985. The Munich Hailstorm of July 12, 1984: a discussion of the synoptic situation. Beitr. Phys. Atmos. 58, 528–544
20. Höller, H., Reinhardt, M.E., 1986. The Munich hailstorm of July 12, 1984 — convective development and preliminary hailstone analysis. Beitr. Phys. Atmos. 59, 1–12.
21. Kaspar, M.,Müller, M., Kakos, V., Rezacova, D., Sokol, Z., 2009. Severe storm in Bavaria, the Czech Republic and Poland on 12–13 July 1984: a statistic- and model-based analysis.Atmos. Res. 93 (1-3), 99–110.

22. Kemény, G., Varga, T., Fogarasi, J., Tóth, K., Tóth, O., 2012. Problemkreis and Schadenkalkulation einer Mehrgefahrenversicherung in dem ungarischen Ackerbau.
23. Ökosystemdienstleistungen und Landwirtschaft. Österreichischen Gesellschaft für Agrarökonomie, pp. 41–42 (in German).
24. Koontz, Jim D., *"A Comparative Study of Dynamic Impactand Static Loading of One-Ply Roofing Assemblies,"*Special Technical Publication 959, American Society for Testing and Materials, 1988, page 24.
25. Koontz, Jim D., "The Effects of Hail on Residential Roofing Products, "*Proceedings of the Third InternationalSymposium on Roofing Technology,*" *NRCA/NIST*, 1991, pg. 206.
26. Kunkel, K. E., R. A. Pielke Jr., and S. A. Changnon, 1999: Temporal fluctuations in weather and climate extremes that cause economic and human health impacts: A review. *Bull. Amer. Meteor. Soc.,* **80,** 1077–1098.
27. Laurie, J.A. P., "Hail and its Effects on Buildings," *ResearchReport No. 176, NBRI, Pretoria, South Africa,* August 1960.
28. Marshall, Timothy P. and Richard F. Herzog, 1999: Protocol for assessment of hail-damaged roofing, North American Conf. on Roofing Technology, Toronto, Canada, 40-46.
29. Mathey, Robert C., "Hail Resistance Tests of Aluminum Skin Honeycomb Panels for the Relocatable Lewis Building, Phase II," 1970, page 20.
30. Matson, R. J., and A. W. Huggins, 1980: The direct measurement of the sizes, shapes, and kinematics of falling hailstones. *J. Atmos. Sci.*, **37**, 1107–1125
31. Morrison, Scott J., "Long-Term Effects of Hail Impacts-An Interim Report", *North American Conference on Roofing Technology,* NRCA, 1999, pg. 30.
32. National Climatic Data Center, lwf.ncdc.noaa.gov /oa/climate /severeweather/ extremes.html.
33. Rigby, Charles A. and Steyn, A.K., The Hail Resistance of South African Roofing Materials, *South African Architectural Record, Vol. 37, No. 4,* April 1952.
34. Roos, D. S., 1972: A giant hailstone from Kansas in free fall. *J. Appl. Meteor.*, **11**, 1008–1011.
35. Saha, S.K.(1961) IJMG.

36. Schleusener, Richard A. and Jennings, Paul C., "An Energy Method for Relative Estimates of Hail Intensity," *Bull. Amer. Meteorol. Soc. Vol. 41, No. 7,* July 1960.

37. Somasundaram, Visvarajah. http://www.slideshare.net/visvarajah/the-characterisation-of-hail-and-fraudulent-impacts-to-vehicle-body-panels

38. Stucki, M., Egli, T., 2007. Synthesebericht Elementarschutzregister Hagel. Tech.

    a. rep.Präventionsstiftung der kantonalen Gebäudeversicherungen (in German)

39. SwissRe, 2014. Sigma: Natural- and Man-made atastrophes 2013. Tech. Rep.Swiss Re Economic Research and Consulting

40. "Specification Test Protocol for Impact Resistance Testing of Rigid Roofing materials by Impacting with Freezer Ice Balls," *Factory Mutual Research Corporation, Class Number 4473,* February 2000.

41. "Susceptibility to Hail Damage, Test Standard for Class 1 Roof Covers," *Factory Mutual Research Corporation, Class Number 4470, Class 1 Roof Covers,* Revised August 29, 1982.

    "Standard Test Method for Impact Resistance of Bituminous Roofing Systems," *ASTM D-3746.*

APPENDIX - A

Damage due to Hailstorm in India during 1986-91

Damage due to Hailstorm during 1986-91								
Sr. No.	Date	(States of India) Name of Region/District	Damages					Total Loss in financial terms
			Human		Houses	Cattle	Crop	
			Killed	Inju-red				
1	7-14 Feb 86	(Maharashtra) Nagpur, Bhandara, Chandrapur, Wardha, Amaraoti, Akola, Yeotmal					Heavy losses to standing crop	
2	7-13 Feb 86	(Maharashtra) Jalgaon, Dhule, Nanded Parbhani	2	10				205 Lakh

Contd...

3	6th Apr 86	(West Bengal) 24 pargana, Bardwan	2			Several		
4	9-13 Feb. 86	(Utter Pradesh) Lalitpur and Manpuri					Extensive Damage	
5	19th Feb. 86	(Utter Pradesh) Mathura					Extensive Damage	
6	9th Feb. 86	(Rajasthan) Chittorgarh, Jalawar					Extensive Damage	
7	10th Feb. 86	(Andhra Paradesh) Warngal					Extensive Damage	
8	12th Feb. 86	(Karnataka) Gulbarga					Extensive Damage	
9	13th Apr. 86	(Tripura) Kamalapur		50				
10	16th May 86	(Rajasthan) Baharatpur & Dholpur	17			512		25 Lakh 125 Lakh
11	25th Dec. 86	(Maharashtra) Nagpur					Considerable damage to Rabi Crops	
12	26th Dec. 86	(Maharashtra) Maratwada Pune	7			175		5.50 Cr in 80 villages in Pune
13	27th Dec. 86	(Maharashtra) Madhya Maharashtra *Nasik, ** Ahmednagar						*3.5 Lakh ** 80 Lakh
14	10-11 Jan. 87	(Madhya Pradesh) Sehore, Ujjain, Ratlam, Shajapur & Dewas					Considerable Rabi crops damaged	
15	13th Mar. 87	(Andhra Pradesh) Nellore	4		200	40	Extensive damage to standing crop	1 Crore

Contd...

16	16th Feb. 88	(Madhya Pradesh.) Chandrapur	6					
17	19th Feb. 88	(Chattisgarh) Raipur					Heavy damage to crops	
18	21st Feb. 88	(Maharashtra) Kamtee tahsil 10 villages in Bhandara Dist	1					
19	1st Mar. 88	(West Bengal) Howrah)	1	4				
20	24th Mar. 88	(Assam) Gotaghat, Kamprup, Nowgungm sonitpur, barpet, Lakhinpur	3		100			
21	22nd Apr. 88	(Maharashtra) Panchgani & Mahabaleswar	3				Extensive damage to raspberi crop	
22	24th Apr. 88	(Karnataka) Kolar Dist				Considerable Damage	Considerable damage	
23	9th Mar. 89	(Andhra Pradesh) 15 villages of Ananthpur dist 8 villages of mehabubnager Dist				Widespread Damage	Widespread Damage	1.04 Cr.
24	25th Mar. 89	(Orissa) Jangam Dist	1	4				
25	29TH Mar. 89	(Maharashtra and Karnataka) North Madhya, Maharashtra, Maratwad & in North Interior Karnataka, Gadag & Dharwar			Widespread	Many	Widespread	

Contd...

26	18th Apr. 89	(Assam) Sibsagar	4		800 houses			
27	26th Apr. 89	(Karnataka) Shimoga,South Interior Karnataka				Extensive damage to Cocunut and banana gardens		
28	2nd May. 89	(Maharashtra) Nasik City	2					
29	17-18 Feb. 90	(Assam) Dibrugarh & Lakhiapur Ditt				Extensive damaged to Rabi crops		
30	20th Feb. 90	(west Bengal) 24-Pargana Dist	1			Several Houses		
31	29th Mar. 90	(Bihar) 400 villages of vaisali Dist						1 Crore
32	2 & 5 May.90	(Himachal Pradesh) Simla					Horticulture crops damaged	
33	11 Feb. 91	Delhi						30 Lakhs
34	27th Feb. 91	(Assam) Lakhimpur Dist				Extensive Damage to power supply		
35	28th Feb. 91	(Assam) Karimnagar Dist				Considera ble No. of livestock perished		10 Crore
36	8th Mar. 91	(Assam) Tinsukhia Dist					Tea Garden, Vegetable crops damaged	
37	2 & 12 Apr. 91	(Orissa) Puri, Balasoore, Cuttack, Kerojhar	5	28			Extensive damage to Vegetable crops	

Contd...

38	24th Apr. 91	(West Bengal) Ranaghat-Nadia Dist	4	21				

Note:

1. Blank in the last column means reliable financial estimate of damage not available.

2. Blank in other columns means "Nil".

3. Damage valuation is as per 1991 Rupee- rate.

Appendix - B

Meteorological Definitions and Indices Related to Thunderstorm

I. Skew-t– Log P Diagram

The skew-t–Log P-diagram is the commonly used thermodynamic diagram. A large number of meteorological variables, indices, and atmospheric conditions can be found directly or through simple analytical procedures. Typically, the environmental temperature, dew point temperature, wind speed and wind direction at various pressure levels are plotted on the diagram. This plot is commonly called a sounding. Sounding data come from weather balloons that are launched around the country at 00Z and 12Z, as well as various special situations in which they are used in field experiments and other campaigns. Figure1and 2are examples skew-t-log P diagram.

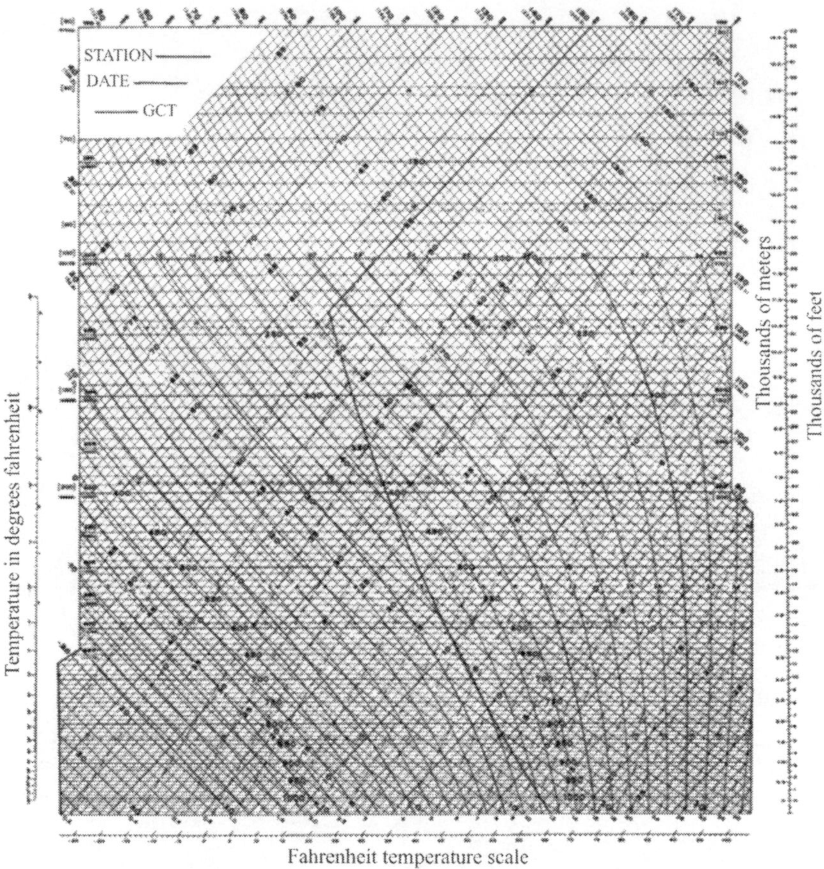

Fig. 1 Skew-t-Log P Thermodynamic Diagram.

Let's take a closer look at Figure 1 and identify the lines on the skew-T diagram. Fig. 2 is a close up of the lower right corner of the diagram in Fig. 1. Each line is labelled accordingly. The solid Diagonal lines are isotherms, lines of constant temperature. Temperatures are in degrees Celsius, and a Fahrenheit scale is also at the bottom of the diagram. The dashed lines are mixing ratios. The dry adiabats on the diagram are the curved lines with the lesser slope and are drawn at 2° intervals. Moist adiabats have a much greater slope and follow much more vertical path on the skew-T diagram. Two height scales are located on the right side of the diagram. The left scale is the height in meters and the right scale is height in thousands of feet. Pressure levels are in millibars (mb)/hectapascals (hPa).

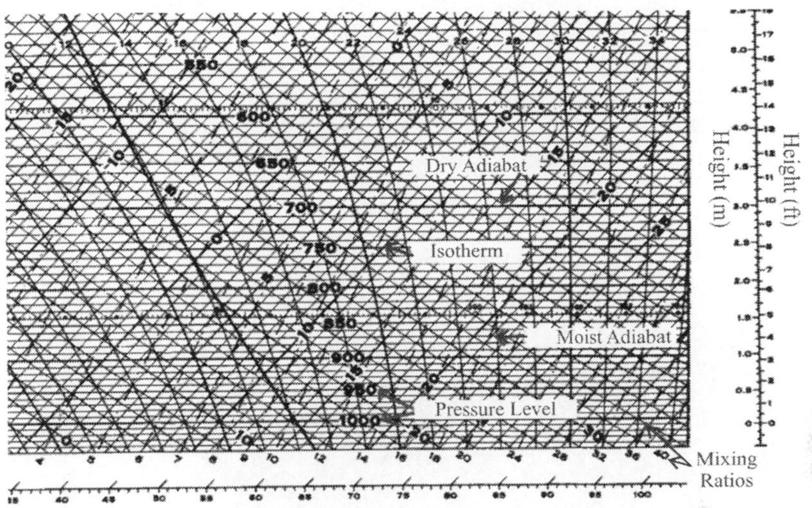

Fig. 2 A close up of a skew – t - Log P diagram presents the various definitions of lines located on the diagram.

II. Levels

(a) ***Lifting Condensation Level (LCL)*:** The level at which a parcel of air first becomes saturated when lifted dry adiabatically. This level can be found by finding the inter section of the dry adiabat through the temperature at the pressure level of interest, and the mixing ratio through the dew point temperature at the pressure level of interest (Figure 3).

(b) ***Convective Condensation Level (CCL)*:** The level that a parcel, if heated sufficiently from below, will rise adiabatically until it is saturated. This is a good estimate for a cumuliform cloud base from surface heating. To find the convective condensation level, find the intersection of the mixing ratio through the dewpoint temperature at the pressure level of interest and the temperature sounding (Fig. 3).

(c) ***Level of Free Convection (LFC)*:** The level in which a parcel first becomes positively buoyant. To find the level of free convection, find the lifting condensation for the level of interest, and find the intersection of the moist adiabat that goes through the LCL, and the temperature curve (Figure 3).

(d) ***Equilibrium Level (EL)*:** The point at which a positively buoyant parcel becomes negatively buoyant, which typically will occur in the upper troposphere. To find this level, find the level of free convection, follow the moist adiabat through this level of free convection up until it intersects the temperature sounding again. This point is the equilibrium level (Fig. 3).

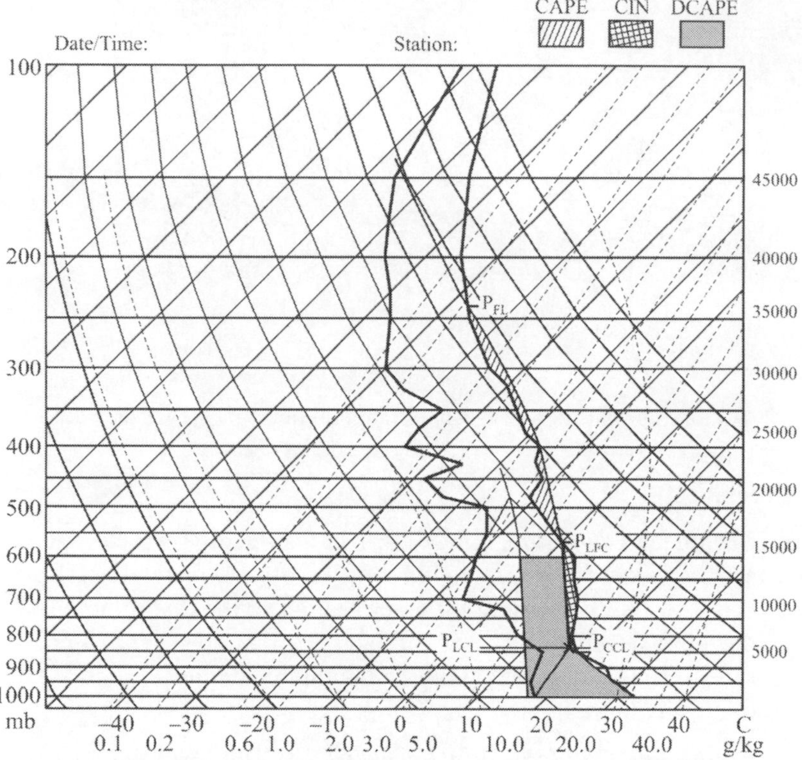

Fig. 3 An example skew-t showing the levels and energies defined in section II.

III. Meteorological Variables

As stated before, a sounding can allow the user determine values of many meteorological variables, making it one of the most useful resources for meteorologists. A variety of temperatures, mixing ratios, vapor pressures, stability indices, and conditions can be derived from temperature and dew point temperature soundings on a skew-t.

(a) Temperatures

1. ***Potential Temperature*** **(Θ):** Potential temperature is the temperature a parcel of air would have if it were lifted (expanded)

or sunk (compressed) adiabatically to 1000 mb. The value of the potential temperature is the temperature of the dry adiabat that runs through the temperature at the pressure level of interest, at 1000 mb (Figure 4).

2. ***Equivalent Temperature*** **(T_e):** The equivalent temperature is the temperature of a parcel if, via a moist adiabatic process, all moisture was condensed into the parcel. Finding the equivalent temperature is slightly more difficult. To find T_e, follow the moist adiabat that runs through the lifting condensation level at the pressure level of interest to a pressure level in which the moist adiabat and dry adiabat have similar slopes, then go down the dry adiabatat this point back down to the original pressure level of interest; this temperature is the equivalent temperature. If the dry adiabat continues beyond the boundary of the skew-t in which it cannot be determined, an alternative is to read off the temperature scale that runs diagonally in the middle of the skew-t (Figure 4).

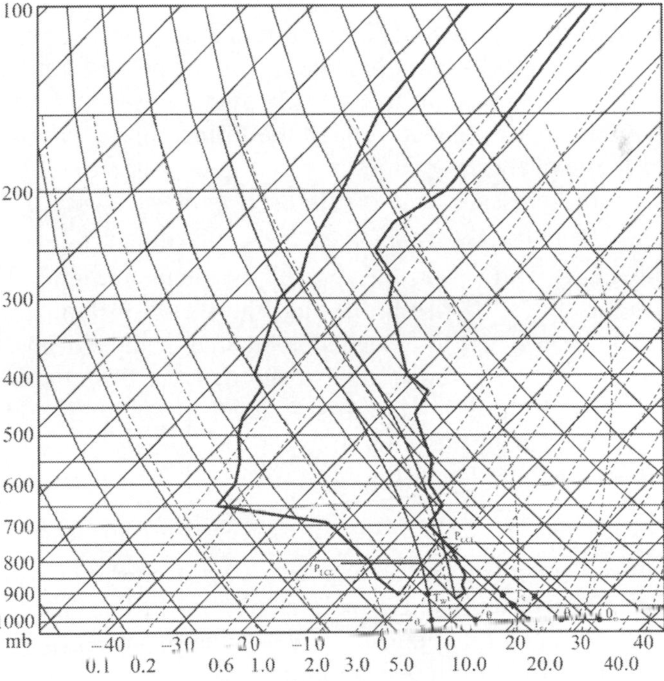

Fig. 4 This figure shows the methods of finding the temperatures mentioned in section III (a).

3. ***Equivalent Potential Temperature*** **(Θ_e):** The equivalent potential temperature is similar to the equivalent temperature however after the moisture has been condensed out of the parcel, the parcel is brought down dry adiabatically to 1000 mb. The process to find the equivalent potential temperature is the same as the regular equivalent temperature however when the parcel is brought down the dry adiabat it continues past the original pressure level and is brought down to1000 mb. The temperature at this intersection is the equivalent potential temperature.
4. ***Saturated Equivalent Potential Temperature*** **(Θ_{es}):** The temperature at which an unsaturated parcel would have if it were saturated. To find the saturated equivalent potential temperature, use a similar process used for determining the equivalent potential temperature however one must follow the moist adiabat through the environmental temperature at the necessary pressure level, unlike using the lifting condensation level for equivalent potential temperature (Fig. 4).
5. ***Wet-bulb Temperature*** **(T_w):** The minimum temperature at which a parcel of air can obtain by cooling via the process of evaporating water into it at constant pressure. To find the wet-bulb temperature follow the moist adiabat through the lifting condensation level and find the temperature of the intersection of this moist adiabat with the original pressure level of interest (Figure 4).
6. ***Wet-bulb Potential Temperature*** **(Θ_w):** Similar to the wet-bulb temperature, however the parcel is then brought down dry adiabatically to 1000 mb. To find the wet-bulb temperature, use similar means as the regular potential temperature, however continue down the moist adiabat through the lifting condensation level through the original pressure level to 1000 mb and read the temperature at the intersection of the moist adiabat and the 1000 mb pressure level (Fig. 4).

(b) Vapor Pressures

1. ***Vapor Pressure (e)*:** The amount of atmospheric pressure that is are sult of the pressure from water vapor in the atmosphere. To find the vapor pressure follow an isotherm (a line parallel to an isotherm) through the dewpoint temperature at the pressure level of interest, up to 622 mb. The value of the mixing ratio at this intersection is the vapor pressure in millibars.
2. ***Saturated Vapor Pressure (e_s)*:** The amount of atmospheric pressure that is are sult of the pressure of water vapor in saturated

air. This quantity can be found using similar means as the vapor pressure however one must follow a parallel isotherm through the temperature at the pressure level of interest.

(c) Mixing Ratios

1. ***Mixing Ratio* (*w*):** The mixing ratio is the ratio of the mass of water vapor in the air over the mass of dry air. This quantity is found by reading the mixing ratio line that goes through the dewpoint temperature at the pressure level of interest.

2. ***Saturated Mixing Ratio* (w_s):** A similar mixing ratio as above, however it is the mixing ratio of a saturated parcel of air at a given temperature and pressure. It can be found by finding the value of the mixing ratio through the temperature at a pressure level of interest.

IV. Stability Indices

(a) ***K-index:*** Used for determining what the probability and spatial coverage of ordinary thunderstorms would be based on temperature and dew point temperature.

$$K = T_{850} + T_{d850} + T_{d700} - T_{700} - T_{500}$$

$T_{d700} - T_{700}$ is also known as dew point depression at 700 hPa (dd_{700}).

For K > 35, widespread thunderstorms (> 75%) are likely. For K values between 31 and 35, fairly widespread thunderstorms (50-75%) may occur. For K values between 26 and 30 scattered (25-50%) thunderstorms are probable. For K values between 20 and 25, isolated thunderstorms (< 25%) are probable, and below 20, thunderstorm will only have a small chance (< 10%) to develop. A summary of these values are in Table B1.

(b) ***Lifted Index* (*LI*):** If storms form, this is an index that indicates the severity of the storms.

$$L_I = T_{500} - T_{p850}$$

T_p is the temperature of a parcel of air lifted to 500 mb moist adiabatically from the surface lifting condensation level. T500 is the environmental temperature at this level. LI > –2 is only as light severity, LI from –3 to –5 has a much strong severity, and the strongest severity are values with LI < 5. A summary of these values are in Table B1.

(c) ***Showalter Index (SI)*:** This is similar to the LI however the level of interest is the 850 mb pressure level.

$$SI = T850 - T_{p850}$$

This index is good for indicating elevated thunderstorms which are not picked up by the lifted index.

(d) ***Total Totals Index (TT)*:** This give the probability of severe thunderstorm activity.

$$TT = T_{850} + T_{d850} - 2T_{500}$$

Values of TT above 52 indicate a high probability of thunderstorms, in which many of the thunderstorms will become severe. Values between 48 and 52 indicate the possibility exists for severe thunderstorms. Values between 44 and 48 indicate the probability for scattered thunderstorms with only a low probability of severe thunderstorms. Finally, values lower than that 44 indicate that only normal thunderstorms will occur and there would not be any severe thunderstorm.

(e) ***Convective Available Potential Energy (CAPE)*:** The amount of potential energy that a parcel can obtain from environmental conditions. Mathematically, it is the area between the level of free convection and the equilibrium level. Inorder to identify CAPE on a sounding, find the level of free convection and follow the moist adiabat through this level up to the equilibrium level. The area between this curve and the temperature curve is positive area, CAPE (Fig. 3).

(f) ***Convective Inhibition (CIN)*:** Amount of energy from the environmental conditions that are required for a parcel to reach the level off reconvection. This is considered to be opposite that of the CAPE, where as it is called negative area. It is a requirement for strong thunderstorms to occur. The CIN is proportional to the area between the temperature curve and a parcels ascent via both a dry and moist adiabatically lapse rates (Figure 3).

(g) ***Downdraft Convective Available Potential Energy (DCAPE):*** This is proportional to the amount of energy that a saturated downdraft would have while falling to the surface. It is found by finding the lifting condensation level at 600 mb, descending the moist adiabat through this level down to the surface, and is proportional to the area that is between this line and the temperature curve (Fig. 3).

(h) ***Cap Strength*:**The cap is a stable region of the lower troposphere that impedes convection in the PBL and is caused by the inversion layer. The cap is determined as the maximum temperature difference between a parcel of air and the surrounding actual temperature in the lower troposphere. The cap will always be a positive number since a cap region has a parcel temperature that is colder than the surrounding environment. Cap is given in units of Celsius. This temperature difference is the amount of warming required to remove cap. Values 2 to 4 °C would be considered as moderate cap and > 4 °C is strong cap and any thunderstorm initiation unlikely.

Cap = Environmental temperature - parcel temperature (both in region with max temperature difference)

(i) ***Summary of Index Values***

Table BI Summaryof K-Index, Lifted Index and TT Index

Index	**Lower Limit**	**Upper Limit**	**Prediction for**	**Prediction***
K-Index	< 20		Thunderstorm Coverage	Rare (less than 10%)
	20	25		Isolated (less than 25%)
	26	30		Scattered (25-50%)
	31	35		Fairley Widespread (50-7 5%)
	>35			Widespread (more than 75%)
Lifted Index(LI)	> –2		Thunderstorm Strength	Weak
	–3	–5		Strong
	< –5			Very Strong
TT Index	< 44		Probability of Severe Thunderstorm	Unlikely
	44	48		Scattered , Non severe
	48	52		Few Severe
	> 52			Many Severe

*Meaning of the terms for strength and probability of severity of thunderstorm are relative to each other.

V. Cloud Layers

Skew-t-Log P diagrams readily allow the user to determine possible cloud layers in the atmosphere. Cloud layers are indicated by locations in which the dew point temperature is very close to the regular temperature. This means that the air is near saturation and a cloud may form.

(a) A shallow cloud layer is seen when the two temperature curves are only near each other for a short vertical layer.

(b) Deep cloud layers can be identified by locations in which the dew point temperature and regular temperature curves are near the same magnitude for deep vertical layers in the atmosphere. Furthermore, one can identify cloud cover by noticing a sudden drop in the dew point temperature. This represents the condensation of water vapour into cloud drops.

Another way to look at it is that a cloud is located in a region in which there is a significant drop in mixing ratio. One will recall that the mixing ratio is defined to be the ratio of then mass of water vapor to the mass of dry air. As water condenses into cloud drops, then mass of water vapor decreases, and thus the mixing ratio decreases.

VI. Atmospheric Mixing

In the morning, when solar radiation warms the Earth, warm pockets of air become positively buoyant and rise, and upon reaching a level aloft, will eventually become negatively buoyant and fall. This process is called turbulence and one of the consequences of turbulence in the atmosphere is mixing. During the afternoon, typically mixing will reach a peak, and when this occurs many meteorological variables become constant. Using a skew-t Log P diagram, one can determine just how well mixed the atmosphere is using two methods. When the atmosphere has reached a point in which mixing is at a maximum, the mixing ratio and potential temperature will be relatively constant with height. A dew point curve that runs parallel to a mixing ratio line (typically dashed on a skew-t) is considered constant mixing ratio. Constant potential temperature occurs when the temperature sounding is close to that of slope of the dry adiabat. An example of a well-mixed atmosphere can be seen in Figure 5. Notice the constant potential temperature (the temperature sounding almost parallels the dry adiabat) and the mixing ratio is constant with height. This indicates that the atmosphere is very well mixed, particularly the atmosphere up to

650 hPa. This type of profile is often observed during the summer months when strong daytime heating causes a very large amount of turbulence in the atmosphere.

A second approach to determining how well mixed the atmosphere is for a particular sounding is to find the lifting condensation level for the surface temperature and surface dewpoint temperature and compare it to the actual observed cloud base at that time. If these two levels compare to each other, then the atmosphere is very well mixed. Large variations in these two levels indicate only a moderate mixing in the atmosphere.

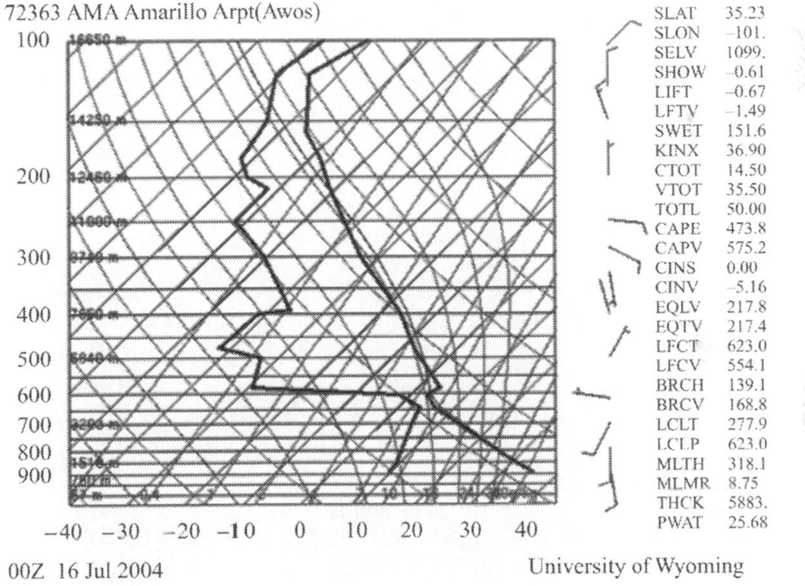

Fig. 5 A well-mixed atmosphere in the levels from the surface up to 650 mb.

VII. Atmospheric Static Stability

Identification of layers that are unstable, stable, or neutrally stable is important for identifying how high buoyant parcels of air will rise in the atmosphere. Unstable layers will promote and enhance positive buoyancy and stable layers will tend to cease lifting and cause parcels to become negatively buoyant. A neutrally stable layer will have no net effect on the buoyancy of a parcel of air. A conditionally unstable layer, depending on the environment can have a positive or negative affect on a parcels buoyancy. There are a few methods for determining the stability of a layer in the atmosphere, using lapse rates, using potential temperature, and or the saturated equivalent potential temperature.

(a) ***Using Lapse Rates***: The necessary lapse rates for determining stability are the moist adiabatic lapse rate, the dry adiabatic lapse rate, and the environmental lapse rate of the layer of interest. The following are the conditions for stability by comparing the environmental lapse rates to the dry and moist adiabatic lapse rate.

$\Gamma < \Gamma_w$	Absolutely Stable
$\Gamma_d > \Gamma > \Gamma_w$	Conditionally Instable
$\Gamma > \Gamma_d$	Absolutely Instable

(b) ***Using Potential Temperature***: The conditions for stability using the change in potential temperature with height are as follows:

$$\frac{\partial\Theta}{\partial z} > 0 \text{ Stable,} \frac{\partial\Theta}{\partial z} = 0 \text{ Neutral, } \frac{\partial\Theta}{\partial z} < 0 \text{ Unstable}$$

(c) ***Using the Saturation Equivalent Potential Temperature***: The conditions for static stability using the change in saturation equivalent potential temperature with height are the following:

$$\frac{\partial\Theta_{es}}{\partial z} > 0 \text{ Stable, } \frac{\partial\Theta_{es}}{\partial z} < 0 \text{ Conditionally Unstable}$$

VIII. Wind Shear

Often times the wind speed and direction are plotted on a skew-t diagram on the left edge of the diagram, near the height axis. This plot can be used to determine levels of strong wind shear. Windshear has two components: speed shear and directional shear. Speed shear is defined as an abrupt increase or decrease in wind speed in a relatively short distance, and directional shear is defined as an abrupt change in wind direction over a relatively short distance. One determines these conditions by interpreting the wind plots on the skew-t diagram. The amount of shear in the atmosphere is critical for the development of strong thunderstorms, and is producers of waves in the atmosphere.

IX. Remarks on skew-t

The use of the skew-t is a very important resource to understand the current state of the atmosphere and, in addition, the forecasted state of the atmosphere. A forecasted skew-t can be just as good as a spatial model that is typically viewed when making a forecast. Various critical meteorological variables can be determined, cloud layer depth and height can be found, and severe weather probabilities and coverage can be derived from the skew-t diagram. Likewise, wind shear and static stabilities can be calculated from the skew-t diagram.

X. Damage Scale

(a) Fujita Tornado Damage Scale

This scale was developed in 1971 by T. Theodore Fujita of the University of Chicago. It is described in the table B2, mentioned below.

Table B2 Fujita Tornado Damage Scale

Scale	Wind Estimate		Typical Damage
	Kmph	mph	
F0	< 117	< 73	Light damage. Some damage to chimneys; branches broken off trees;shallow-rooted trees pushed over; sign boards damaged.
F1	117-180	73-112	Moderate damage. Peels surface off roofs; mobile homes pushed off foundations or overturned; moving autos blown off roads.
F2	181-253	113-157	Considerable damage. Roofs torn off frame houses; mobile homes demolished; boxcars overturned; large trees snapped or uprooted; light-object missiles generated; cars lifted off ground.
F3	254-333	158-207	Severe damage. Roofs and some walls torn off well-constructed houses;trains overturned; most trees in forest uprooted; heavy cars lifted off the ground and thrown.
F4	334-418	208-260	Devastating damage. Well-constructed houses levelled; structures with weak foundations blown away some distance; cars thrown and large missiles generated.
F5	419-512	261-318	Incredible damage. Strong frame houses levelled off foundations and swept away; automobile-sized missiles fly through the air in excess of 100 meters (109 yds.); trees debarked; incredible phenomena will occur.

Practical application of Fujita's F-scale winds had several difficulties. These precise wind speed numbers are actually guesses and had never been scientifically verified. Different wind speeds may cause similar-looking damages from place to place - even from building to building. Without a thorough engineering analysis of tornado damage in any event, the actual wind speeds needed to cause that damage remained unknown under this system. Hence since February 2007 an Enhanced F-scale was introduced in U.S. It is as per Table-B3, mentioned below.

(b) Enhanced F Scale for Tornado Damage

Table B3 Enhanced F Scale for Tornado Damage

	Fujita Scale					Derived EF Scale			**Operational EF Scale**	
F Number	**Max wind**		**3 Second Gust**		**EF Number**	**3 Second Gust**		**EF Number**	**3 Second Gust**	
	Kmph	**mph**	**Kmph**	**mph**		**Kmph**	**mph**		**Kmph**	**mph**
0	64-116	40-72	72-125	45-78	0	72-137	65-85	0	72-137	65-85
1	117-180	73-112	126-188	79-117	1	138-176	86-109	1	138-177	86-110
2	181-253	113-157	189-259	118-161	2	177-220	110-137	2	178-217	111-135
3	254-333	158-207	260-336	162-209	3	221-268	138-167	3	218-265	136-165
4	334-418	208-260	337-420	210-261	4	269-320	168-199	4	266-321	166-200
5	419-674	261-318	421-510	262-317	5	321-376	200-234	5	321<	Over 200

The Enhanced F-scale still is a set of wind estimates (not measurements) based on damage. Its uses three-second gusts estimated at the point of damage based on a judgment of 8 levels of damage to the 28 indicators listed in table B4(**http://www.spc.noaa.gov/efscale/**). These estimates vary with height and exposure.

Table B4 Enhanced F Scale Damage Indicators

NUMBER	DAMAGE INDICATOR	ABBREVIATION
1	Small barns, farm outbuildings	SBO
2	One- or two-family residences	FR12
3	Single-wide mobile home (MHSW)	MHSW
4	Double-wide mobile home	MHDW
5	Apt, condo, townhouse (3 stories or less)	ACT
6	Motel	M
7	Masonry apt. or motel	MAM
8	Small retail bldg. (fast food)	SRB
9	Small professional (doctor office, branch bank)	SPB
10	Strip mall	SM
11	Large shopping mall	LSM
12	Large, isolated ("big box") retail bldg.	LIRB
13	Automobile showroom	ASR
14	Automotive service building	ASB
15	School - 1-story elementary (interior or exterior halls)	ES
16	School - jr. or sr. high school	JHSH
17	Low-rise (1-4 story) bldg.	LRB
18	Mid-rise (5-20 story) bldg.	MRB
19	High-rise (over 20 stories)	HRB
20	Institutional bldg. (hospital, govt. or university)	IB
21	Metal building system	MBS
22	Service station canopy	SSC
23	Warehouse (tilt-up walls or heavy timber)	WHB
24	Transmission line tower	TLT
25	Free-standing tower	FST
26	Free standing pole (light, flag, luminary)	FSP
27	Tree - hardwood	TH
28	Tree - softwood	TS

For more details with respect to the development and makeup of the Enhanced F-scale readers may refer the Storm Prediction Centre (SPC) website www.spc.noaa.gov/faq/tornado/EFScale.pdf.

XI. Brightness Temperature

The brightness temperature (or TB) is a measurement of the radiance of the microwave radiation traveling upward from the top of the atmosphere to the satellite, expressed in units of the temperature of an equivalent black body. It is the fundamental parameter measured by passive microwave radiometers.

Satellite passive microwave radiometers measure raw antenna counts from which the antenna temperature is determined and then the brightness temperature of the Earth is calculated. Large antennas are used for the various channels of the radiometer, and during operation, each antenna feedhorn passes a hot and cold target in order to provide consistently calibrated raw counts. The conversion from radiometer counts to top-of-the-atmosphere TB is called the calibration process. Several calibration processing steps are required to derive the TB values. Microwave radiometer TB are considered a fundamental climate data record and are the values from which the ocean measurements of wind speed, water vapour, cloud liquid water, rain rate, and sea surface temperature are derived.

Colour Plates

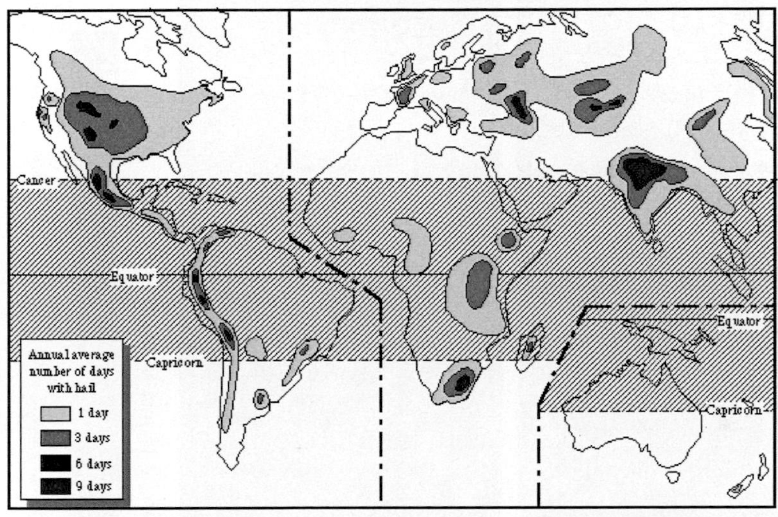

Fig. 1.1 Global map of annual hail days, from Williams (1973), based largely on Frisby and Sansom (1967) based on surface observations.

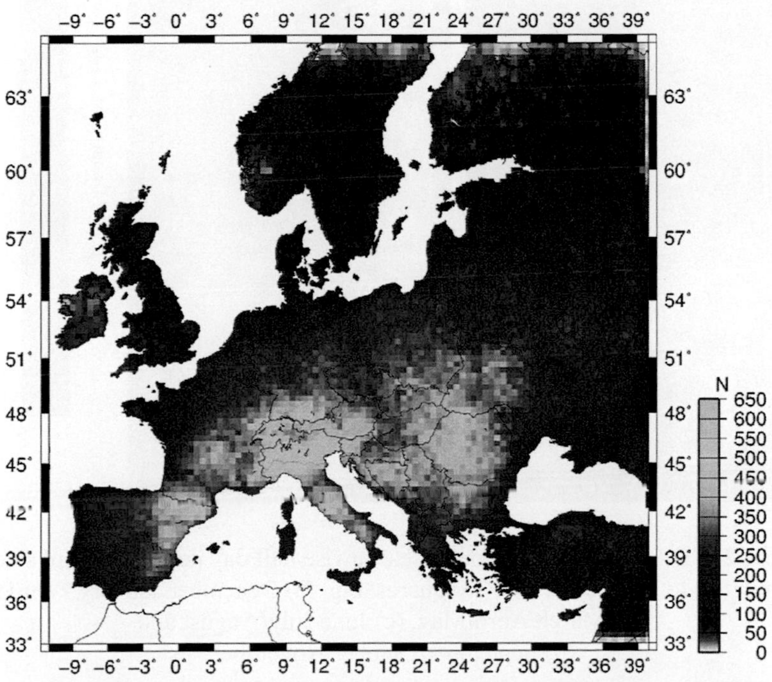

Fig. 1.13 Hail event frequency as estimated by Punge et al. (2014).

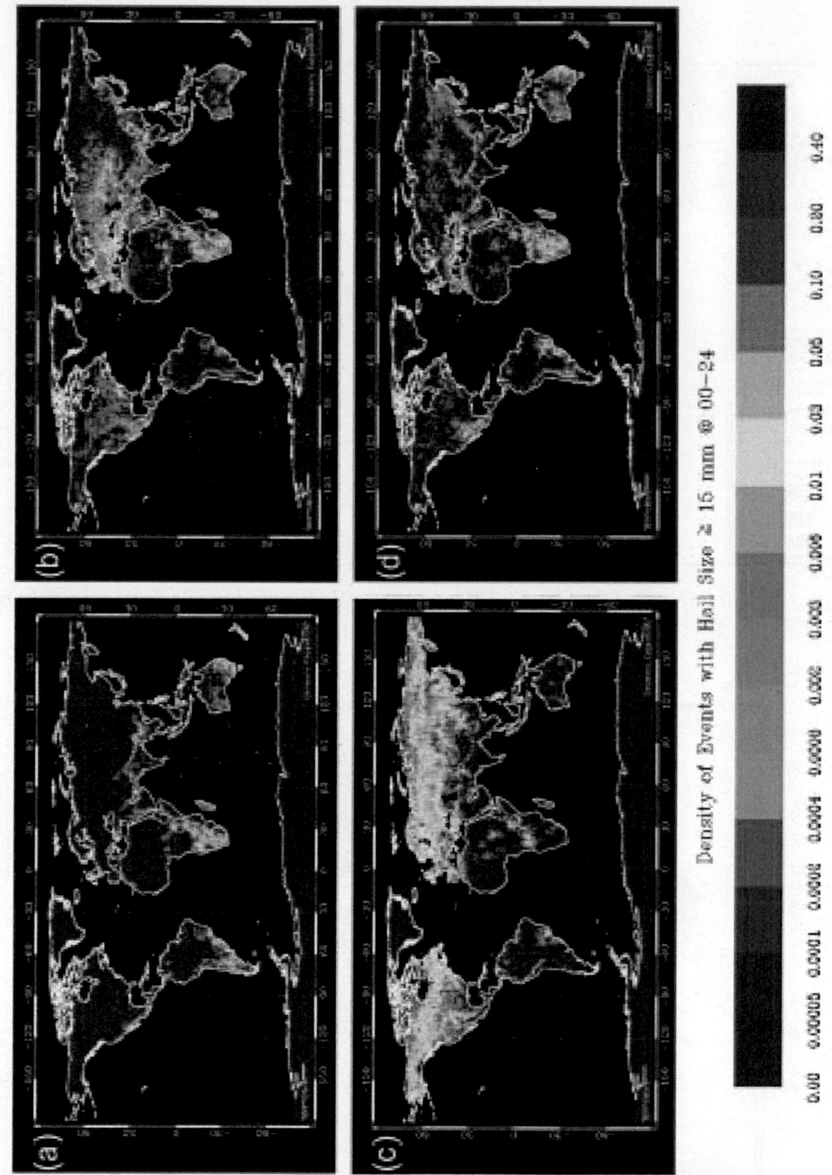

Fig. 1.2 CDP global distribution of seasonwise hail day density for hail size ≥ 15 mm (00 to 24UTC) in 1° × 1° squares. Fig. (a) December/January/February, (b) March/April/May, (c) June/July/August and (d) September/October/November.

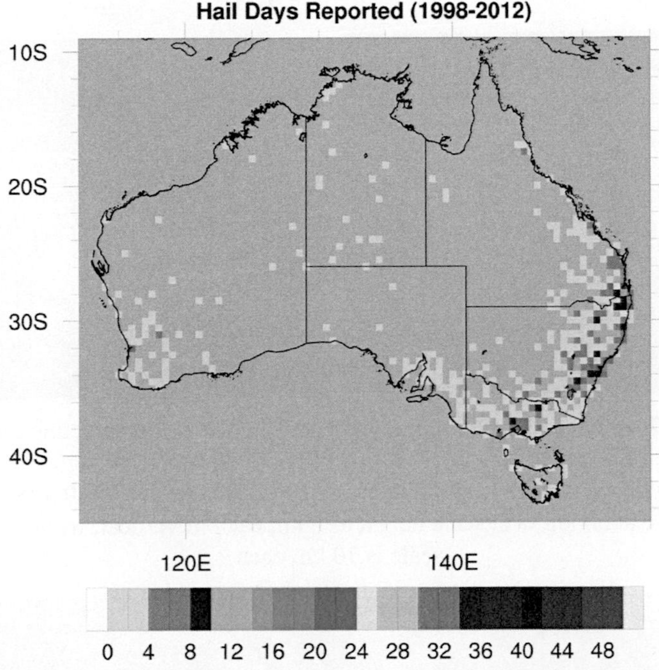

Fig. 1.15 Total number of reported hail days per grid cell (1998-2012). Australia Bureau of Meteorology: Severe Storms Archive. [http://www.bom.gov.au/Australia/stormarchive/]

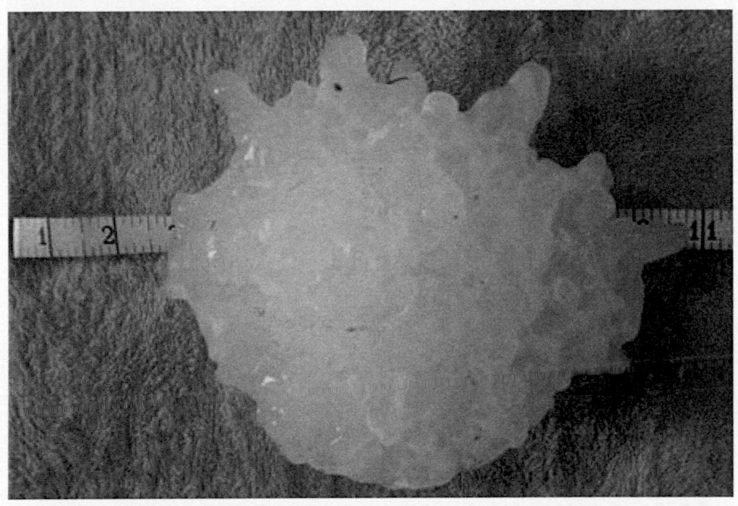

Fig. 2.9 The largest official hailstone ever collected in the U.S. An eight-inch monster that fell at Vivian, South Dakota on July 23, 2010

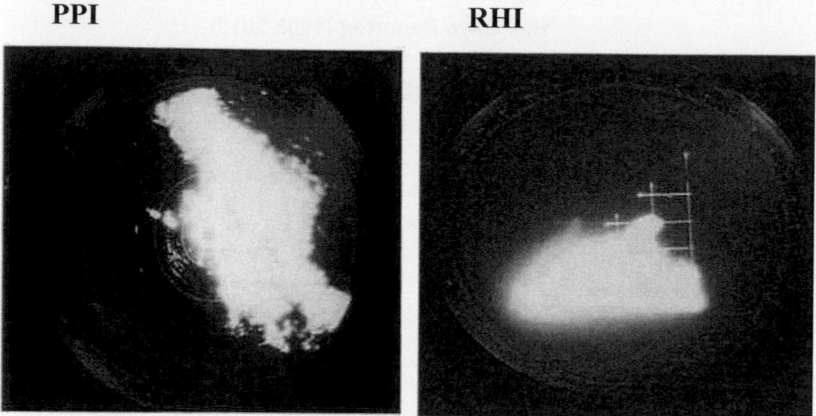

Fig. 5.7 PPI and RHI cloud photographs from the storm detecting radar at Mumbai (India) on 26 July 2005. (Left) PPI at 1130 h IST having radar range at 100 km with each range ring of 20 km scale. (Right) RHI at 1430 h IST having radar range at 50 km with scale height as 5 km each in vertical, while horizontal scale is 10 km each

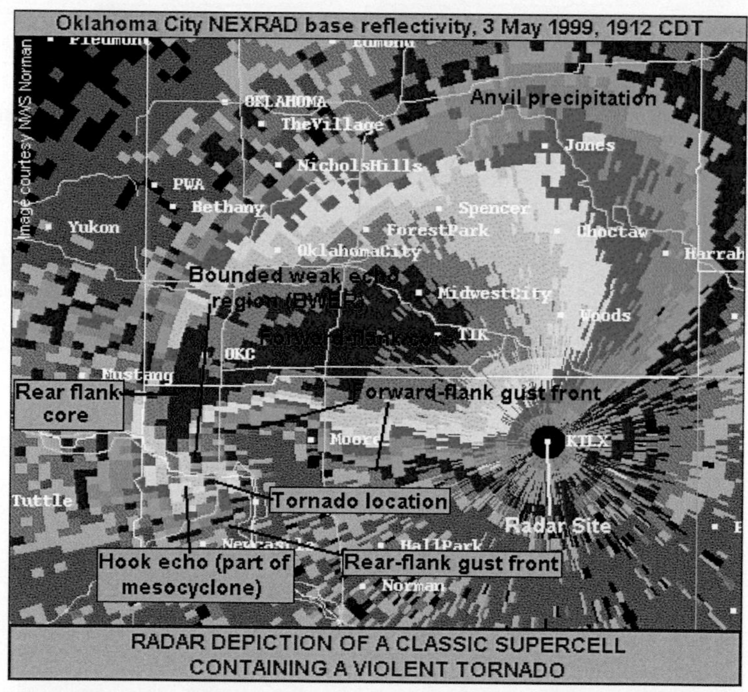

Fig. 5.8 Typical Hook echo and Bounded Weak Eecho Region (BWER)

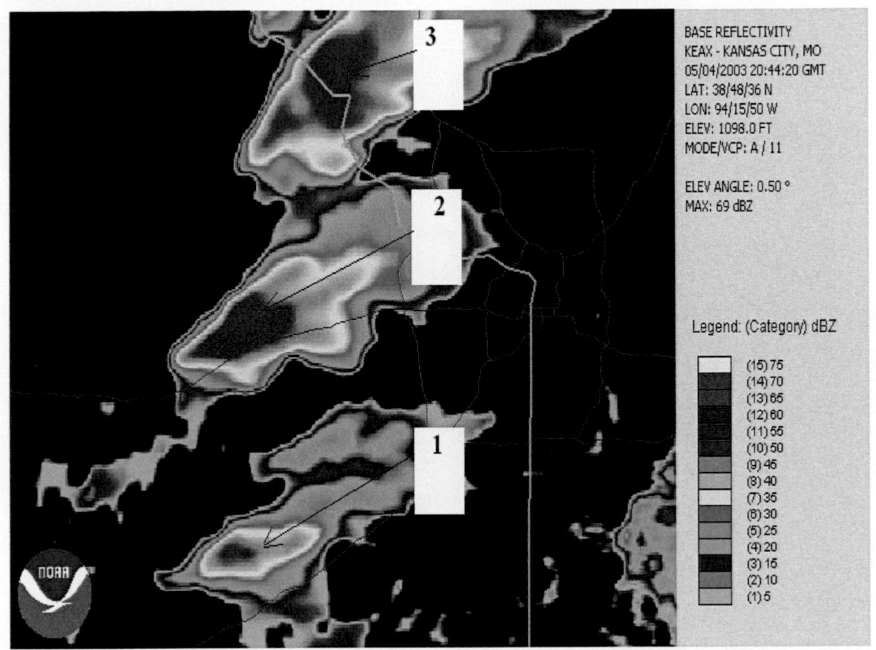

Fig. 5.9 Forming (1), formed (2) and dissipating (3) hook echoes from bottom to top.

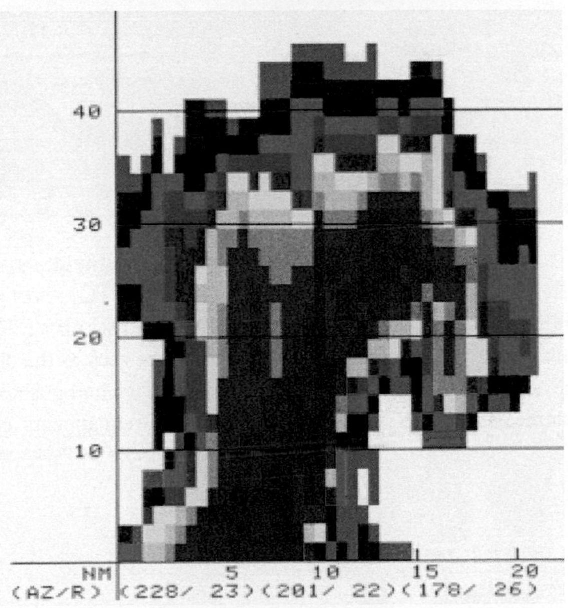

Fig. 5.10 Bounded weak echo region (BWER)

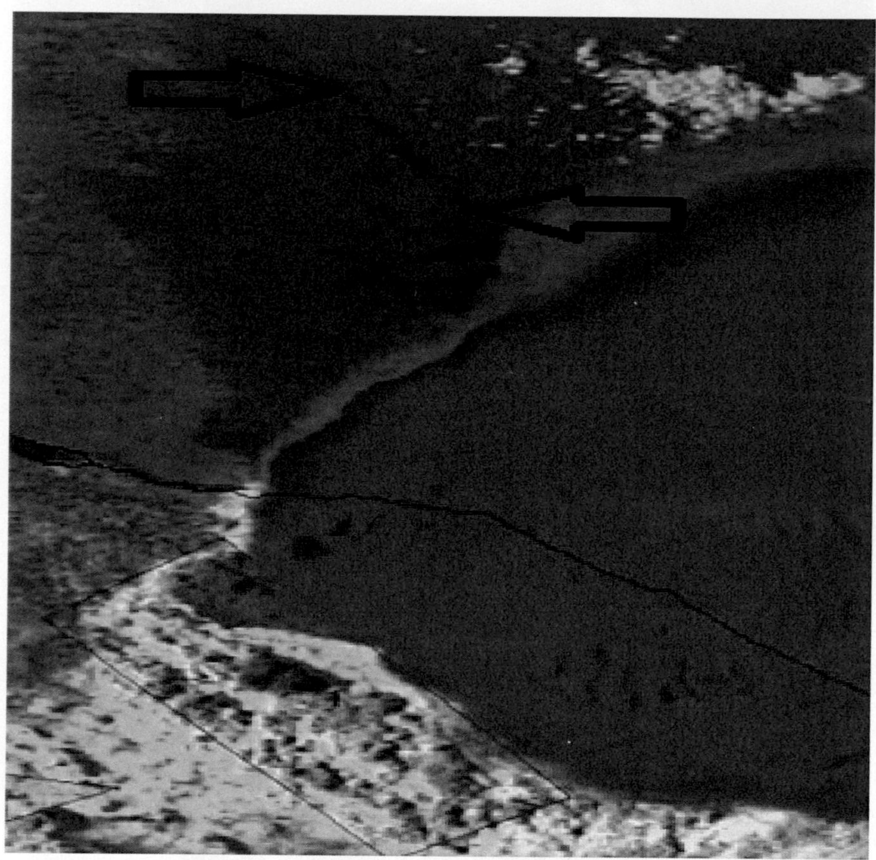

Fig. 5.11 A tornadic storm with 4.5 inch hail. The image is based on the NOAA-AVHRR overpass on 29 June 2000, 2221 UTC, over a domain of 282 264 AVHRR 1-km pixels. The cloud occurred in southwestern Nebraska.

A hail swath (marked with arrow) on the ground can be seen as the dark purple line emerging off the north flank of the storm, oriented nw-se. Two hail gushes are evident on the swath near the edge of the storm. The precipitation swath appears as darker blue

Author Index

Index

E

F

N

O

P

Q

R

S

T

U